VOLUME SEVENTY SIX

ADVANCES IN BOTANICAL RESEARCH

Plant Cyclotides

ADVANCES IN BOTANICAL RESEARCH

Series Editors

VOLUME SEVENTY SIX

Advances in
BOTANICAL RESEARCH
Plant Cyclotides

Volume Editor

DAVID J. CRAIK

Institute for Molecular Bioscience,
The University of Queensland,
Brisbane, QLD, Australia

ELSEVIER

AMSTERDAM • BOSTON • HEIDELBERG • LONDON
NEW YORK • OXFORD • PARIS • SAN DIEGO
SAN FRANCISCO • SINGAPORE • SYDNEY • TOKYO

Academic Press is an imprint of Elsevier

AP

Academic Press is an imprint of Elsevier
125 London Wall, London EC2Y 5AS, UK
The Boulevard, Langford Lane, Kidlington, Oxford OX5 1GB, UK
225 Wyman Street, Waltham, MA 02451, USA
525 B Street, Suite 1800, San Diego, CA 92101-4495, USA

First edition 2015

ISBN: 978-0-12-800030-4
ISSN: 0065-2296

For information on all Academic Press publications
visit our website at http://store.elsevier.com

CONTENTS

CONTRIBUTORS

Marilyn Anderson
Department of Biochemistry and Genetics, La Trobe Institute for Molecular Science, La Trobe University, Melbourne, Vic, Australia

Tao Bi
Department of Pharmacology and Pharmaceutical Sciences, University of Southern California, Los Angeles, CA, USA

Julio A. Camarero
Department of Pharmacology and Pharmaceutical Sciences; Department of Chemistry, University of Southern California, Los Angeles, CA, USA

Michelle L. Colgrave
CSIRO Agriculture Flagship, Brisbane, QLD, Australia

David J. Craik
Institute for Molecular Bioscience, The University of Queensland, Brisbane, QLD, Australia

Norelle L. Daly
Centre for Biodiscovery and Molecular Development of Therapeutics, AITHM, James Cook University, Cairns, QLD, Australia

Edward K. Gilding
Institute for Molecular Bioscience, The University of Queensland, Brisbane, QLD, Australia

Ulf Göransson
Division of Pharmacognosy, Department of Medicinal Chemistry, Uppsala University, Biomedical Center, Uppsala, Sweden

Christian W. Gruber
Center for Physiology and Pharmacology, Medical University of Vienna, Vienna, Austria

Karen Harris
Department of Biochemistry and Genetics, La Trobe Institute for Molecular Science, La Trobe University, Melbourne, Vic, Australia

Wen-Jun He
State Key Laboratory of Phytochemistry and Plant Resources in West China, Kunming Institute of Botany, Chinese Academy of Sciences, Kunming, Yunnan, People's Republic of China

Mark A. Jackson
Institute for Molecular Bioscience, The University of Queensland, Brisbane, QLD, Australia

Meng-Wei Kan
Institute for Molecular Bioscience, The University of Queensland, Brisbane, QLD, Australia

Johannes Koehbach
School of Biomedical Sciences, The University of Queensland, St. Lucia, QLD, Australia

Yilong Li
Department of Pharmacology and Pharmaceutical Sciences, University of Southern California, Los Angeles, CA, USA

Sohaib Malik
Division of Pharmacognosy, Department of Medicinal Chemistry, Uppsala University, Biomedical Center, Uppsala, Sweden

Georgianna Kae Oguis
Institute for Molecular Bioscience, The University of Queensland, Brisbane, QLD, Australia

K. Johan Rosengren
The University of Queensland, School of Biomedical Sciences, Brisbane, QLD, Australia

Thomas Shafee
Department of Biochemistry and Genetics, La Trobe Institute for Molecular Science, La Trobe University, Melbourne, Vic, Australia

Blazej Slazak
Division of Pharmacognosy, Department of Medicinal Chemistry, Uppsala University, Biomedical Center, Uppsala, Sweden; Władysław Szafer Institute of Botany, Polish Academy of Sciences, Cracow, Poland

Ning-Hua Tan
State Key Laboratory of Phytochemistry and Plant Resources in West China, Kunming Institute of Botany, Chinese Academy of Sciences, Kunming, Yunnan, People's Republic of China

PREFACE

Cyclotides are peptides (mini-proteins) that are characterized by their unusual structural features of a head-to-tail cyclized backbone and a knotted arrangement of three cross-linking disulfide bonds. It is thought that their role in plants is as defense agents, a hypothesis based on their observed activity against certain insects, nematodes, and mollusk's, although it is by no means certain that defense is their primary or indeed only function. So far cyclotides have been discovered in five major families of angiosperms, namely the Violaceae, Rubiaceae, Cucurbitaceae, Solanaceae, and Fabaceae and, accordingly are present in some of the most economically important plant families. However, they are sparsely distributed in the plant kingdom, occurring in only a small fraction of species from all of these families, apart from the Violaceae, where they appear to be ubiquitous. One plant may contain from dozens to hundreds of cyclotides.

Cyclotides originally came to notice based on an observation that a plant (*Oldenlandia affinis*, Rubiaceae) containing the prototypical cyclotide, kalata B1, is used in African folk medicine to accelerate child birth. The discovery that kalata B1 has an unusual cyclic and knotted structure was pivotal in explaining the ability of this peptide to survive boiling to make the medicinal tea. It is this unusual structure and associated exceptional stability that sparked interest in cyclotides as potential frameworks in drug design.

Despite more than 25 years of research since the name "cyclotide" was coined in 1999, so far there have been relatively few studies on cyclotides among plant biologists, with most of the cyclotide literature being focused on their applications in drug design. The overall goal of this book is to promote knowledge of cyclotides among plant biologists, with the aim of encouraging the further study of the functions and distribution of this fascinating family of plant proteins. This book comprises 10 chapters arranged to highlight the progression of cyclotide research from discovery to structure to function and applications.

Chapter 1 provides an overview of the history of the discovery and of cyclotides and identifies some of the milestones associated with their structural characterization, synthesis, bioassay, and applications in drug design. In Chapter 2, Ulf Göransson and colleagues from Uppsala University provide an overview of cyclotides in the Violaceae. As noted already, the Violaceae is the only plant family known in which cyclotides occur in every species examined so far. This observation raises interesting questions about the evolution and biosynthesis of cyclotides. What is special about the Violaceae compared to other plant families with respect to cyclotides?

Chapter 3 provides a complementary discussion of cyclotides in the Rubiaceae, where the incidence of cyclotide-bearing species is less than 5%. In this chapter, Gruber and colleagues provide some practical protocols for the isolation of cyclotides from this important plant family, analyze their sequence diversity, discuss their evolution, and summarize their bioactive properties. Chapter 4 focuses on cyclotides from Chinese plants with a particular focus on plants with a history of uses in Chinese Traditional Medicines. In particular, Tan and colleagues describe cost- and time-effective detection and extraction methods to assist in the isolation and characterization of new cyclotides from Chinese plants.

Chapter 5 focuses on methods for the determination of the amino acid sequences of cyclotides. Sequencing studies are challenging for cyclic peptides, which by definition lack termini, and given the resistance of native cyclotides to proteolysis, this was one of the reasons why early studies in the 1970s did not successfully sequence cyclotides. However, the fact that all cyclotides have a conserved glutamic acid in loop 1 and may be cleaved by endo-GluC after reduction, allows linearization of the backbone, and hence subsequent sequence analysis. In early studies this was done by Edman sequencing but this has largely been supplanted by Mass spectrometry MS/MS studies.

Chapter 6 continues on the theme of structural analysis but focuses on the 3D structures of cyclotides. Most such studies have been carried out using NMR spectroscopy because cyclotides, like other small disulfide-rich peptides are notoriously difficult to crystallize. The few crystal structures that have been done have confirmed the NMR analyses and show the cyclic cystine knot structure.

Chapter 7 describes the natural functions and structure–activity relationships of cyclotides. Such studies include screening for pharmaceutical and agricultural (i.e., pesticidal) applications of native cyclotides as well as the chemical modifications of cyclotides to develop structure–activity relationships. The underpinning chemical syntheses for these applications are described in the following two chapters. In particular, Chapter 8 describes the biosynthesis of cyclotides. Nature biosynthesizes these fascinating molecules as conventional linear precursor proteins that are subsequently cyclized via asparaginyl endopeptidases (AEPs), which are common plant enzymes that typically perform proteolytic functions. The "repurposing" of AEPs as ligases is an example of the efficient reuse of an existing enzyme to create a new class of molecules.

Chapter 9 focuses on a range of chemical and biological approaches for the nonnatural synthesis of cyclotides. As noted earlier such technologies have helped to underpin a wide range of structure–activity relationship studies of cyclotides and also a wide range of biotechnological applications as described in Chapter 10. These applications include pesticidal (e.g., insecticidal or nematocidal sprays, or via genetic engineering for insect resistance), pharmaceutical (e.g., for controlling angiogenesis in cancer, antiobesity, multiple sclerosis, antipain, and immunosuppressants), and antifouling and biofilm formation.

Overall, I hope that this book will introduce cyclotides to a new audience and extend research in this area. Although currently there are around 300 sequences published, it has been estimated that the number of cyclotides existing in angiosperms is probably in the tens of thousands. Many questions remain to be studied: Why does one plant produce a suite of many similar cyclotides? Why do different tissues contain different suites of cyclotides? Is pesticidial activity their primary function or might there be other physiological roles for cyclotides? How can these molecules be used in biotechnology applications?

David J. Craik

CHAPTER ONE

Overview on the Discovery and Applications of Cyclotides

David J. Craik
Institute for Molecular Bioscience, The University of Queensland, Brisbane, QLD, Australia
E-mail: d.craik@imb.uq.edu.au

Contents

Abstract

Cyclotides are bioactive peptides from plants that are characterized by their head-to-tail cyclic backbone and knotted arrangement of their three conserved disulfide bonds. Their natural function is thought to be as host defense agents and a single plant can express dozens to hundreds of cyclotides. This article provides a brief overview of the history of the discovery of cyclotides and their applications in drug design and agriculture, the former based on their stability as a peptide scaffold and the latter based on their activity against crop pests. More than 300 papers have been published on cyclotides and they represent a promising field of biotechnological research.

1. INTRODUCTION

The name cyclotide was formally introduced in 1999 (Craik, Daly, Bond, & Waine, 1999) to describe a family of plant-derived peptides that have the unique structural features of a head-to-tail cyclic backbone and a knotted arrangement of three conserved disulfide bonds. However, the origins of the cyclotide field may be traced back much further, to Africa, in the 1960s. Two observations in different parts of Africa on the use of folk medicines from the plant *Oldenlandia affinis* for childbirth and various other medical conditions unknowingly sparked the field (Gran, Sandberg, & Sletten, 2000). In one report (Sandberg, 1965) it was noted that a decoction from *O. affinis* (called Wetegere in the Gbaya language) was used to facilitate labor. In a second report a Red Cross team working at the Central Hospital

Advances in Botanical Research, Volume 76
ISSN 0065-2296
http://dx.doi.org/10.1016/bs.abr.2015.09.008

of Luluabourg in Kananga noted a series of obstetrical complications related to the use of a native drug "kalata–kalata" (Gran, 1970), which prompted an investigation of its properties. In that case, it was known that the medicinal tea was prepared by boiling the aboveground parts of the plant in water, so whatever the active ingredient was, it could survive boiling.

Dr Lorents Gran, one of the physicians on the Norwegian Red Cross team characterized the peptidic nature of one of the uterotonic agents and named it kalata B1 (Gran, 1970). In a series of papers in the early 1970s he characterized the pharmacological properties of the peptide, including a comparison of its uterotonic activity to that of oxytocin, also a peptide and well known in Western obstetric medicine but unrelated in sequence. However, despite the amino acid composition and size of approximately 30 amino acids being largely determined by 1973, the structure of kalata B1 remained unknown, with attempts to characterize it stymied by what turned out to be some quite unusual structural features that had not been encountered earlier in peptides of this size.

The field went silent for some 20 years, with no papers reported on kalata B1 until our 1995 publication on its three-dimensional structure (Saether et al., 1995), which revealed in detail the macrocyclic backbone and proposed knotted arrangement of disulfide bonds. We had actually determined the structure in 1991 using NMR spectroscopy, but because it was so unusual we did not publish it until we had finally convinced ourselves that such a knotted structure within a macrocyclic backbone was possible. However, we had discussed the structure with colleagues in the intervening period, including revealing the proposed cystine knot and the first literature reference to the structure of kalata B1 is noted in a 1993 Cell paper (McDonald & Hendrickson, 1993) in which kalata B1 is compared to human cystine knot proteins that had just been structurally elucidated. We also compared it to some inhibitory cystine knot peptides that were beginning to be reported at the time (Pallaghy, Nielsen, Craik, & Norton, 1994).

Figure 1 shows the sequence and three-dimensional structure of kalata B1, highlighting its circular topology and compact folded topography. The lack of termini in the mature peptide makes any numbering scheme for the amino acids somewhat arbitrary, but the one indicated here fits with what we now know about the biosynthetic route. Before this though, we had gone through several iterations of alternative numbering schemes. Although perhaps a trivial point, this ambiguity highlights the unconventional nature of circular proteins and the fact that the seamless circle of peptide bonds gives no insight in itself as to the origin of the linear precursor protein.

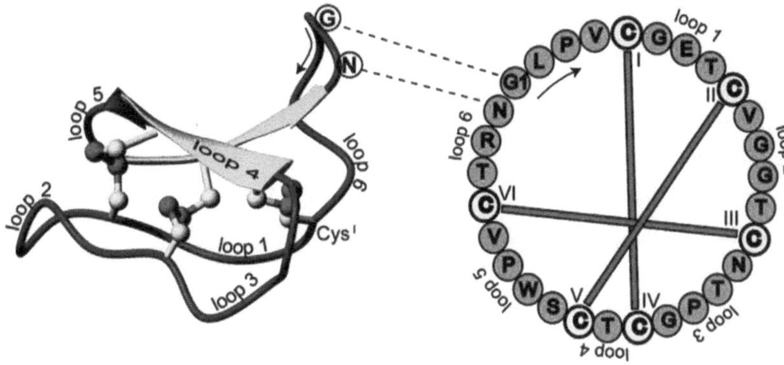

Figure 1 ***Sequence and structure of kalata B1.*** On the left is the NMR structure (PDB ID code 1NB1). On the right is the amino acid sequence, with the circular backbone shown using the one-letter amino acid code. Arrows indicate the direction of the peptide chain from N to C direction. Backbone loops between successive cysteine residues are numbered loops 1 to 6. Cysteine residues are numbered I to VI. (See color plate)

For me the unique structural features of the cystine knot and head-to-tail cyclic backbone immediately sparked interest in this peptide as a possible tool in drug design, as they explained how kalata B1 might have survived boiling to make the medicinal tea as well as surviving the human digestive system to deliver its pharmacological actions. At that time we had no idea what the natural function of the peptide was, nor how it might be biosynthesized, or indeed whether other related peptides might exist.

Coincidentally, three other reports of macrocyclic peptides of similar size (approximately 30 amino acids) appeared in 1993 and 1994, the first on violapeptide 1 (Schöpke, Hasan Agha, Kraft, Otto, & Hiller, 1993) from *Viola arvensis* of the Violaceae plant family and two others from the Rubiaceae plants *Psychotria longipes* and *Chassalia parvifolia*, describing anti-HIV and neurotensin antagonistic properties, respectively (Gustafson et al., 1994; Witherup et al., 1994). The violapeptide report, published in German in an obscure journal, went essentially unnoticed, but the JACS paper by Gustafson et al. from the U.S. National Cancer Institute (NCI) on the sequences and bioactivities of the anti-HIV peptides circulin A and B created much interest for us, and we recognized that kalata B1, might not be alone but instead could be part of a larger family of similar peptides. This was confirmed when we collaborated with the NCI group and determined that circulin A also had a cyclic cystine knot structure (Daly, Koltay, et al., 1999). This prompted us to initiate a systematic discovery effort to look for related peptides in plants, which culminated in the discovery of around a dozen new sequences and the definition of the name "cyclotides"

(Craik et al. 1999). As described in Chapter 2 of this book, Ulf Göransson and colleagues in Uppsala were conducting parallel studies on Violaceae family plants (Claeson, Göransson, Johansson, Luijendijk, & Bohlin, 1998; Göransson, Luijendijk, Johansson, Bohlin, & Claeson, 1999). More recently, a number of other groups have initiated cyclotide discovery research and some of these studies are reviewed in Chapters 3 and 4. Chapters 5 and 6 outline the structural investigations by mass spectrometry and NMR spectroscopy, respectively, that have complemented these discovery efforts.

From these early beginnings, the cyclotide field has expanded significantly so that there are now more than 300 research papers published on cyclotides or closely related peptides. Indeed it is now clear that macrocyclic peptides are present in bacteria, fungi, plants, and animals. The database CyBase (http://cybase.org.au) has been established to catalogue and curate sequences of circular proteins (Kaas & Craik, 2010; Mulvenna, Wang, & Craik, 2006; Wang, Kaas, Chiche, & Craik, 2008), and the reader is referred there for updates on the current numbers of sequences. Around 300 cyclotide sequences have been published but it is estimated that the size of the family is around 50,000 members based on their currently known distribution in flowering plants (Gruber et al., 2008). Cyclotides have been screened for a diverse range of bioactivities that range from antifouling to anti-HIV, as described in more detail in Chapter 7.

Figure 2 provides a timeline of cyclotide research since the discovery of kalata B1 in 1970 until today and Table 1 provides some key references associated with major milestones. Of particular note were the development of methods to chemically synthesize cyclotides in the late 1990s (Daly, Love, Alewood, & Craik, 1999; Tam & Lu, 1997, 1998; Tam, Lu, Yang, & Chiu, 1999) and the discovery of the first cyclotide gene in 2001 (Jennings, West, Waine, Craik, & Anderson, 2001). The Jennings paper provided the first clues on cyclotide biosynthesis. Although it seems obvious now, before then it was not clear whether cyclotides were ribosomally or nonribosomally synthesized, with the latter being the case for other smaller naturally occurring cyclic peptides known then, such as cyclosporin. The presumed natural function of cyclotides as plant defense agents was confirmed in the same report that we described the gene encoding kalata B1 (Jennings et al., 2001). Kalata B1 was shown to be a potent insecticidal agent against the larvae of the lepidopteran species *Helicoverpa punctigera*.

In, perhaps, an indication of just how intriguing the structure was, a paper that queried the cystine knot structure of kalata B1 was published in 2002 (Skjeldal, Gran, Sletten, & Volkman, 2002). The new proposal for

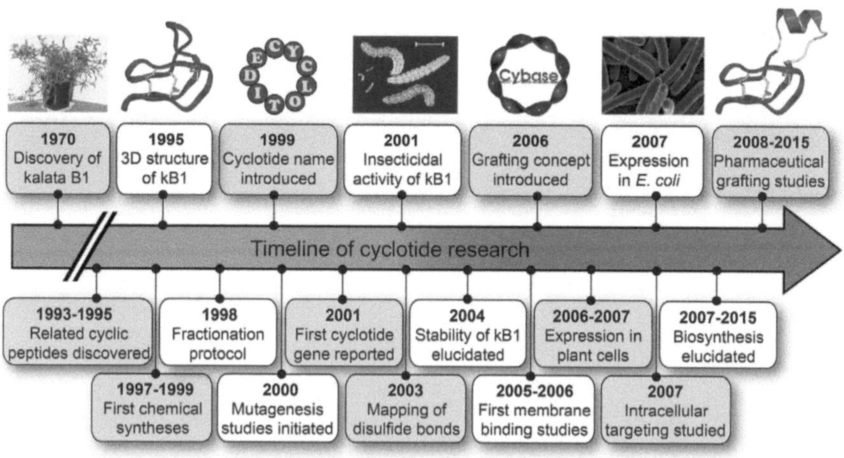

Figure 2 ***Timeline illustrating the history of cyclotide research.*** The field may be traced to the discovery of kalata B1 in 1970 based on its use in an African folk medicine. Several other macrocyclic peptides were discovered in the late 1990s, and with the elucidation of the structure of kalata B1 in 1995 and fractionation protocols for the isolation of peptides from plants, the name cyclotides was introduced in 1999. In 2011, it became clear that cyclotides are ribosomally synthesized (i.e., directly gene-encoded). The exceptional stability of kalata B1 was elucidated in 2004 and the concept that because of this stability it could be used as a framework in drug design emerged shortly thereafter (2006). Mechanistic studies of the natural function of cyclotides in 2005 and 2006 demonstrated membrane binding by using surface plasmon resonance and NMR, and approaches for the biosynthesis of cyclotides was developed in 2007. Over the last decade much of the activity in cyclotide research has focused on delineating biosynthetic pathways and applying cyclotides as templates in drug design particularly for the grafting of epitope sequences onto them to stabilize them.

the disulfide connectivity turned out to be incorrect, but it nevertheless sparked efforts to absolutely confirm the cystine knot structure, which was done by chemical disulfide mapping in 2003 (Göransson & Craik, 2003) and then reconfirmed using NMR in 2004 (Rosengren, Daly, Plan, Waine, & Craik, 2003). Interestingly, most structures of cyclotides have been determined using NMR as small disulfide-rich peptides are notoriously difficult to crystallize (see Chapter 6). Very recently, though, racemic crystallography has been demonstrated to be a valuable tool to facilitate the crystallization of cyclotides and other cyclic peptides (Wang, King, Northfield, Ojeda, & Craik, 2014). A vast array of NMR, crystallography and chemical evidence now supports the cystine knot structure of kalata B1 and other cyclotides.

This cyclic cystine knot structure is integral to the stability of cyclotides and a key paper in which the thermal, chemical, and enzymatic stability of

Table 1 Milestones in Cyclotide Research

Year(s)	Milestone	References
1970	Discovery of kalata B1 as a peptidic uterotonic agent	Gran (1970)
1993–1995	Discovery of related macrocyclic peptides in Rubiaceae and Violaceae	Schöpke et al. (1993), Gustafson et al. (1994), and Witherup et al. (1994)
1994	Anti-HIV activity of circulins reported	Gustafson et al. (1994)
1995	3D structure of kalata B1 reported	Saether et al. (1995)
1997–1999	Chemical syntheses of cyclotides developed	Daly, Koltay, Gustafson, Boyd, Casas-Finet, & Craik (1999), Tam and Lu (1997, 1998), and Tam, Lu, and Yu (1999)
1998	Fractionation protocol developed for peptides from plant biomass	Claeson et al. (1998)
1999	The name "cyclotide" is introduced	Craik et al. (1999)
1999	The antimicrobial activity of cyclotides is reported	Tam et al. (1999)
2000–2001	MCoTI trypsin inhibitors (Cucurbitaceae) discovered and structurally characterized	Hernandez et al. (2000), Felizmenio-Quimio, Daly, and Craik (2001), and Heitz et al. (2001)
2001	The first cyclotide gene is reported	Jennings et al. (2001)
2001	The insecticidal activity of kalata B1 is reported	Jennings et al. (2001)
2003	The disulfide bonding pattern of kalata B1 is chemically confirmed	Göransson and Craik (2003)
2004	Factors underlying the exceptional stability of kalata B1 are elucidated	Colgrave and Craik (2004)
2005	First membrane binding study	Kamimori et al. (2005)
2005–2006	Prediction of cyclotide-like sequences in the Poaceae	Basse (2005) and Mulvenna, Mylne, et al. (2006)

2006	Grafting concept introduced	Clark et al. (2006)
2006–2009	Chemo-enzymatic cyclization of MCoTI cyclotides	Thongyoo et al. (2006, 2007, 2008, 2009)
2006–2010	CyBase created and updated	Mulvenna, Wang, et al. (2006), Wang et al. (2008), and Kaas and Craik (2010)
2006–2007	Expression of cyclotides in plant cells	Seydel and Dörnenburg (2006) and Dörnenburg (2008)
2007	Expression of cyclotides in *Escherichia coli*	Camarero, Kimura, Woo, Shekhtman, et al. (2007)
2007	Cell-penetrating ability of MCoTI-II reported	Greenwood et al. (2007)
2008	Mechanism of insecticidal activity established	Barbeta et al. (2008)
2011	Discovery of cyclotides in the Fabaceae	Poth, Colgrave, Lyons, Daly, and Craik (2011) and Nguyen et al. (2011)
2012	Discovery of cyclotides in the Solanaceae	Poth et al. (2012)
2012	MALDI imaging of cyclotides *ex planta*	Poth et al. (2012)
2013	Discovery of linear cyclotides in the Poaceae	Nguyen et al. (2013)
2013	Intracellular targeting of a protein–protein interaction	Ji et al. (2013)
2014	Isolation of butelase-1-cyclizing enzyme	Nguyen et al. (2014)

kalata B1 and related cystine knot analogues was examined in detail was reported in 2004 (Colgrave & Craik, 2004). This study provided the quantitative data to confirm what had been noted by Gran in 1973, i.e., that kalata B1 was exceptionally stable (Sletten & Gran, 1973). This exceptional stability presents some challenges in mass spectrometry sequencing studies, as described in more detail in Chapter 5.

The beginnings of our mechanistic understanding of how cyclotides exert their pesticidal activity came in 2005 with a report using surface plasmon resonance to measure the membrane binding capability of kalata B1 (Kamimori, Hall, Craik, & Aguilar, 2005). It appears that cyclotides act by disrupting the midgut membranes of insect pests (Barbeta, Marshall, Gillon, Craik, & Anderson, 2008), based on the ability to target particular phsopholipids (Henriques, Huang, Chaousis, Wang, & Craik, 2014). Numerous studies examining the specificity of binding of kalata B1 and other cyclotides to membranes is further described in Chapter 7 and in several recent articles (Burman, Stromstedt, Malmsten, & Göransson, 2011; Henriques & Craik, 2012; Henriques et al., 2012).

It was recognized early on in cyclotide research that the main benefit of kalata B1 and related cyclotides was probably not from their uterotonic activity, but from the fact that cyclotides were so stable and could be used as scaffolds in drug design. The first paper providing proof of concept for cyclotide grafting was reported in 2006 in which we modified certain residues in loop 5 of kalata B1 and showed that the framework was tolerant to substitution (Clark, Daly, & Craik, 2006). This paved the way for a series of grafting studies which are described in detail in Chapters 9 and 10, and in recent reviews (Jagadish & Camarero, 2010; Poth, Chan, & Craik, 2013).

Until 2006, cyclotides were either isolated directly from plant sources or synthesized chemically using solid phase peptide synthesis. A significant breakthrough was made with methods developed for the biological expression of cyclotides, using intein-based systems (Garcia et al., 2011), and in cellular systems, including expression in plant cells (Dörnenburg, 2008, 2009) and *Escherichia coli*, the latter using modified inteins (Garcia et al., 2011). Such studies paved the way for the generation of cyclotide libraries (Camarero, Kimura, Woo, Cantor, & Steenblock, 2007). A chemo-enzymatic approach for the cyclization of a linear cyclotide precursor was developed around the same time and applied to make modified cyclotides for pharmaceutical applications (Thongyoo, Bonomelli, Leatherbarrow, & Tate, 2009; Thongyoo, Jaulent, Tate, & Leatherbarrow, 2007; Thongyoo, Roque-Rosell, Leatherbarrow, & Tate, 2008; Thongyoo, Tate, & Leatherbarrow, 2006).

The use of cyclotides as drug scaffolds was initially focused on extracellular targets such as enzymes or cell surface receptors, but a new possibility arose in 2007 with the demonstration that the trypsin inhibitor cyclotide MCoTI-II could penetrate cells (Greenwood, Daly, Brown, Stow, & Craik, 2007). Subsequent studies by ourselves (Cascales et al., 2011) and the group of Camarero (Contreras, Elnagar, Hamm-Alvarez, & Camarero, 2011) confirmed this observation and culminated in the development of a grafted cyclotide to block a protein–protein interaction implicated in cancer (Ji et al., 2013). Over recent years there has been continuing activity in the field of grafting of cyclotides and now there are more than a dozen examples of grafted cyclotides for a wide range of therapeutic targets. These are described in more detail in Chapter 7.

The idea that cyclotides might be cyclized by asparaginyl endopeptidase enzymes was first proposed in 2001 in our paper reporting the gene structure of kalata B1 (Jennings et al., 2001). However, firm evidence for this only began to emerge in 2007 (Gillon et al., 2008; Saska et al., 2007). Further biosynthetic studies in the period 2007–2015 have led to a much increased understanding of cyclotide biosynthesis, as described herein in Chapters 8 and 9. A recent milestone was the identification of butelase-1, an asparaginyl endoproteinase, as a key enzyme implicated in the biosynthesis of cyclotides from butterfly pea (*Clitoria ternatea*) (Nguyen et al., 2014).

Finally, although cyclotide research is a relatively narrowly defined field, focused on plant peptides of around 30 amino acids, the discovery that cyclotides are a large family of proteins and that there are to be other classes of ribosomally synthesized cyclic peptides in all kingdoms of life may change our perception of protein structure more generally (i.e., not all proteins need to have ends) and lead to ways of stabilizing completely unrelated proteins. In an example of this application, we reengineered the natural conotoxin peptide Vc1.1 present in the venom of marine cone snails but known to have analgesic properties (Clark et al., 2010). We cyclized this peptide in an attempt to stabilize it and found in the process that the cyclic peptide was orally active in contrast to the linear peptide which was not orally active. It has been estimated that up to a quarter of all proteins have their termini within a distance of approximately 10 Å and hence potentially many more proteins are amenable to cyclization via linkers of a few amino acids. In a sense, plants can teach us great lessons in pharmaceutical design. Plant peptides are highly potent and stable and the technology for producing them biosynthetically is now available.

In conclusion, the cyclotide field is now more than 20 years old. No grafted cyclotide has yet reached human clinical trials but the typical pathway for drug development takes many years, so there remains hope that a modified cyclotide may reach clinical trials in the near future. Hopefully this book will provide newcomers to the field an overview of developments to this point and stimulate further work in this area.

ACKNOWLEDGMENTS

Work on cyclotides in our laboratory has been supported by the Australian Research Council (ARC) (DP150100443, DP0984390, LP130100550) and the National Health and Medical Research Council (NHMRC) (APP1047857, APP1009267, APP1028509). DJC is a National Health and Medical Research Council (NHMRC) Professorial Fellow (APP1026501).

REFERENCES

Barbeta, B. L., Marshall, A. T., Gillon, A. D., Craik, D. J., & Anderson, M. A. (2008). Plant cyclotides disrupt epithelial cells in the midgut of lepidopteran larvae. *Proceedings of the National Academy of Sciences of the United States of America, 105*, 1221–1225.

Basse, C. W. (2005). Dissecting defense-related and developmental transcriptional responses of maize during *Ustilago maydis* infection and subsequent tumor formation. *Plant Physiology, 138*, 1774–1784.

Burman, R., Stromstedt, A. A., Malmsten, M., & Göransson, U. (2011). Cyclotide-membrane interactions: defining factors of membrane binding, depletion and disruption. *Biochimica et Biophysica Acta-Biomembranes, 1808*, 2665–2673.

Camarero, J. A., Kimura, R. H., Woo, Y. H., Cantor, J., & Steenblock, E. (2007). A cell-based approach for the biosynthesis/screening of cyclic peptide libraries against bacterial toxins. *Chimica Oggi, 25*, 20–23.

Camarero, J. A., Kimura, R. H., Woo, Y.-H., Shekhtman, A., & Cantor, J. (2007). Biosynthesis of a fully functional cyclotide inside living bacterial cells. *ChemBioChem, 8*, 1363–1366.

Cascales, L., Henriques, S. T., Kerr, M. C., Huang, Y. H., Sweet, M. J., Daly, N. L., et al. (2011). Identification and characterization of a new family of cell-penetrating peptides. Cyclic cell-penetrating peptides. *The Journal of Biological Chemistry, 286*, 36932–36943.

Claeson, P., Göransson, U., Johansson, S., Luijendijk, T., & Bohlin, L. (1998). Fractionation protocol for the isolation of polypeptides from plant biomass. *Journal of Natural Products, 61*, 77–81.

Clark, R. J., Daly, N. L., & Craik, D. J. (2006). Structural plasticity of the cyclic-cystine-knot framework: implications for biological activity and drug design. *Biochemistry Journal, 394*, 85–93.

Clark, R. J., Jensen, J., Nevin, S. T., Callaghan, B. P., Adams, D. J., & Craik, D. J. (2010). The engineering of an orally active conotoxin for the treatment of neuropathic pain. *Angewandte Chemie International Edition, 49*, 6545–6548.

Colgrave, M. L., & Craik, D. J. (2004). Thermal, chemical, and enzymatic stability of the cyclotide kalata B1: the importance of the cyclic cystine knot. *Biochemistry, 43*, 5965–5975.

Contreras, J., Elnagar, A. Y. O., Hamm-Alvarez, S. F., & Camarero, J. A. (2011). Cellular uptake of cyclotide MCoTI-I follows multiple endocytic pathways. *Journal of Controlled Release, 155*, 134–143.

Craik, D. J., Daly, N. L., Bond, T., & Waine, C. (1999). Plant cyclotides: a unique family of cyclic and knotted proteins that defines the cyclic cystine knot structural motif. *Journal of Molecular Biology, 294*, 1327–1336.

Daly, N. L., Koltay, A., Gustafson, K. R., Boyd, M. R., Casas-Finet, J. R., & Craik, D. J. (1999). Solution structure by NMR of circulin A: a macrocyclic knotted peptide having anti-HIV activity. *Journal of Molecular Biology, 285*, 333–345.

Daly, N. L., Love, S., Alewood, P. F., & Craik, D. J. (1999). Chemical synthesis and folding pathways of large cyclic polypeptides: studies of the cystine knot polypeptide kalata B1. *Biochemistry, 38*, 10606–10614.

Dörnenburg, H. (2008). Plant cell culture technology—harnessing a biological approach for competitive cyclotides production. *Biotechnology Letters, 30*, 1311–1321.

Dörnenburg, H. (2009). Progress in kalata peptide production via plant cell bioprocessing. *Biotechnology Journal, 4*, 632–645.

Felizmenio-Quimio, M. E., Daly, N. L., & Craik, D. J. (2001). Circular proteins in plants: solution structure of a novel macrocyclic trypsin inhibitor from *Momordica cochinchinensis*. *Journal of Biological Chemistry, 276*, 22875–22882.

Garcia, A. E., Tai, K. P., Puttamadappa, S. S., Shekhtman, A., Ouellette, A. J., & Camarero, J. A. (2011). Biosynthesis and antimicrobial evaluation of backbone-cyclized α-defensins. *Biochemistry, 50*, 10508–10519.

Gillon, A. D., Saska, I., Jennings, C. V., Guarino, R. F., Craik, D. J., & Anderson, M. A. (2008). Biosynthesis of circular proteins in plants. *Plant Journal, 53*, 505–515.

Göransson, U., & Craik, D. J. (2003). Disulfide mapping of the cyclotide kalata B1. Chemical proof of the cystic cystine knot motif. *Journal of Biological Chemistry, 278*, 48188–48196.

Göransson, U., Luijendijk, T., Johansson, S., Bohlin, L., & Claeson, P. (1999). Seven novel macrocyclic polypeptides from *Viola arvensis*. *Journal of Natural Products, 62*, 283–286.

Gran, L. (1970). An oxytocic principle found in *Oldenlandia affinis* DC. *Meddelelser fra Norsk Farmaceutisk Selskap, 12*, 173–180.

Gran, L., Sandberg, F., & Sletten, K. (2000). *Oldenlandia affinis* (R&S) DC. A plant containing uteroactive peptides used in African traditional medicine. *Journal of Ethnopharmacology, 70*, 197–203.

Greenwood, K. P., Daly, N. L., Brown, D. L., Stow, J. L., & Craik, D. J. (2007). The cyclic cystine knot miniprotein MCoTI-II is internalized into cells by macropinocytosis. *International Journal of Biochemistry and Cell Biology, 39*, 2252–2264.

Gruber, C. W., Elliott, A. G., Ireland, D. C., Delprete, P. G., Dessein, S., Goransson, U., et al. (2008). Distribution and evolution of circular miniproteins in flowering plants. *Plant Cell, 20*, 2471–2483.

Gustafson, K. R., Sowder, R. C. I., Henderson, L. E., Parsons, I. C., Kashman, Y., Cardellina, J. H. I., et al. (1994). Circulins A and B: novel HIV-inhibitory macrocyclic peptides from the tropical tree *Chassalia parvifolia*. *Journal of the American Chemical Society, 116*, 9337–9338.

Heitz, A., Hernandez, J. F., Gagnon, J., Hong, T. T., Pham, T. T., Nguyen, T. M., et al. (2001). Solution structure of the squash trypsin inhibitor MCoTI-II. A new family for cyclic knottins. *Biochemistry, 40*, 7973–7983.

Henriques, S. T., & Craik, D. J. (2012). Importance of the cell membrane on the mechanism of action of cyclotides. *ACS Chemical Biology, 7*, 626–636.

Henriques, S. T., Huang, Y. H., Castanho, M. A. R. B., Bagatolli, L. A., Sonza, S., Tachedjian, G., et al. (2012). Phosphatidylethanolamine binding is a conserved feature of cyclotide-membrane interactions. *The Journal of Biological Chemistry, 287*, 33629–33643.

Henriques, S. T., Huang, Y.-H., Chaousis, S., Wang, C. K., & Craik, D. J. (2014). Anticancer and toxic properties of cyclotides are dependent on phosphatidylethanolamine phospholipid targeting. *ChemBioChem, 15*, 1956–1965.

Hernandez, J. F., Gagnon, J., Chiche, L., Nguyen, T. M., Andrieu, J. P., Heitz, A., et al. (2000). Squash trypsin inhibitors from *Momordica cochinchinensis* exhibit an atypical macrocyclic structure. *Biochemistry, 39*, 5722–5730.

Jagadish, K., & Camarero, J. A. (2010). Cyclotides, a promising molecular scaffold for peptide-based therapeutics. *Biopolymers, 94*, 611–616.

Jennings, C., West, J., Waine, C., Craik, D., & Anderson, M. (2001). Biosynthesis and insecticidal properties of plant cyclotides: the cyclic knotted proteins from *Oldenlandia affinis*. *Proceedings of the National Academy of Sciences of the United States of America, 98*, 10614–10619.

Ji, Y., Majumder, S., Millard, M., Borra, R., Bi, T., Elnagar, A. Y., et al. (2013). In vivo activation of the p53 tumor suppressor pathway by an engineered cyclotide. *Journal of the American Chemical Society, 135*, 11623–11633.

Kaas, Q., & Craik, D. J. (2010). Analysis and classification of circular proteins in CyBase. *Biopolymers: Peptide Science, 94*, 584–591.

Kamimori, H., Hall, K., Craik, D. J., & Aguilar, M. I. (2005). Studies on the membrane interactions of the cyclotides kalata B1 and kalata B6 on model membrane systems by surface plasmon resonance. *Analytical Biochemistry, 337*, 149–153.

McDonald, N. Q., & Hendrickson, W. A. (1993). A structural superfamily of growth factors containing a cystine knot motif. *Cell, 73*, 421–424.

Mulvenna, J. P., Mylne, J. S., Bharathi, R., Burton, R. A., Shirley, N. J., Fincher, G. B., et al. (2006). Discovery of cyclotide-like protein sequences in graminaceous crop plants: ancestral precursors of circular proteins? *Plant Cell, 18*, 2134–2144.

Mulvenna, J. P., Wang, C., & Craik, D. J. (2006). CyBase: a database of cyclic protein sequence and structure. *Nucleic Acids Research, 34*, D192–D194.

Nguyen, G. K., Lian, Y. L., Pang, E. W. H., Phuong, Q. T. N., Tran, T. D., & Tam, J. P. (2013). Discovery of linear cyclotides in monocot plant *Panicum laxum* of Poaceae family provides new insights into evolution and distribution of cyclotides in plants. *Journal of Biological Chemistry, 288*, 3370–3380.

Nguyen, G. K. T., Wang, S., Qiu, Y., Hemu, X., Lian, Y., & Tam, J. P. (2014). Butelase 1 is an Asx-specific ligase enabling peptide macrocyclization and synthesis. *Nature Chemical Biology, 10*, 732–738.

Nguyen, G. K. T., Zhang, S., Ngan, T. K. N., Phuong, Q. T. N., Chiu, M. S., Hardjojo, A., et al. (2011). Discovery and characterization of novel cyclotides originated from chimeric precursors consisting of albumin-1 chain a and cyclotide domains in the Fabaceae family. *Journal of Biological Chemistry, 286*, 24275–24287.

Pallaghy, P. K., Nielsen, K. J., Craik, D. J., & Norton, R. S. (1994). A common structural motif incorporating a cystine knot and a triple-stranded b-sheet in toxic and inhibitory polypeptides. *Protein Science, 3*, 1833–1839.

Poth, A. G., Chan, L. Y., & Craik, D. J. (2013). Cyclotides as grafting frameworks for protein engineering and drug design applications. *Biopolymers: Peptide Science, 100*, 480–491.

Poth, A. G., Colgrave, M. L., Lyons, R. E., Daly, N. L., & Craik, D. J. (2011). Discovery of an unusual biosynthetic origin for circular proteins in legumes. *Proceedings of the National Academy of Sciences of the United States of America, 108*, 10127–10132.

Poth, A. G., Mylne, J. S., Grassl, J., Lyons, R. E., Millar, A. H., Colgrave, M. L., et al. (2012). Cyclotides associate with leaf vasculature and are the products of a novel precursor in Petunia (Solanaceae). *Journal of Biological Chemistry, 287*, 27033–27046.

Rosengren, K. J., Daly, N. L., Plan, M. R., Waine, C., & Craik, D. J. (2003). Twists, knots, and rings in proteins. Structural definition of the cyclotide framework. *The Journal of Biological Chemistry, 278*, 8606–8616.

Saether, O., Craik, D. J., Campbell, I. D., Sletten, K., Juul, J., & Norman, D. G. (1995). Elucidation of the primary and three-dimensional structure of the uterotonic polypeptide kalata B1. *Biochemistry, 34*, 4147–4158.

Sandberg, F. (1965). Etude sur les plantes medicinales et toxiques d-Afrique equatoriale. *Cahiers de la Moboke*, 5–49.

Saska, I., Gillon, A. D., Hatsugai, N., Dietzgen, R. G., Hara-Nishimura, I., Anderson, M. A., et al. (2007). An asparaginyl endopeptidase mediates in vivo protein backbone cyclisation. *Journal of Biological Chemistry, 282*, 29721–29728.

Schöpke, T., Hasan Agha, M. I., Kraft, R., Otto, A., & Hiller, K. (1993). Hämolytisch aktive komponenten aus *Viola tricolor* L. und *Viola arvensis* Murray. *Scientia Pharmaceutica, 61*, 145–153.

Seydel, P., & Dörnenburg, H. (2006). Establishment of in vitro plants, cell and tissue cultures from *Oldenlandia affinis* for the production of cyclic peptides. *Plant Cell Tissue Organ Culture, 85*, 247–255.

Skjeldal, L., Gran, L., Sletten, K., & Volkman, B. F. (2002). Refined structure and metal binding site of the kalata B1 peptide. *Archives of Biochemistry and Biophysics, 399*, 142–148.

Sletten, K., & Gran, L. (1973). Some molecular properties of kalatapeptide B-1. A uterotonic polypeptide isolated from *Oldenlandia affinis* DC. *Meddelelser fra Norsk farmaceutisk selskap, 7–8*, 69–82.

Tam, J. P., & Lu, Y.-A. (1997). Synthesis of large cyclic cystine-knot peptide by orthogonal coupling strategy using unprotected peptide precursors. *Tetrahedron Letters, 38*, 5599–5602.

Tam, J. P., & Lu, Y.-A. (1998). A biomimetic strategy in the synthesis and fragmentation of cyclic protein. *Protein Science,* 7, 1583–1592.

Tam, J. P., Lu, Y.-A., Yang, J. L., & Chiu, K. W. (1999). An unusual structural motif of antimicrobial peptides containing end-to-end macrocycle and cystine-knot disulfides. *Proceedings of the National Academy of Sciences of the United States of America, 96*, 8913–8918.

Tam, J. P., Lu, Y.-A., & Yu, Q. (1999). Thia zip reaction for synthesis of large cyclic peptides: Mechanisms and applications. *Journal of the American Chemical Society, 121*, 4316–4324.

Thongyoo, P., Bonomelli, C., Leatherbarrow, R. J., & Tate, E. W. (2009). Potent inhibitors of β-tryptase and human leukocyte elastase based on the MCoTI-II scaffold. *Journal of Medicinal Chemistry, 52*, 6197–6200.

Thongyoo, P., Jaulent, A. M., Tate, E. W., & Leatherbarrow, R. J. (2007). Immobilized protease-assisted synthesis of engineered cysteine-knot microproteins. *ChemBioChem, 8*, 1107–1109.

Thongyoo, P., Roque-Rosell, N., Leatherbarrow, R. J., & Tate, E. W. (2008). Chemical and biomimetic total syntheses of natural and engineered MCoTI cyclotides. *Organic and Biomolecular Chemistry, 6*, 1462–1470.

Thongyoo, P., Tate, E. W., & Leatherbarrow, R. J. (2006). Total synthesis of the macrocyclic cysteine knot microprotein MCoTI-II. *Chemical Communications (Cambridge, England)*, 2848–2850.

Wang, C. K., Kaas, Q., Chiche, L., & Craik, D. J. (2008). CyBase: a database of cyclic protein sequences and structures, with applications in protein discovery and engineering. *Nucleic Acids Research, 36*, D206–D210.

Wang, C. K., King, G. J., Northfield, S. E., Ojeda, P. G., & Craik, D. J. (2014). Racemic and quasi-racemic x-ray structures of cyclic disulfide-rich peptide drug scaffolds. *Angewandte Chemie International Edition, 53*, 11236–11241.

Witherup, K. M., Bogusky, M. J., Anderson, P. S., Ramjit, H., Ransom, R. W., Wood, T., et al. (1994). Cyclopsychotride A, A biologically active, 31-residue cyclic peptide isolated from *Psychotria longipes*. *Journal of Natural Products, 57*, 1619–1625.

CHAPTER TWO

Cyclotides in the Violaceae

Ulf Göransson[*,1], **Sohaib Malik**[*], **Blazej Slazak**[*,§]
[*]Division of Pharmacognosy, Department of Medicinal Chemistry, Uppsala University, Biomedical Center, Uppsala, Sweden
[§]Władysław Szafer Institute of Botany, Polish Academy of Sciences, Cracow, Poland
[1]Corresponding author: E-mail: ulf.goransson@fkog.uu.se

Contents

Abstract

It has been 20 years since the cyclic cystine-knotted compounds, called cyclotides, tied our knot with the violets. This chapter, above all, is a lucid summary of the twists, turns, knots, and rings in the relationship that followed.

The chapter begins by giving an account of how a little known scientific article facilitated the discovery of cyclotides in *Violaceae*, how it captured our imagination, and how it was realized that the discovery of a series of cyclotides was within a touching distance. The processes of extraction, isolation, and characterization as well as chemical synthesis of cyclotides used in our laboratory have been described, and in doing so, the solvents, materials, conditions, instrumentation, and techniques involved have been pointed out. An introduction to *Violaceae* and a description of these plants in diverse cultural settings for various therapeutic purposes is provided. A section follows it on the discovery of cycloviolacin O2 from *Viola odorata* as a potent antimicrobial peptide, and how its interaction with bacteria in terms of mechanism of action and resistance development was investigated. It is followed by a reflection on how the recent innovations in biotechnology and bioinformatics have helped out along the way: the use of novel strategies for cyclization and the use of transcriptomics data in the discovery of new cyclotides. Finally, the question about biological role of cyclotides in *Violaceae* has been asked, and an attempt to answer this question has been presented.

Advances in Botanical Research, Volume 76
ISSN 0065-2296
http://dx.doi.org/10.1016/bs.abr.2015.09.001

1. THE DISCOVERY OF CYCLOTIDES IN VIOLACEAE

We set out to explore a wide set of plant species in the summer of 1995 in pursuit of new natural product chemistry in the form of peptides. At that time, only a few plant polypeptides, of any category, had been described. The thesis describing one of them, kalata B1, was mentioned as a blueprint of research in pharmacognosy, that is, the science of the natural products that can be used as drugs or for the production of drugs. Discovery of that prototypic cyclotide, is a schoolbook example of pharmacognostic research: the thesis of Lorents Gran contains a fair share of ethnopharmacology, botany, chemistry, medicine, and biology; spiced up with the exciting story of personal experiences and observations of Gran working as a Red Cross physician during the civil war in Luluabourg, Democratic Republic of the Congo (then Zaire) (Gran, 1973c; Gran, Sandberg, & Sletten, 2000). Because this work was carried out in the early days of modern biotechnology, neither the circular structure nor the complete sequence of this prototypic cyclotide was shown at the time.

The first isolation of kalata B1 was made from a Rubiaceae plant (*Oldenlandia affinis*), not from Violaceae (Gran, 1973a, 1973b; Sletten & Gran, 1973). (However today, kalata B1 has been found in several species of Violaceae too.) The big leap in cyclotide research coincides with their discovery in Violaceae around 25 years later. In 1993, Schöpke, and coworkers stumbled across violapeptide I during their search for saponins in *Viola arvensis* and *Viola tricolor* (Schöpke, Hasan, Kraft, Otto, & Hiller, 1993) and reported the sequence of a 29-amino acid long peptide from *V. arvensis*. They observed that the termini were blocked and the structure was cyclic, but they most likely made a mistake in their sequence analyses because of unspecific cleavage of the enzyme Arg-C and because they missed a tryptophan residue in the quantitative amino acid analysis (tryptophan is degraded in the hydrolysis to release amino acid residues from the peptide, and its presence is usually determined by ultraviolet spectroscopy). No mass spectrum was reported either. However, Edman sequencing helped to reveal 28 out of 29 residues, and 27 of those match with the cyclotide of highest abundance in *V. arvensis*, which we reported 5 years later (Claeson, Göransson, Johansson, Luijendijk, & Bohlin, 1998). Most likely, they had isolated varv peptide A [or kalata S, as it was named later when the same sequence was rediscovered in *O. affinis* (Craik, Daly, Bond, & Waine, 1999)].

At first, the discovery of violapeptide I, the first cyclotide from Violaceae, escaped unnoticed. The work of Schöpke and coworkers was not cited in the two independent discoveries of three other cyclic peptides that were reported in 1994, all from Rubiaceae: circulins A and B (Gustafson et al., 1994) and cyclopsychotride A (Witherup et al., 1994). It was not noticed in the first articles disclosing the full sequence and structure of kalata B1 either (Pallaghy, Nielsen, Craik, & Norton, 1994; Saether et al., 1995). Perhaps the reason for this is that Schöpke and coworkers published in the German language and in a relatively small journal of the Austrian Pharmaceutical Society, *Scientia Pharmaceutica*. In a way, this proved to be fortunate for our group in Uppsala: in our hunt for polypeptides, we "discovered" violapeptide I by the data mining of the time—manually going through the volumes of the journal, Natural Products Updates. Having the sequence of kalata B1 in hand, and being aware of the cyclic structure, it was a moment of epiphany to turn the wheel of sequences around and see the almost complete match to violapeptide I. (These two cyclotides differ in only one residue.)

As such, during the summer of 1995, we knew about existence of the peptides which would be known as cyclotides later on. Although this knowledge triggered us to quickly collect *V. arvensis*, our first aim was to use these peptides to validate a protocol that we had developed for the fractionation and detection of a peptide-enriched fraction from plants (Claeson et al., 1998). However, this screen did not produce any hits as clear as the ones we saw in the violets, and soon we realized the untapped source of chemistry that we were sitting on. We discovered that the varv peptide A was not expressed alone, rather it was one of a library of compounds (Göransson, Luijendijk, Johansson, Bohlin, & Claeson, 1999) and that they could be discovered in other Violaceae species too, e.g., *Viola odorata*. The presence of cyclotides in that particular species was however to be publicly disclosed by Craik et al. (1999), who had come on the same track independent of us. These two species and sequences and structure of cyclotides are shown in Figure 1.

2. UNLOCKING THE CYCLOTIDE LIBRARY

As it is clear from the preceeding section, cyclotides, and their sequences and structures, were not easily accessible. First of all, their peptidic nature and size combined with their origin in plants probably explains some of these difficulties: to a large extent, the pharmacognosists or natural product chemists still appear to avoid working with aqueous solutions and

Figure 1 ***Origin of cyclotides in Violaceae.*** The species that started the cyclotide story in Violaceae, *Viola arvensis* (A), through the serendipitous discovery of violapeptide I. The varv peptides A-H confined the first "library" of cyclotides (Claeson et al., 1998; Göransson et al., 1999). Varv peptide A is most likely identical to the violapeptide I, which sequence was not complete when reported. (B) *Viola odorata*, the main source of cycloviolacin O2 (cyO2) and other cyclotides that formed the bracelet subfamily (Craik et al., 1999). The sequence alignment below exemplifies cyclotide sequences: the prototype kalata B1, varv peptide A, and cycloviolacin O2. Brackets indicate the three disulfide bonds and the cyclic backbone. (See color plate)

biomolecules, and the research area of plant peptides in biochemistry can still be considered to be in its infancy. For extraction two main approaches have been used. The first approach is similar to the protocols commonly used in the laboratory of a natural product chemist, using a 1 to 1 (v/v) mixture of dichloromethane and methanol. The second approach, based on the protocol developed in our laboratory, uses hydroalcoholic mixtures. Each protocol normally contains a step of defattening: if the extraction is done using dichloromethane/methanol defattening is done after the main extraction by adding water to the extract. Cyclotides will then partition to the upper phase, containing water and methanol, and give the phase a yellow to brown color, whereas lipophilic compounds move to the lower phase, constituted by dichloromethane. The lower phase is normally dark greenish from the photosynthetic pigments present (e.g., chlorophylls). If water/alcohol is used for extraction, defattening is done before the main extraction, normally by using dichloromethane. The plant material is then dried before the main extraction. Aqueous mixtures containing either ethanol or methanol have been used, targeted extraction may be done by changing the ratio of water to alcohol. 50% ethanol, as used in our original protocol, extracts a load of compounds making the extract viscous for some Violaceae species (e.g., *V. odorata*), which can make the following separation of the extract from the plant material debris (usually by filtration) less than straightforward. Using methanol instead avoids this problem. In our laboratory, using 60% methanol in water has become standard. However, neither the possibility to target the extraction toward more hydrophilic or more hydrophobic cyclotides (Yeshak, Burman, Eriksson, & Göransson, 2012), nor the possibility to choose other solvents should be neglected (Mahatmanto, Poth, Mylne, & Craik, 2014).

For example, in small-scale analysis of, e.g., herbaria samples, the extraction is often done using acidic mixtures of acetonitrile and water, i.e., combinations of the same solvents that are later used for analyses using high-performance liquid chromatography and mass spectrometry. This kind of extraction is suitable to adapt toward cyclotides of varying hydrophobicity, as quite often, the aim in such an extraction is to get a full profile of the cyclotide content. In a standard procedure we perform the first small-scale extraction using 60% (v/v) acetonitrile, 0.1% formic acid, in water, followed by a second and third round of extraction while decreasing the concentration of acetonitrile to 30% and 10%, respectively. This protocol has been used at a volume of 4 ml for amounts of (dry) plant material of 10–250 mg. In large-scale extractions, the ratio of plant material to volume is normally kept to ~1:10 (w/v), and the extractions are repeated three times.

Most extractions are done using dried and powdered plant material. If fresh plant material is used, the ratio and concentration of solvents should be adjusted accordingly to compensate for the amount of water. The herbal violets, i.e., *Viola* spp., are relatively dense with a dry weight of 20–25% of the fresh material. Because of the extreme stability of the cyclotide structure, collected samples are stable over long periods of time if dried and stored properly. This has been exploited by sampling plants for chemical screens of cyclotide content from herbaria samples from both outside (Gruber et al., 2008) and inside Violaceae (Simonsen et al., 2005). The most extreme example of cyclotide stability over time to our knowledge is the analysis of a herbarium sample of *V. odorata* from 1820: the liquid chromatography—mass spectrometry analysis was identical to a fresh collect (Burman et al., 2015). In fact, nearly identical profiles were obtained from the samples from 1849, 1886, and 1948.

Before liquid chromatographic and mass spectrometric analyses, the crude extracts are usually subjected to some kind of fractionation or sample preparation. Most commonly used method is solid phase extraction, using reverse phase material. This means that the concentration of organic solvent in the crude extract must be decreased before processing the sample; this can be achieved by concentrating/drying the sample (using a rotary evaporator, speedvac, freeze dryer, or by drying under a stream of nitrogen). Diluting the sample may also decrease the concentration of organic solvent. However, it is crucial to keep the concentration under control because the retention of cyclotides on reverse phase material is extremely sensitive to the percentage of organic modifier in both sample and eluents. That sensitivity can also be exploited: it is, for example, possible to load large volumes (liters) of diluted samples on relatively small columns. Besides reverse phase, fractionations by ion exchange and size exclusion chromatography have also been used, see, e.g., Claeson et al. (1998), Göransson, Sjögren, Svangård, Claeson, and Bohlin (2004). In our laboratory, strong cation exchange resins are regularly used to capture positively charged cyclotides, and, e.g., Sephadex columns are used to desalt (i.e., separate small natural products from peptides) in an analytical scale. In those separations, the content of organic solvent (acetonitrile) is usually kept at 30%.

For final analysis and/or purification, reverse phase is again the prevailing and most powerful method. The scale of this final separation depends on the purpose for screening using state-of-the-art analytical equipment in liquid chromatography—mass spectrometry, extract corresponding to the fractions of a microgram of plant material is consumed. That is, sampling a single leaf or pistil may be enough for analysis. Figure 2 shows an example chromatogram using this technique, in this case of leafs of *Viola uliginosa*.

In comparison to those miniaturized systems, which use columns with inner diameter of 0.075 mm and are operated at flow rates of 250 nl/min, laboratory scale preparative columns have inner diameters of 20 mm and flow rates of 20 ml/min, and the extract from 100 g of plant material is usually loaded. In total, the number of species screened for their cyclotide content is approaching ~200 and represents most genera (see CyBase (www.cybase.org.au) for current updates of species and cyclotides).

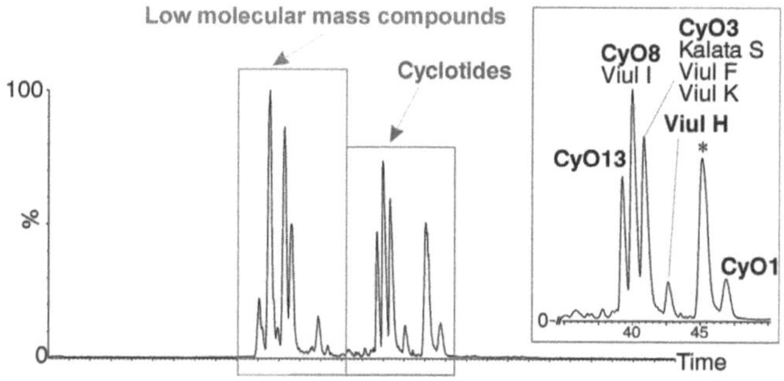

Figure 2 ***Liquid chromatography–mass spectrometry analysis of* Viola uliginosa.** The base peak ion mass chromatograms of *Viola uliginosa* leaf extract. Even after SPE purification plant extracts are complex mixtures of cyclotides and other compounds. Some of the cyclotides are inseparable in HPLC, and mass spectrometry is needed to characterize different cyclotides eluting in one peak (main constituents in **bold**). Low molecular compounds in Violaceae include glycosylated flavones (e.g., the ubiquitous peak with an m/z of 579) and anthocyanidins.

Cyclotide sequences range typically between 28 and 37 amino acid residues, and they all contain a head-to-tail cyclic backbone and the so-called cystine knot, which is formed by three disulfide bonds: CysI-CysIV, CysII-CysV, and CysIII-CysVI. Together, it is these features that render the cyclotides their extreme stability, which makes them resilient to heat, enzymes, and chemical degradation (Colgrave & Craik, 2004). Understandably, these features present a challenge for their structure determination. As mentioned above, the first isolation of cyclotides both in Violaceae and Rubiaceae resulted in incomplete sequences and connectivities of disulfide bonds (Gran, 1973b; Schöpke et al., 1993; Sletten & Gran, 1973). The basic strategy they used, however, is the same as the strategy of today: the three disulfide bonds forming the cystine knot need to be reduced, preferably alkylated, and followed by a subsequent enzymatic digestion. The resulting product is then subjected to enzymatic cleavage, to generate peptide fragments amenable for sequencing. Most often more than one enzyme is needed. Endoproteinase GluC has proven to be very useful because it generates only one linear product for most cyclotides (i.e., they contain only one glutamic acid cleavage point). Trypsin is used next, cleaving C-terminal to positively charged residues (arginine and lysine), and in some cases chymotrypsin, cleaving after hydrophobic residues (e.g., tryptophan, tyrosine, phenylalanine, and at slower rate, leucine). Incomplete sequences from one enzymatic digest can often be found in another, and by combining the data the full sequence is revealed. In most cases, cleavage using GluC and trypsin is enough; the use of chymotrypsin has essentially been limited to disclose the position of the isobaric residues leucine or isoleucine when mass spectrometry is used for sequencing.

Much has changed in the field of analytical biochemistry since the sequencing of the first cyclotides in the Violaceae, and the then prevailing Edman technique for protein sequencing has now largely been exceeded by *de novo* sequencing by mass spectrometry. Both techniques however need support in the form of, e.g., quantitative amino acid analyses or nucleotide sequences generated from cDNA libraries. During the last years, the number of cyclotide sequences has increased rapidly as a result of the quick development of nucleotide sequencing techniques. Discovery rate is increased by the continuous monitoring of new genetic data added to public databases (e.g., GenBank (Benson et al., 2013)), as well as more specialized databases such as the one making the results of the 1KP project available (www.onekp.com). In this project, the transcriptomes (the total RNA expression) of more than 1000 plants have been sequenced, including the transcriptomes of *V. tricolor* (LPGY, 154526 contigs) and *Viola canadensis* (NJLF, 59763 contigs). During 2015, the first articles describing the use of transcriptomes specifically sequenced for cyclotide studies have been published, on *Viola baoshanensis* (Zhang, Li, et al., 2015) and *V. uliginosa* (yielding 95899 contigs from the callus culture and 75716 from the wild type) (Slazak, Jacobsson, Kuta, & Göransson, 2015). The latter study can serve as an example of what can be obtained using this approach.

As described above, the rapid development of next-generation DNA and RNA sequencing techniques provides a tool giving insight into both cyclotides *and* their precursor's sequences. The recent study of the whole transcriptome of *V. uliginosa*, complemented with analysis on peptide level using liquid chromatography—mass spectrometry, yielded 24 precursor peptides (12 complete and 12 partial) and 19 complete cyclotides sequences as shown in Table 1 and Figure 3. Of these, 12 were new and named viul A-L according to the nomenclature proposed by Broussalis et al., (2001). Others had been identified before in other species with the most similarities with *V. odorata* and most of them belong to the bracelet type. Interestingly, the one Möbius-type cyclotide (varv peptide A/kalata S) was identical to the peptide found in a plant belonging to different family (*O. affinis*, Rubiaceae).

The transcriptome sequencing was used for two purposes: firstly, to sequence the cyclotides produced by this species and secondly, to assess the changes in cyclotide gene expression under the prolonged influence of plant growth regulators. Moreover, the availability of whole transcriptomes allows

Table 1 Alignment of *Viola uliginosa* Cyclotides Amino Acid Sequences Isolated and/or Predicted from Transcriptome

Cyclotide	Monoisotopic Mass [Da]	Sequence/Consensus	Peptide Existence
		g e i p C - g E s C v w i p C - i t s a i g C s C k s w k v C y r n - -	
kalata S[b]	2876	[1 G L P V C - G E T C V G G T C - - - N T P G C S C S W - P V C T R N - - 29]	M_1, Tr
viul A[a]	3045	[1 G - I P C - G E S C V W I P C - I S S L I G C S C R G - K V C Y H - - - 29]	Tr
viul B[a]	3051	[1 G - V P C - G E S C V W I P C - L T G A I G C S C S N - K V C Y L N - - 30]	M_1, Tr
mram 8[a]	3113	[1 G - I P C - G E S C V F I P C - L T S A I G C S C K S - K V C Y R N - - 30]	M_1, Tr
CyO1[a]	3114	[1 G - I P C - A E S C V Y I P C T V T A L L G C S C S N - R V C Y N - - - 30]	M_1, Tr
viul C	3118	[1 G - T F C - G E T C V M F P C - F S S A R G C G C H N - L G C E L N - - 30]	M_2, Tr
CyO13[a]	3122	[1 G - I P C - G E S C V W I P C - I S A A I G C S C K S - K V C Y R N - - 30]	M_1, Ms
viul D[a]	3124	[1 G - I P C - G E S C V W I P C - L T S A I G C S C K S - K V C Y K N - - 30]	Tr
CyO2[a]	3138	[1 G - I P C - G E S C V W I P C - I S S A I G C S C K S - K V C Y R N - - 30]	M_1, Tr
cycloviolin D[a]	3144	[1 G - F P C - G E S C V F I P C - I S A A I G C S C K N - K V C Y R N - - 30]	Tr
CyO3[a]	3152	[1 G - I P C - G E S C V W I P C - L T S A I G C S C K S - K V C Y R N - - 30]	M_1, Ms, Tr
viul E	3171	[1 G - G H C - G E S C M L L P C - F T A R I G C S C S R - S I C Y K N - - 30]	M_2, Tr
viul F	3199	[1 G - R F C - G E I C S R G F C - - - S N P R C T C N A S R Q C V R N - - 29]	M_2, Tr
CyO8[a]	3225	[1 G T L P C - G E S C V W I P C - I S S V V G C S C K S - K V C Y K N - - 31]	M_1, Ms, Tr
viul G	3230	[1 G R A V C - G E T C F A G I C - - - Y T P V C V C G K W D L C R M N - - 30]	M_1, Tr
viul H[a]	3238	[1 S E L P C - G E S C V F I P C - I T S I A G C S C S H - K V C Y L N - - 31]	M_2, Tr
viul I[a]	3246	[1 G P T P C - G E T C I W I S C - V T A V M G C S C K N - S I C Y M N - - 31]	M_1, Tr
viul J	3326	[1 G E V T C N G E T C F T G K C - - - N A K G C N C K N W P L C T R N - - 31]	Tr
viul K	3343	[1 S - I F C - S E T C R T F P C - F T K A V G C S C V S - K R C Y K N - - 30]	M_2, Tr
viul L[a]	3407	[1 G - I P C - A E T C L W R P C - R T A I M G C S C E Y - N F C Y K N - - 30]	M_2, Tr

[a]cyclotide sequence fits the pattern for a bracelet structure.

[b]cyclotide sequence fits the pattern for a Möbius structure.

CyO, cycloviolacin; M_1, mass fingerprint with ΔM ($M_{\text{predicted from sequence}} - M_{\text{seen}}$) < 0.25u; M_2, mass fingerprint with ΔM < 2u (low abundance ions); Ms, sequenced by shotgun LC-MSMS;Tr, predicted from transcript; viul A-L, new sequences unique to *Viola uliginosa*.

Table reproduced from Slazak, Jacobsson, et al. (2015) with permission from Elsevier.

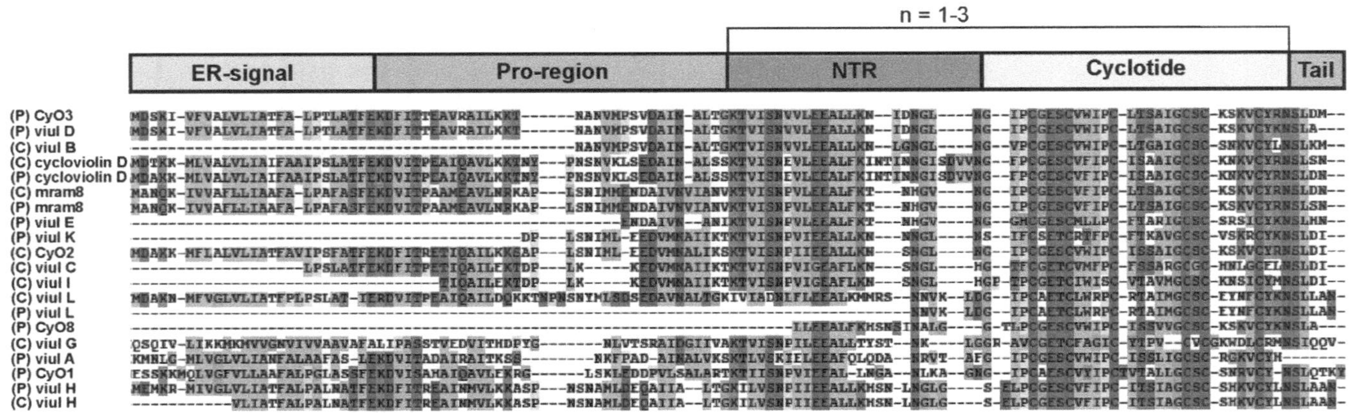

Figure 3 ***Example of structures of the cyclotide precursors.*** Peptide and alignment of sequences predicted from two *Viola uliginosa* transcripts: the native plant (P) and callus from in vitro culture (C) (Slazak, Sliwinska, et al., 2015). Similarly to precursors produced by plants from other families, five regions can be distinguished: ER-signal, pro region, N-terminal repeat (NTR), the cyclotide sequence, and a short C-terminal tail (characteristic for Violaceae). (See color plate)

searching for enzymes probably involved in cyclotide biosynthesis. Thus, the transcriptomes were searched for homology with *Clitoria ternatea* cyclization enzyme butelase 1 (accession: GenBank: AIB06797.1), possibly related to cyclotide biosynthesis (Nguyen et al., 2014). The closest sequence had a similarity of 68% to the active site of the enzyme. In the active region, the catalytic triad (N59, H165, and C207) was conserved. The butelase 1 belongs to asparaginyl endopeptidases (AEPs) and shares 71% of the sequence and the catalytic triad with other enzymes from this family (Nguyen et al., 2014). As AEPs play important roles in metabolic and regulatory pathways, it can be expected that such an enzyme would be expressed. Nevertheless, the butelase 1 homolog, found in the transcript, shared 65% similarity to the catalytic region of the enzyme whereas other AEPs share only 35% similarities in this region (Nguyen et al., 2014; Slazak, Jacobsson, et al., 2015). Though, more studies focused on enzymatic activity would be needed to confirm the identity and role of the sequence found in the transcript.

Without doubt, we have only seen the first of many studies to come that will take advantage of next-generation sequencing techniques. However, the massive amount of data that has suddenly become accessible in the form of nucleotide sequences creates new problems, and it is problematic to fully exploit bioinformatics for relatively short peptide sequences. For example, it has proven difficult to streamline the sequencing process into a peptidomic approach, i.e., by directly combining mass spectrometry data, usually from fragmentation, with a nucleotide database. The methods used are all optimized for the identification of large proteins, but there have been some attempts to solve the problem (Colgrave, Poth, Kaas, & Craik, 2010; Göransson, Broussalis, & Claeson, 2003; Hashempour, Koehbach, Daly, Ghassempour, & Gruber, 2013; Mohimani et al., 2011). Similar difficulties appear in pure data mining, and the powerful BLAST tool looses some of its edge when query sequences are too short. Recently two Blast-independent tools, CyPerl and CyExcel, were developed to significantly expand the distribution of possible cyclotide occurrence as compared to using BLAST (Zhang, Hua, et al., 2015). Notably, cyclotides are also included in a dedicated database: CyBase, the database of cyclic proteins (www.cybase.org.au) (Kaas & Craik, 2010; Mulvenna, Wang, & Craik, 2006). However surprisingly, not much is yet known of the arrangement and sequences of cyclotides on the genomic level. Despite the progress in sequencing techniques and bioinformatics, to assemble and annotate a full plant genome still represents a challenge and a substantial investment. We foresee that the sequencing of full genomes will be preceded by targeted genetic sequencing using techniques developed for long reads, e.g., Pacific Biosciences (PacBio) Sequencing. Genetic sequencing will be crucial to answer questions of where and how genes are localized and organized; it will give further clues for cyclotide biosynthesis and evolution.

As described above, and in a dedicated chapter of this volume, the methods for determination of cyclotide sequences have evolved during the years since their discovery. In other aspects, methods for structure determination have progressed at a slower pace. To specifically determine the connectivities of disulfide bonds still presents a major challenge. However, the cystine knot has been experimentally determined for cyclotides of different subfamilies and origin and using different techniques. Examples include circulins A and B (Derua, Gustafson, & Pannell, 1996) and kalata B1 (Göransson & Craik, 2003) by mass spectrometry studies of native peptides, stepwise synthesis of circulin B and cyclopsychotride A (Tam & Lu, 1998), kalata B1 and cycloviolacin O1 by NMR (Rosengren, Daly, Plan, Waine, & Craik, 2003), and varv peptide F by X-ray (Wang et al., 2009). As such, there is little doubt about the I-IV, II-V, and III-VI connectivity. (The numbering of the connectivity is true whichever cysteine is chosen as the starting point because the peptide is cyclic.)

The resulting cyclotide structure is a fortified framework. It owes its exceptional strength to the cyclic cystine knot, and the network of hydrogen bonds adding another layer of firmness to the scaffold, producing the unyielding complex. Most of the cyclotides fall into one of the two subfamilies: the Möbius and bracelet, based on the presence or absence of a *cis*-proline residue, respectively, in the loop 5 (Figure 4 and c.f. Figure 1). Hence, there is a conceptual twist of 180° in the Möbius cyclotides. In addition to the *cis*-proline residue, other general distinguishing features in the two subfamilies include presence of a short 3_{10} helix (Rosengren et al., 2003) and a cluster of positively charged residues in bracelets (Göransson, Herrmann, Burman, Haugaard-Jonsson, & Rosengren, 2009). On the other hand, some conserved structural features are present across the subfamilies. These include the absolutely conserved six cysteine residues involved in CCK, a glutamic acid residue in the loop 1 (with the exception of kB12) that has an important role in stabilizing the structure by contributing to hydrogen bonding, and a β-hairpin and a third β-sheet (which may be distorted in some cyclotides) centered on loop 5. The three-dimensional structures of cyclotides are described in detail in a separate chapter of this book. So the natural question is then: Is cyclotide diversity connected to plant species diversity of the Violaceae?

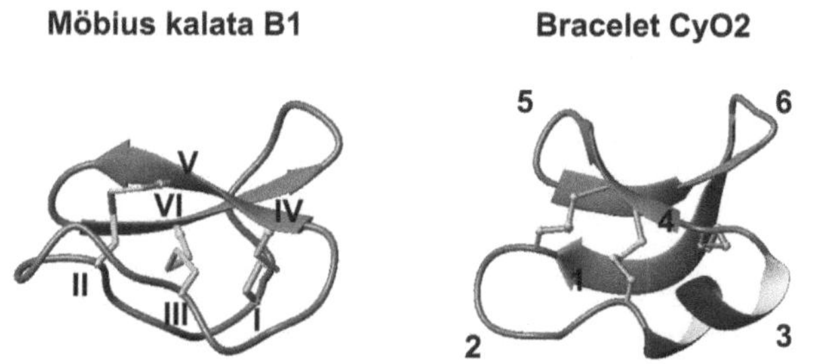

Figure 4 ***Structures of Möbius and bracelet cyclotides.*** Structures highlight the fact that the cyclotides fall into two major subfamilies, which are called bracelet and Möbius (so named because of a twist in the circular backbone). The cyclotides feature six loops (marked 1–6) between the cysteines, two of which lie in the core of the protein and four that are presented on the surface. The size and sequence of the surface-exposed loops can vary considerably; the number of combinations of natural loop sequences counts in millions (Craik, Cemazar, Wang, & Daly, 2006), which provides an excellent opportunity for protein engineering, e.g., for grafting of pharmacologically active sequence epitopes. The conserved cysteine residues forming three disulfide bonds in a knotted arrangement (i.e., I–IV, II–V, III–VI) are marked. CyO2, cycloviolacin O2.

3. THE VIOLACEAE

Historically, the classification of Violaceae is largely based on floral symmetry (Hekking, 1984, 1988; Melchior, 1925). Recently, however, analysis of genetic markers (trnL/trnL-F and rbcL plastid sequences of 102 species) has shown that the phylogeny of Violaceae is in need of a major revision (Wahlert, Marcussen, de Paula-Souza, Feng, & Ballard, 2014). Currently, the family Violaceae may be considered to consist of ~1050 species, divided into 33 genera, as summarized in Table 2. The Table is adapted from Burman et al. (2015) and based on de Paula-Souza and Ballard (2014), de Paula-Souza and Pirani (2014), Tokuoka (2008), Wahlert, Ballard, and de Paula-Souza (2015), Wahlert et al. (2014). The analysis of Wahlert and coworkers showed *Fusispermum* and *Rinorea crenata* being successive sister groups to the remaining Violaceae, in which *Rinorea sensu strictu* formed one clade and other genera form four clades with strong support. Within these clades, however, there is little support for the classical division of genera; instead the family appears to be even more diverse than previously recognized.

The family demonstrates great morphological diversity, and besides the herbal species of the genus *Viola* that comes first into mind when we think of violets, most genera include trees, shrubs, and lianas. *Viola* is by far the largest genus, comprising ~600 species, followed by *Rinorea* (~250 species)

Table 2 The Family Violaceae

Genus	Number of Species (1002–1098)	Clade/Comment
Fusispermum	3	Sister to all species
Rinorea crenata group	3	
Rinorea s.s.	225–275	
Rinoreocarpus	1	Unplaced
Decorsella	1	Unplaced
Allexis	2–4	Clade 1
Noisettia	1	
Schweiggeria	1	
Viola	580–620	
Paypayrola	8	Clade 2
Hekkingia	1	
Leonia	6	Clade 3
Amphirrhox	1–3	
Gloeospermum	~12	
Pigea	10	
Cubelium	1	
Hybanthus s.s.	2	
Mayanaea	1	
Orthion	6	
Hybanthus guanacastensis group	3	Clade 4
Dioryktandra (Scyphellandra)	2	
Afrohybanthus	25	
Mexion (proposed)	2	
Hybanthus thiemii group	4–6	
Ixchelia (proposed)	2	
Isodendrion	4	
Pombalia	~65	
Melicytus	~10	
Anchietea	6	
Hybanthopsis	1	
Corynostylis/Calyptrion	3	
Agatea	10	

and *Hybanthus* and *Pombalia*, according to the current classification. With these major genera constituting ~90% of the number of species, the other genera are poor in species and several are monotypic containing only one species (e.g., *Hybanthopsis*, *Noisettia*, and *Decorsella*). The major genera are widely distributed throughout the world.

It is clear that species diversity is reflected in cyclotide expression, but to date, we have not been able to use this connection to match expression to phylogeny other than by intuition. Neither have the analytical methods used so far, i.e., studies on peptide level using liquid chromatography and mass spectrometry, been able to generate enough marker compounds to help build new phylogenies. Some peptides appear in seemingly distantly related species (c.f. the occurrence at peptide level of kalata B1 in *O. affinis* and in *V. tricolor* but not in *V. arvensis*). However, it should be pointed out that to date, analyses of RNA expression have shown big discrepancies when compared to the expression on peptides level. Cyclotide RNA that is found in relatively high abundance is not always detected as peptides, and no RNA can be detected for some peptides occurring at high levels. The new knowledge that we currently acquire, in the form of transcriptomes revealing sequences and arrangements of additional peptides *and* their precursor sequences, can be expected to substantially increase our chances to make that connection. Table 3 shows an example of the *V. uliginosa* cyclotides that are coexpressed in other species.

Clearly, there are many cyclotides to be found in Violaceae, and the family stands out in the context of cyclotide production: all species in Violaceae that have been investigated for presence of cyclotides have shown positive results. In comparison, only a fraction of the species in other known cyclotide-containing families (*Rubiaceae, Cucurbitaceae, Fabaceae, Solanaceae*, and *Poaceae*) have been shown to contain cyclotides. Individual plant species have been shown to contain >50 different cyclotides, and the total number has been estimated in thousands (>9000 (Simonsen et al., 2005); 5000–25,000 (Burman et al., 2015) based on screens on peptide level; and >30,000 based on transcriptome sequencing (Zhang, Li, et al., 2015)).

Table 3 *Viola uliginosa* Cyclotides Already Found in Other Species

Cyclotide	Species
Cycloviolacin O1	*Viola odorata*
Cycloviolacin O2	*V. odorata, Viola biflora, Viola philippica*
Cycloviolacin O3	*V. odorata*
Cycloviolacin O13	*V. odorata*
Cycloviolin D	*Leonia cymosa*
Varv A (syn. kalata S)	*V. odorata, Oldenlandia affinis, Viola tricolor, Viola arvensis, Viola baoshanensis, Viola yedoensis, V. biflora, V. philippica*
Mram 8	*Melicytus ramiflorus, V. philippica*

4. VIOLACEAE—THE PHARMACOPOEIA

It was the uterus-contracting properties of a decoction of *Oldenlandia affinis* that led to isolation of the first cyclotide, kalata B1. Hence, first question in the quest for pharmacological activities of cyclotide-bearing violets could be as follows: Are there any similar observations (on effects or use) of plants from the family of Violaceae? A search in the literature and relevant databases, including the largest database of natural products Napralert, which includes data on extract level, their biological activities, and use in ethnomedicine, reveals a poignant answer: Yes! Several reports from geographically distant places demonstrate that plants from Violaceae have been used for similar or related gynecological and obstetrical purposes, summarized in Table 4. Of the species found all belong to genera that have been shown experimentally to contain cyclotides. Maybe it is time to revisit some of these plants, particularly when considering the fact that Gruber and coworkers recently demonstrated that the cyclotide kalata B7 does indeed interact with the oxytocin receptor (Koehbach et al., 2013).

Plants from the Violaceae have a long history as medicinal plants: *V. tricolor*—its common name in English "heartsease" implies its importance in traditional medicine—and *V. odorata* have been used to treat, e.g., skin disorders such as scabs, itching, ulcers, eczema, or psoriasis. Indeed, preparations of violets were included in the first Pharmacopoeias, e.g., they are mentioned in the texts of Nicholas Culpeper from 1653, and *V. odorata* was included in the first Swedish Pharmacopoeia *Pharmacopoeia Holmiensis Galeno-Chymica* from 1686. In Sweden, *V. tricolor* and *V. odorata* have been traditionally used against skin disorders (Lindgren, 1927), and other indications include respiratory ailments and digestive problems. Today, *V. tricolor* and/or subspecies *V. arvensis* and *Viola vulgaris* (under the herbal drug name *Violae herba cum flore*) are included in the European Pharmacopoeia and in the list of herbal medicines for human use by the European Medicines Agency (www.ema.europa.eu). Plants of the genus *Viola* have also been used against cancer; both European and Chinese folk medicine have made use of them for this purpose (Perry, 1980). Several members of the Violaceae plant family are key herbs used in traditional Chinese medicine. One such herb and the activity based on its cyclotide content have been studied in detail, namely *Viola yedoensis* (Wang et al., 2008). Other studies, e.g., Witkowska-Banaszczak, Bylka, Matlawska, Goslinska, and Muszynski (2005), Zarrabi, Dalirfardouei, Sepehrizade, and Kermanshahi (2013), have reported the use of fractions that most likely do contain

Table 4 Ethnopharmacological Observations of the Use of Violaceae Plants for Gynecological and Obstetrical Indications

Plant Species	Plant Part/Formulation	Indication	Place	References
Viola species (unnamed)	Used in the diet	Facilitate childbirth	New Guinea	Stopp (1962)
Viola odorata	Hot water extract of flowers	Abortifacient	Argentina	Martinez-Crovetto (1981)
Viola humboldtii	Juice of the whole plant	To increase lactation	Northwestern Amazonia	Schultes and Raffauf (1990)
Viola adunca	Roots and leaves	Chewed during labor	USA	Gunther (1973)
Rinorea anguifera	Root decoction	Protective after childbirth	Malaysia	Burkhill (1966)
Hybanthus brevis	Decoction of entire plant	Treat menorrhagia and hemorrhage between menstrual periods	Mexico	Browner (1985)
Hybanthus enneaspermus	—	Tonic before childbirth and after delivery	Africa	Iwu (1993)
Hybanthus concolor	Infusion of roots and stems of mixed with feed	Veterinary aid for mares to expel injured fetuses	USA	Herrick (1977)

Connected to this list are then the two ethnopharmocological observations of the use of *Oldenlandia affinis*, to facilitate childbirth (Gran, 1973b; Sandberg, 1965).

cyclotides. In those studies, anti-infective properties are in focus to raise the question: Can cyclotides contribute to the development of new antibiotics?

No plant-derived antibiotics—not even plant-inspired antibiotics—are available in the market today. This is perplexing because of two obvious reasons: Firstly, plants, just like human beings, need to protect themselves from pathogens. They have developed a number of secondary metabolites to do that. Although a large number of extracts of plants with an antimicrobial activity have been reported, not many promising compounds have been pursued. Secondly, plant-based anticancer drugs such as *Vinca* alkaloids, taxol, and podophyllotoxins have been developed. It is only natural to assume that from an ecological perspective, plants would produce antibiotic substances, in addition to being a source of anticancer substances (Gibbons, 2004).

Violets have been used in diverse cultural settings in a number of indications implicating their anti-infective properties. In India, *Ionidium suffruticosum* (Vern. Ratanpurs) has been used as a demulcent in gonorrhea, and *V. odorata* (Vern. Banafsha) has been used in fever (Dymock, Hooper, & Warden, 1892). In addition, *Viola biflora*, *Viola cinerea*, and *Viola diffusa* have been used in similar indications (Chopra, Chopra, & Nayer, 1956). In the Philippines, different preparations of violets have been used in the treatment of cancers (Quisumbing, 1951). In North Carolina, a soup of *Viola palmata* and *Viola septemloba* together with *Sassafras* has been used to "bring out measles" (Duke, 2002).

We know that cyclotides are among the antimicrobial principles from violets. Pränting, Lööv, Burman, Göransson, and Andersson (2010) used two violets to look for possible cyclotide antibiotics. The authors used cycloviolacin O2 from *V. odorata* and vaby A and vaby D from *Viola abyssinica*. They found out that cycloviolacin O2 is a potent and broad-spectrum inhibitor of Gram-negative bacteria. In addition to killing wild-type bacteria, cycloviolacin O2 was effective in killing the capsule producing as well as multiresistant strains. By using the model membranes, Burman, Strömstedt, Malmsten, and Göransson (2011) have demonstrated that cyclotide activity owes to a preferential binding and extraction of the phosphatidyl ethanolamine phospholipids from the membrane, which is supported by other studies (Henriques & Craik, 2012; Henriques et al., 2011). The work by Pränting et al. (2010) demonstrated that all the charged amino acid residues are important for antimicrobial activity of cycloviolacin O2. Ongoing investigations in our lab have shown that this is not an isolated example; instead, a range of cyclotides isolated from *V. odorata*, *V. arvensis*, and *V. uliginosa* have potent antimicrobial properties. In another study, resistance development to cyO2 in Gram-negative bacteria has been explored, and the genes involved in low-level resistance have been identified (personal observations).

Answering the following two questions asked in this context is important in explaining the presence of cyclotide cocktails in plants as well as in developing a rationale for roles of different subclasses of cyclotides during evolution: first question is about cross-resistance and the second concerns spectrum of activity. (1) Whether bacteria resistant to one bracelet cyclotide would be cross-resistant to other bracelet cyclotides, would they be cross-resistant to Möbius cyclotides as well (if there are any active Möbius cyclotides)? Would different cross-resistant profiles suggest that suites of cyclotides in the same plant help in overcoming cross-resistance. (2) Is there a correlation between the bacterial organisms found in the ecological niche of a cyclotide-producing plant and the spectrum of activity of cyclotides it produces? Would all antimicrobial cyclotides target similar bacterial species? What if they have a complementary or overlapping target spectrum? Would complementary spectrum of activity imply that different cyclotide sequences evolved for different organisms?

Presence of antimicrobial peptides (AMPs) as a family of similar sequences is a recurring theme in plants as well as other organisms. Families of plant AMPs include thionins, defensins, lipid transfer proteins, and snakins (Segura, Moreno, Madueno, Molina, & Garcia-Olmedo, 1999). These are structurally diverse but share some features, for example, variable number of cysteine residues forming disulfide bonds, thereby conserving their intricate scaffolds (Padovan, Scocchi, & Tossi, 2010). Observations in line with the role of these peptides in plant defense include *in vitro* inhibition of bacterial growth, wide distribution of peptide-coding genes, and their induced expression after exposure to pathogen and constitutive expression in flowers and seeds, which are particularly sensitive to infection (Maroti, Kereszt, Kondorosi, & Mergaert, 2011). However, cyclotides stand out as a family of AMPs because of their exceptional stability, structural plasticity, unique biochemical target, and Gram-negative selective antimicrobial action. These features together with recent advancements in the methods of production of cyclotides make them an intriguing prospect from a drug development perspective.

5. GROWING NEEDS AND SOURCES OF CYCLOTIDES

Since the complete description of the cyclotide structure, and the discovery that the cyclotides form a peptide library of a considerable size, the interest in using this chemical library has mushroomed. Applications may include the medical, biotechnological, or agricultural use of native peptides, or mutants in which residues have been changed to increase the effect of native peptides or mutants in which other bioactive sequences have been inserted. To date, native cyclotides have been reported to have oxytocic/uterotonic (Gran, 1973a; Koehbach et al., 2013), insecticidal (Jennings, West, Waine, Craik, & Anderson, 2001), molluscicidal (Plan, Saska, Cagauan, & Craik, 2008), anticancer (Lindholm et al., 2002), antimicrobial (Pränting et al., 2010; Tam, Lu, Yang, & Chiu, 1999), antifouling (Göransson, Sjögren, et al., 2004), immunosuppressive (Hellinger et al., 2014), and anti-HIV (Gerlach et al., 2013; Gustafson, McKee, & Bokesch, 2004) activities, as well as membrane disruptive properties (Burman et al., 2011; Henriques & Craik, 2012; Svangård et al., 2007). Cyclotides with these effects originate from plants outside of the Violaceae family as well. Most studies have been done on kalata B1, the prototypic cyclotide that was first discovered in Rubiaceae, but which also occurs in high relative concentrations in many Violaceae species (e.g., in *V. odorata, V. baoshanensis, and V. yedoensis*). Kalata B1, together with the atypical cyclotide *Momordica cochinchinensis* trypsin inhibitor I and II (MCoTI-I and -II) from Cucurbitaceae, are the cyclotides that have been subjected to most structure activity/mutation studies. For example, kalata B1 has been a subject for extensive alanine and lysine scans (Huang, Colgrave, Clark, Kotze, & Craik, 2010; Simonsen et al., 2008), and even the complete peptide has been synthesized using amino acid residues of D-configurations to create a mirror image of the three-dimensional peptide structure (Sando et al., 2011)! The most well-studied effects of cyclotides, in terms of the number of peptides tested, are their hemolytic and cytotoxic effects and their insecticidal activity. For this reason, these activities were also used to build the first model for their quantitative structure—activity relationship (Park, Strömstedt, & Göransson, 2014).

So far in this chapter, only isolation from plants has been discussed as a source of cyclotides. However, it should be pointed out that chemical synthesis of cyclotides was done at an early stage (Daly, Love, Alewood, & Craik, 1999; Tam & Lu, 1997, 1998; Tam et al., 1999), and that methods for synthesis have been developed over the years. The current protocols (e.g., Blanco-Canosa, Nardone, Albericio, & Dawson, 2015; Gunasekera, Aboye, Madian, El-Seedi, & Göransson, 2013) take advantage of native chemical ligation (Dawson, Muir, Clark-Lewis, & Kent, 1994; Hackeng, Griffin, & Dawson, 1999) to join the head-to-tail cyclic backbone.

Although peptide synthesis is now routinely performed in many labs (and can be commercially outsourced if not available in-house) backbone cyclization adds to the complexity of the synthesis and greatly affects both yields and costs. One alternative to this would be to use the natural machinery, the biosynthetic route of cyclotides.

This can be done in several ways; if the catalytic enzymes were known, they could be used for the final step of cyclization for precursors obtained from synthesis. These enzyme systems could also be used to cyclize precursors expressed in microorganisms, plant cells and tissues, or even whole plants, in processes taking place in vivo or ex vivo. Leatherbarrow and coworkers were the first to demonstrate the possibility to use the reverse function of immobilized proteases for the synthesis of cyclotides (*M. cochinchinensis* trypsin inhibitor I and II) (Thongyoo, Jaulent, Tate, & Leatherbarrow, 2007; Thongyoo, Roque-Rosell, Leatherbarrow, & Tate, 2008). However, utilizing this reverse direction of protease activity has not been able to compete with native chemical ligation. Lately, sortase A has been demonstrated to be useful for cyclization of cyclotides (Jia et al., 2014; Stanger et al., 2014), including kalata B1, but they leave a scar—the remains of the signature in the finished sequence—and sortases are consumed in the process. We are currently exploring the use of the macrocyclase domain of PatG (Koehnke et al., 2012) from *Prochloron* sp. of bacteria, and the patellamide pathway, and preliminary results indicate that kalata B1 is cyclized (Muhammad, T., Hussain, W., Gunasekera, S., Jaspars M., Göransson U., personal observations).

The most promising route so far appears to be the use of butelase 1, the first asparagine/aspartate peptide ligase to be reported (Nguyen et al., 2014). This enzyme is isolated from a cyclotide-producing plant, *Clitoria ternatea* (Fabaceae), and demonstrates impressive turn over rates, yields, and broad activity on various peptides of plant and animal origins. It was discovered by an ingenious screening strategy, exploiting a short conserved C-terminal sequence. However, the natural function of butelase has not yet been demonstrated, i.e., Is butelase involved in cyclotides biosynthesis per se or Is its function just another example of serendipity? The successful cyclization of a full precursor peptide would help answering these questions. In this context it is interesting to note that the first such enzyme in plants has been discovered, i.e., an enzyme whose proven biosynthetic function is to ligate proteogenic peptide sequences head to tail (Barber et al., 2013). However, peptide cyclase 1 does not produce cyclotides, but the smaller (5–12-amino acid residue) cyclic peptides found in plants from the family Caryophyllaceae (Arnison et al., 2013; Barber et al., 2013; Condie et al., 2011). However, this enzyme has not yet been in focus to aid the production of cyclotides.

All methods mentioned above aim to yield a cyclic sequence. However, cyclotides include a cystine knot motif also. The oxidative folding into the cystine knot is unpredictable; some peptides primarily those of the Möbius subfamily (e.g., kalata B1) together with the *M. cochinchinensis* trypsin inhibitor I and II (MCoTI-I and -II) fold quickly (minutes to hours) and in high yields (>95%), into native structures. This is also the main reason for why these peptides occur in most grafting studies. Bracelet cyclotides seem to present real problems for folding, and the syntheses and folding of cycloviolacin O2 (first isolated from *V. odorata*) (Hashempour et al., 2013; Leta Aboye, Clark, Craik, & Göransson, 2008), and hedyotide B1 (Wong, Taichi, Nishio, Nishiuchi, & Tam, 2011), are so far the only attempts ending with reasonable yield. As such, ~2/3 of all cyclotides (i.e., the bracelet subfamily) is currently intractable for synthesis.

Biotechnological tools are currently exploited to express cyclotides in linear and cyclic forms, e.g., in bacterial displays (Getz, Cheneval, Craik, & Daugherty, 2013) and in cell screening in *Escherichia coli* (Jagadish et al., 2013) and *Saccharomyces cerevisiae* (Jagadish et al., 2015). (For specifics, see the chapter dedicated to the topic in this volume). However, none of these methods have been used for any large-scale expression of cyclotides, and the prospects seem slim. In addition, most studies and attempts of expression in plant cells have been done using model plants, e.g., *Arabidopsis thaliana* that do not naturally produce cyclotides (Mylne et al., 2012), and only a fraction of the expressed peptide appears to be cyclized. The next step must be to use cyclotide-expressing plants, for both expression and to build our understanding of cyclotide biosynthesis. Although our knowledge of the latter is increasing (Craik & Malik, 2013), we still lack basic knowledge about the expression and fate of the cyclotide precursor.

Native cyclotides have been expressed in plant cell cultures, as pioneered by the group of Dörnenburg in their work on *O. affinis* (Dörnenburg, 2010; Seydel & Dörnenburg, 2006). Recently, we established cell cultures of *V. uliginosa* (Slazak, Jacobsson, et al., 2015) that provided a yield of more than 4 mg of individual cyclotides (i.e., cycloviolacin O13) per gram dry mass. In addition, we demonstrated that the expression is, and can be, controlled by the choice of plant hormones added (media containing only a cytokinin gave the highest yield), and that cyclotide expression correlates with accumulation of storage material. The expression is maintained in regenerated plants from callus cultures, and is dependent on ploidy level (the number of sets of chromosomes in a cell) and organogenesis (Slazak, Sliwinska, et al., 2015). The latter supports the tissue-specific expression (Göransson et al., 2003; Trabi & Craik, 2004). Figure 5 summarizes possibilities of in vitro culturing; and the stage is set for moving into expression systems of transgenic cultures and to establish expression systems in plants for protein farming.

6. COMPLETING THE CIRCLE

It has been 20 years since the Eureka moment in Uppsala when we turned the violapeptide I sequence around to fit with kalata B1, and our research went in the direction of chemistry of plants of the Violaceae species. We have learned a lot since then, as shown in this chapter and others in this volume as well as in a series of recent reviews (Burman, Gunasekera, Strömstedt, & Göransson, 2014; Craik, Henriques, Mylne, & Wang, 2012; Craik & Malik, 2013; Göransson, Svangård, Claeson, & Bohlin, 2004; Northfield et al., 2014; Schroeder, Swedberg, & Craik, 2013). Still many opportunities are lurking around and many challenges are awaiting us. For drug discovery, based on native cyclotides or mutants or derivatives, there is a lack of fundamental pharmaceutical studies such as bioavailability, biodistribution, and biodegradation. Where do these peptides end up in nature (Ovesen, Brandt, et al., 2011; Ovesen, Nielsen, & Hansen, 2011)? And are there any targets other than the cellular membrane? Some results suggest that we need to reconsider membrane disruption as the mode of action, a fact revealed recently. For example, Gruber and coworkers recently demonstrated that the oxytocin receptor is indeed in play (Koehbach et al., 2013), and cyclotides have demonstrated activity far below the cytotoxic concentration (Yeshak, Göransson, Burman, & Hellman, 2012). In addition, the immunosuppressive effect (Grundemann et al., 2013) and the antifouling activity (Göransson, Sjögren, et al., 2004) are difficult to explain with the membrane model. However, perhaps the main question that is still unanswered is the following: What is the role of cyclotides *in planta*? Their insecticidal and antimicrobial activities suggest that they are defense molecules, but these studies as most other studies are done using cyclotides outside their natural setting in the plant. To conclude this chapter, we show an example on how we have started to reveal the role of cyclotides *in planta* using immunolocalization.

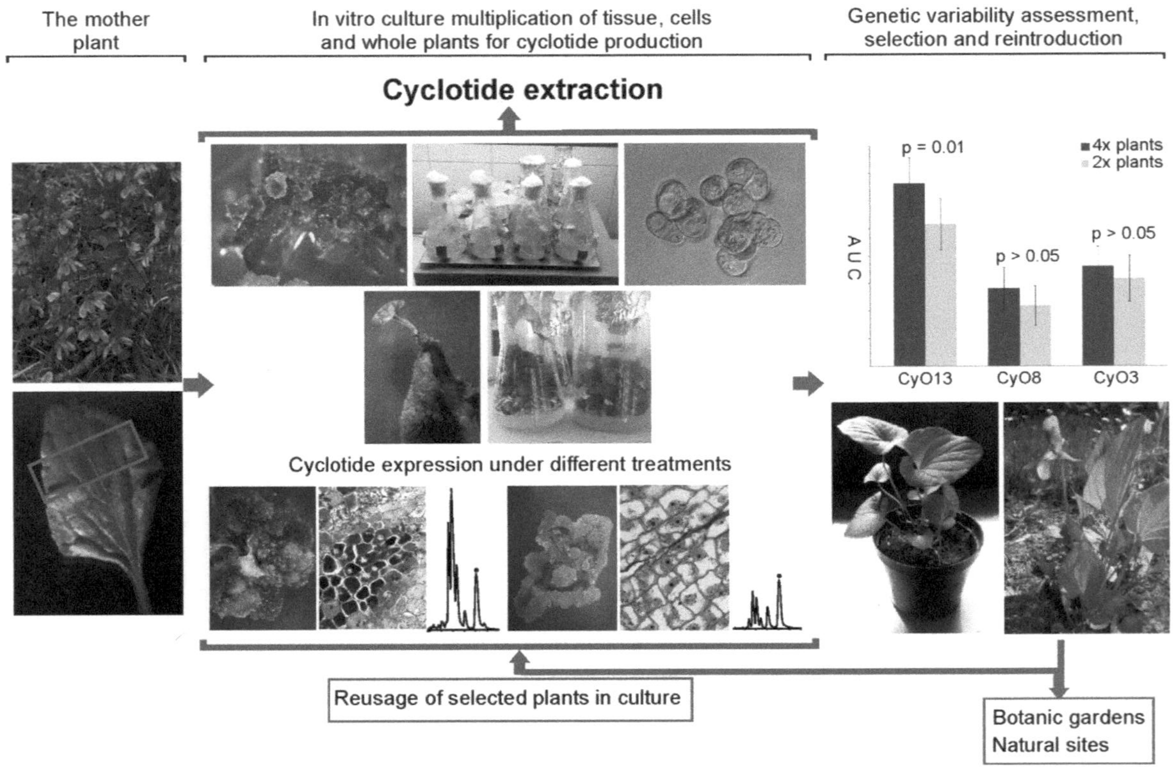

Figure 5 ***In vitro cultures for micropropagation of* Viola uliginosa *and cyclotide production.*** By using in vitro culture techniques, multiplication of maternal plant, its tissues and cells are achieved in relatively short time. Regenerated plantlets can serve as a source of material for reintroduction, whereas tissues, especially fast-growing suspension cultures, can be utilized for cyclotide production. Culture conditions induce somaclonal variation leading to changes in genomes and cause genetic variation of regenerated plants. The changes may be beneficial as demonstrated for *V. uliginosa*, where tetraploid (4×)-regenerated plants produced more cyclotides than plants with standard genome size (diploid, 2×). The selected specimens can be reused in *in vitro* culture as initial material to obtain higher yields (Slazak, Jacobsson, et al., 2015; Slazak, Sliwinska, et al., 2015). *Parts of Figure 5 reproduced from Slazak, Jacobsson, et al. (2015), Slazak, Sliwinska, et al. (2015) with permissions from Elsevier and Springer.* (See color plate)

Previously, MALDI imaging has been used to depict cyclotides distribution in *Petunia* (Solanaceae) leaf (Poth et al., 2012), and the vacuole has been implicated to be the final deposition and storage place of cyclotides in *Nicotiana benthamiana* using cyclotide precursor incorporating green fluorescent protein (GFP) (Conlan, Gillon, Barbeta, & Anderson, 2011). The methods exploited in those studies have certain advantages and disadvantages: MALDI imaging shows distribution of molecules with particular masses, but it has only a resolution of 20–25 μm (Cillero-Pastor & Heeren, 2014; Walch, Rauser, Deininger, & Hofler, 2008). This is not sufficient to distinguish tissues and, in particular, does not allow the visualization of cyclotides at the subcellular level. On the other hand, the biotechnological tools used are indirect and the influence of GFP label and choice of model organism is unclear.

Taking these limitations into consideration, it seems that the most appropriate method to depict cyclotide distribution in cells, tissues, and organs would be to grow antibodies and apply immunohistochemistry. Recently, we successfully raised antibodies against the bracelet cyclotide cycloviolacin O2 (Slazak, B., Kapusta, M., Malik, S., Bohdanowicz, J., Kuta, E., Malec, P., Göransson, U., unpublished results). These polyclonal antibodies were shown to bind to cycloviolacin O2 very efficiently, but also to other bracelet cyclotides differing from cycloviolacin O2 by one or two amino acid residues. Binding decreased substantially for a bracelet differing in four residues and they did not bind at all to Möbius cyclotides. The lack of complete specificity can be considered as an advantage, which allows us to picture cyclotides as a class of compounds in many species of Violaceae.

Immunofluorescence was then used to visualize cyclotides in cell, tissue, and organs, using secondary antibodies conjugated with a fluorochrome as shown in Figure 6. The antibodies were first tested in a series of western blot, dot blot and negative control experiments. All these experiments further confirmed the specificity of antibodies, showing neither binding of primary antibodies nor secondary antibodies alone. In the cross section of *V. odorata* leafs, cyclotides were found to form big clumps in the lower epidermis. Cyclotides were also found, although in smaller amounts, in all mesophyll cells and in the cells of upper epidermis. In the petiole, cyclotides were evenly distributed but also placed in vascular bundles. Further comparison of the images with handmade cross sections, showing anatomical features, showed that the cyclotides in vascular bundle are placed in xylem parenchyma, collenchyma, and the phloem.

This study showed that the cyclotides are stored in the vacuole using a direct method. (The cytoplasm was additionally stained with antibodies

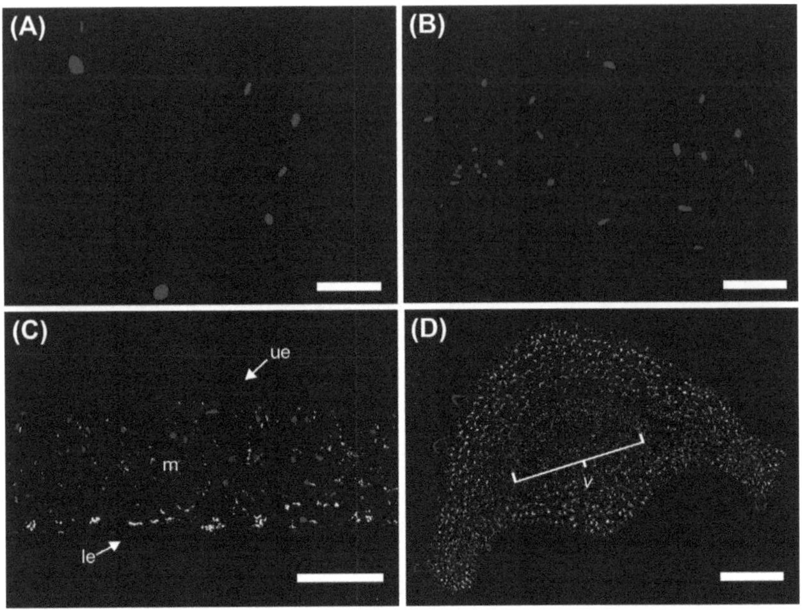

Figure 6 Immunohistochemistry techniques in depicting cyclotides localization in plant tissues and organs. (A) and (B), negative controls proving antibodies specificity: *Nicotiana benthamiana* (A), plant known for not producing cyclotides; *Viola odorata* leaf cross section (B) stained without using primary antibody against CyO2. Only blue autofluorescence of cell walls, nuclei, and chloroplasts (stained with DAPI) could be seen. (C) and (D), *V. odorata* leaf (C) and petiole (D) cross sections with stained cyclotides (red fluorescence). In the leaf (C) big clumps of cyclotides were found in lower epidermis region (le) and smaller amounts in leaf mezofile (m) and upper epidermis (ue). In the petiole (D) cyclotides are evenly distributed and found in the vascular bundle (v). The clumps of cyclotides are most probably an artifact of the fixation procedure (proteins and peptides, e.g., very concentrated cyclotides, may cross-link with paraformaldehyde and glutaraldehyde and become insoluble in water (Mannich & Krösche, 1912; Richards & Knowles, 1968)). Bars represent 100 μm in (A–C), and 250 μm in (D). (See color plate)

against tubulin to show the vacuole.) This knowledge of distribution of cyclotides gives food for thought on cyclotide role in plant physiology and more broadly, biology. And with new knowledge, new questions emerge: What could be the role of cyclotides in the phloem? The sieve tubes in the phloem tissue are composed of cells that do not contain nuclei or other organelles, including vacuoles to store cyclotides. The role of the phloem is to transport assimilates. To get to the phloem sap inside the sieve tubes, cyclotides would have to be excreted from vacuoles and outside cells producing them. There is some evidence from in vitro studies that this can take place (Dörnenburg, 2008; Slazak, Jacobsson, et al., 2015). Another possibility is that cyclotides are produced in the embryonic tissue, or in meristematic tissues in later stages of development, from which the phloem is formed: vacuoles are decomposed during the course of differentiation into mature sieve elements and possibly releasing cyclotides into the phloem sap. A similar mechanism has been proposed to explain the presence of *p*-proteins in sieve elements (Heo, Roszak, Furuta, & Helariutta, 2014). In fact, when we applied the immunohistochemistry in embryos of *V. odorata*, large quantities of cyclotides were found exactly in this kind of tissue (Slazak, B., Kapusta, M., Malik, S., Bohdanowicz, J., Kuta, E., Malec, P., Göransson, U., unpublished observations). The analogy of cyclotides with *p*-proteins goes beyond their origin in the phloem sap, because they too are considered to play a role in plant defense.

Knowledge of distribution of a particular defense molecule helps to develop a hypothesis about its target organism. Compounds protecting the plant from microorganisms or small herbivores, like aphids or spider mites, tend to be placed in the epidermis whereas compounds protecting from larger herbivores feeding on whole organs seem to require defenses deeper in the tissue, e.g., in the leaf mesophyll (Nuringtyas, Choi, Verpoorte, Klinkhamer, & Leiss, 2012). In the light of the current results, it is tempting to speculate that the large amount of cyclotides in the lower epidermis is there to protect against spider mites (Tetranychidae). Spider mites live on the lower epidermis, feeding by punctuating plant cells and sucking its contents; the inability of those organisms to digest cyclotides may be a factor limiting their feeding. And is it possible that the cyclotides are present in the phloem and collenchyma to protect the whole vascular bundle? The fact that spider mites often attack violets lends further support to this hypothesis (Slazak, B., personal observations). Or are cyclotides there to protect against infections by microbial and viral pathogens that are spreading through the plant phloem? The antimicrobial and antiviral properties of cyclotides support this idea. We are currently trying to answer these questions by studies of cell ultrastructures.

So the circle is closing up—the plant-inspired chemical adventure is turning back to its origin. We can only access the full potential of cyclotides when we understand the details in their natural setting, *in planta*.

ACKNOWLEDGMENTS

Past and present members of the cyclotide group in Uppsala, and our collaborators, are acknowledged. In particular, we thank Dr Per Claeson and Prof. Lars Bohlin for backing the first ventures into cyclotide research; Prof. Elżbieta Kuta and Andrzej Malobędzki for photos of *Viola odorata* and *Viola uliginosa*, respectively; Dr Sonny Larsson for support with systematics; and Dr Małgorzata Kapusta for help with microscopy. The Swedish Research Council (#2012-5063), the Strategic Foundation for Strategic Research (F06-0058), and the Władysław Szafer Institute of Botany, Polish Academy of Sciences, are acknowledged for their financial support.

REFERENCES

Arnison, P. G., Bibb, M. J., Bierbaum, G., Bowers, A. A., Bugni, T. S., Bulaj, G., et al. (2013). Ribosomally synthesized and post-translationally modified peptide natural products: overview and recommendations for a universal nomenclature. *Natural Product Reports, 30*(1), 108–160. http://dx.doi.org/10.1039/c2np20085f.

Barber, C. J., Pujara, P. T., Reed, D. W., Chiwocha, S., Zhang, H., & Covello, P. S. (2013). The two-step biosynthesis of cyclic peptides from linear precursors in a member of the plant family Caryophyllaceae involves cyclization by a serine protease-like enzyme. *Journal of Biological Chemistry, 288*(18), 12500–12510. http://dx.doi.org/10.1074/jbc.M112.437947.

Benson, D. A., Cavanaugh, M., Clark, K., Karsch-Mizrachi, I., Lipman, D. J., Ostell, J., et al. (2013). GenBank. *Nucleic Acids Research, 41*(Database issue), D36–D42. http://dx.doi.org/10.1093/nar/gks1195.

Blanco-Canosa, J. B., Nardone, B., Albericio, F., & Dawson, P. E. (2015). Chemical protein synthesis using a second-generation N-acylurea linker for the preparation of peptide-thioester precursors. *Journal of the American Chemical Society, 137*(22), 7197–7209. http://dx.doi.org/10.1021/jacs.5b03504.

Broussalis, A. M., Göransson, U., Coussio, J. D., Ferraro, G., Martino, V., & Claeson, P. (2001). First cyclotide from Hybanthus (Violaceae). *Phytochemistry, 58*(1), 47–51.

Browner, C. H. (1985). Plants used for reproductive health in Oaxaca, Mexico. *Economic Botany, 39*(4), 482–504.

Burkhill, I. H. (1966). *Dictionary of the economic products of the Malay peninsula* (Vol. 2). Kuala Lumpur: Ministry of Agriculture and Cooperatives.

Burman, R., Gunasekera, S., Strömstedt, A. A., & Göransson, U. (2014). Chemistry and biology of cyclotides: circular plant peptides outside the box. *Journal of Natural Products, 77*(3), 724–736. http://dx.doi.org/10.1021/np401055j.

Burman, R., Strömstedt, A. A., Malmsten, M., & Göransson, U. (2011). Cyclotide-membrane interactions: defining factors of membrane binding, depletion and disruption. *Biochimica et Biophysica Acta, 1808*(11), 2665–2673. S0005-2736(11)00209-4.

Burman, R., Yeshak, M. Y., Larsson, S., Craik, D. J., Rosengren, K. J., & Göransson, U. (2015). Distribution of circular proteins in plants: large-scale mapping of cyclotides in the Violaceae. *Frontiers in Plant Science, 6*, 855. http://dx.doi.org/10.3389/fpls.2015.00855.

Chopra, R. N., Chopra, I. C., & Nayer, S. L. (1956). *Glossary of Indian medicinal plants*. Council of Scientific & Industrial Research.

Cillero-Pastor, B., & Heeren, R. M. (2014). Matrix-assisted laser desorption ionization mass spectrometry imaging for peptide and protein analyses: a critical review of on-tissue digestion. *Journal of Proteome Research, 13*(2), 325–335. http://dx.doi.org/10.1021/pr400743a.

Claeson, P., Göransson, U., Johansson, S., Luijendijk, T., & Bohlin, L. (1998). Fractionation protocol for the isolation of polypeptides from plant biomass. *Journal of Natural Products, 61*(1), 77–81. http://dx.doi.org/10.1021/np970342r.

Colgrave, M. L., & Craik, D. J. (2004). Thermal, chemical, and enzymatic stability of the cyclotide kalata B1: the importance of the cyclic cystine knot. *Biochemistry, 43*(20), 5965–5975. http://dx.doi.org/10.1021/bi049711q.

Colgrave, M. L., Poth, A. G., Kaas, Q., & Craik, D. J. (2010). A new 'era' for cyclotide sequencing. *Biopolymers*. http://dx.doi.org/10.1002/bip.21400.

Condie, J. A., Nowak, G., Reed, D. W., Balsevich, J. J., Reaney, M. J. T., Arnison, P. G., et al. (2011). The biosynthesis of Caryophyllaceae-like cyclic peptides in *Saponaria vaccaria* L. from DNA-encoded precursors. *Plant Journal, 67*(4), 682–690. http://dx.doi.org/10.1111/j.1365-313X.2011.04626.x.

Conlan, B. F., Gillon, A. D., Barbeta, B. L., & Anderson, M. A. (2011). Subcellular targeting and biosynthesis of cyclotides in plant cells. *American Journal of Botany, 98*(12), 2018–2026. http://dx.doi.org/10.3732/ajb.1100382.

Craik, D. J., Cemazar, M., Wang, C. K., & Daly, N. L. (2006). The cyclotide family of circular miniproteins: nature's combinatorial peptide template. *Biopolymers, 84*(3), 250–266. http://dx.doi.org/10.1002/bip.20451.

Craik, D. J., Daly, N. L., Bond, T., & Waine, C. (1999). Plant cyclotides: a unique family of cyclic and knotted proteins that defines the cyclic cystine knot structural motif. *Journal of Molecular Biology, 294*(5), 1327–1336. http://dx.doi.org/10.1006/jmbi.1999.3383.

Craik, D. J., Henriques, S. T., Mylne, J. S., & Wang, C. K. (2012). Cyclotide isolation and characterization. *Methods in Enzymology, 516*, 37–62. http://dx.doi.org/10.1016/B978-0-12-394291-3.00024-1.

Craik, D. J., & Malik, U. (2013). Cyclotide biosynthesis. *Current Opinion in Chemical Biology, 17*(4), 546–554. http://dx.doi.org/10.1016/j.cbpa.2013.05.033.

Daly, N. L., Love, S., Alewood, P. F., & Craik, D. J. (1999). Chemical synthesis and folding pathways of large cyclic polypeptides: studies of the cystine knot polypeptide kalata B1. *Biochemistry, 38*(32), 10606–10614. http://dx.doi.org/10.1021/bi990605b.

Dawson, P. E., Muir, T. W., Clark-Lewis, I., & Kent, S. B. (1994). Synthesis of proteins by native chemical ligation. *Science, 266*(5186), 776–779.

Derua, R., Gustafson, K. R., & Pannell, L. K. (1996). Analysis of the disulfide linkage pattern in circulin A and B, HIV-inhibitory macrocyclic peptides. *Biochemical and Biophysical Research Communications, 228*(2), 632–638. S0006-291X(96)91708-8.

Dörnenburg, H. (2008). Plant cell culture technology-harnessing a biological approach for competitive cyclotides production. *Biotechnology Letters, 30*(8), 1311–1321. http://dx.doi.org/10.1007/s10529-008-9704-7.

Dörnenburg, H. (2010). Cyclotide synthesis and supply: from plant to bioprocess. *Biopolymers.* http://dx.doi.org/10.1002/bip.21466.

Duke, J. A. (2002). *Handbook of medicinal herbs* (2nd ed.). Boca Raton, FL: CRC Press.

Dymock, W., Hooper, D., & Warden, C. J. H. (1892). In W. Dymock, C. J. H. Warden, & D. Hooper (Eds.), *Pharmacographia indica: A history of the principal drugs of vegetable origin, met with in British India* (Vols. 1–3). London: K. Paul, Trench, Trübner & co., ltd.

Gerlach, S. L., Yeshak, M., Goransson, U., Roy, U., Izadpanah, R., & Mondal, D. (2013). Cycloviolacin O2 (CyO2) suppresses productive infection and augments the antiviral efficacy of nelfinavir in HIV-1 infected monocytic cells. *Biopolymers, 100*(5), 471–479. http://dx.doi.org/10.1002/bip.22325.

Getz, J. A., Cheneval, O., Craik, D. J., & Daugherty, P. S. (2013). Design of a cyclotide antagonist of neuropilin-1 and -2 that potently inhibits endothelial cell migration. *ACS Chemical Biology.* http://dx.doi.org/10.1021/cb4000585.

Gibbons, S. (2004). Anti-staphylococcal plant natural products. *Natural Product Reports, 21*(2), 263–277. http://dx.doi.org/10.1039/b212695h.

Göransson, U., Broussalis, A. M., & Claeson, P. (2003). Expression of Viola cyclotides by liquid chromatography-mass spectrometry and tandem mass spectrometry sequencing of intercysteine loops after introduction of charges and cleavage sites by aminoethylation. *Analytical Biochemistry, 318*(1), 107–117. S0003269703001143.

Göransson, U., & Craik, D. J. (2003). Disulfide mapping of the cyclotide kalata B1. Chemical proof of the cystic cystine knot motif. *Journal of Biological Chemistry, 278*(48), 48188–48196. http://dx.doi.org/10.1074/jbc.M308771200.

Göransson, U., Herrmann, A., Burman, R., Haugaard-Jonsson, L. M., & Rosengren, K. J. (2009). The conserved glu in the cyclotide cycloviolacin O2 has a key structural role. *ChemBioChem, 10*(14), 2354–2360. http://dx.doi.org/10.1002/cbic.200900342.
Göransson, U., Luijendijk, T., Johansson, S., Bohlin, L., & Claeson, P. (1999). Seven novel macrocyclic polypeptides from *Viola arvensis*. *Journal of Natural Products, 62*(2), 283–286. http://dx.doi.org/10.1021/np9803878.
Göransson, U., Sjögren, M., Svangård, E., Claeson, P., & Bohlin, L. (2004). Reversible antifouling effect of the cyclotide cycloviolacin O2 against barnacles. *Journal of Natural Products, 67*(8), 1287–1290. http://dx.doi.org/10.1021/np0499719.
Göransson, U., Svangård, E., Claeson, P., & Bohlin, L. (2004). Novel strategies for isolation and characterization of cyclotides: the discovery of bioactive macrocyclic plant polypeptides in the Violaceae. *Current Protein and Peptide Science, 5*(5), 317–329.
Gran, L. (1973a). On the effect of a polypeptide isolated from "Kalata-Kalata" (*Oldenlandia affinis* DC) on the oestrogen dominated uterus. *Acta Pharmacologica et Toxicologica (Copenhagen), 33*(5), 400–408.
Gran, L. (1973b). Oxytocic principles of *Oldenlandia affinis*. *Lloydia, 36*(2), 174–178.
Gran, L. (1973c). *The uteroactive principles of "Kalata-Kalata" (Oldenlandia affinis DC.). (PhD Thesis)*. Oslo: University of Oslo.
Gran, L., Sandberg, F., & Sletten, K. (2000). *Oldenlandia affinis* (R&S) DC. A plant containing uteroactive peptides used in African traditional medicine. *Journal of Ethnopharmacology, 70*(3), 197–203. http://dx.doi.org/10.1016/S0378-8741(99)00175-0.
Gruber, C. W., Elliott, A. G., Ireland, D. C., Delprete, P. G., Dessein, S., Göransson, U., et al. (2008). Distribution and evolution of circular miniproteins in flowering plants. *Plant Cell, 20*(9), 2471–2483. tpc.108.062331.
Grundemann, C., Thell, K., Lengen, K., Garcia-Kaufer, M., Huang, Y. H., Huber, R., et al. (2013). Cyclotides suppress human T-lymphocyte proliferation by an interleukin 2-dependent mechanism. *PLoS One, 8*(6), e68016. http://dx.doi.org/10.1371/journal.pone.0068016.
Gunasekera, S., Aboye, T. L., Madian, W. A., El-Seedi, H. R., & Göransson, U. (2013). Making ends meet: microwave-accelerated synthesis of cyclic and disulfide rich proteins via in situ thioesterification and native chemical ligation. *International Journal of Peptide Research and Therapeutics, 19*(1), 43–54. http://dx.doi.org/10.1007/s10989-012-9331-y.
Gunther, E. (1973). *Ethnobotany of western Washington* (Revised edition). Seattle: University of Washington Press.
Gustafson, K. R., McKee, T. C., & Bokesch, H. R. (2004). Anti-HIV cyclotides. *Current Protein and Peptide Science, 5*(5), 331–340.
Gustafson, K. R., Sowder, R. C., Henderson, L. E., Parsons, I. C., Kashman, Y., Cardellina, J. H., et al. (1994). Circulins A and B. Novel human immunodeficiency virus (HIV)-inhibitory macrocyclic peptides from the tropical tree *Chassalia parvifolia*. *Journal of the American Chemical Society, 116*(20), 9337–9338. http://dx.doi.org/10.1021/ja00099a064.
Hackeng, T. M., Griffin, J. H., & Dawson, P. E. (1999). Protein synthesis by native chemical ligation: expanded scope by using straightforward methodology. *Proceedings of the National Academy of Sciences of the United States of America, 96*(18), 10068–10073.
Hashempour, H., Koehbach, J., Daly, N. L., Ghassempour, A., & Gruber, C. W. (2013). Characterizing circular peptides in mixtures: sequence fragment assembly of cyclotides from a violet plant by MALDI-TOF/TOF mass spectrometry. *Amino Acids, 44*(2), 581–595. http://dx.doi.org/10.1007/s00726-012-1376-x.
Hekking, W. H. A. (1984). Studies on Neotropical violaceae—the genus *Fusispermum*. *Proceedings of the Koninklijke Nederlandse Akademie van Wetenschappen, Series C: Biological and Medical Sciences, 87*(2), 121–130.
Hekking, W. H. A. (1988). Studies on neotropical Violaceae 2. Arrangement of leaves, inflorescences and branchlets in neotropical *Rinorea*. *Flora, 180*(5–6), 345–376.

Hellinger, R., Koehbach, J., Fedchuk, H., Sauer, B., Huber, R., Gruber, C. W., et al. (2014). Immunosuppressive activity of an aqueous *Viola tricolor* herbal extract. *Journal of Ethnopharmacology, 151*(1), 299–306. http://dx.doi.org/10.1016/j.jep.2013.10.044.

Henriques, S. T., & Craik, D. J. (2012). Importance of the cell membrane on the mechanism of action of cyclotides. *ACS Chemical Biology,* 7(4), 626–636. http://dx.doi.org/10.1021/cb200395f.

Henriques, S. T., Huang, Y. H., Rosengren, K. J., Franquelim, H. G., Carvalho, F. A., Johnson, A., et al. (2011). Decoding the membrane activity of the cyclotide kalata B1: the importance of phosphatidylethanolamine phospholipids and lipid organization on hemolytic and anti-HIV activities. *Journal of Biological Chemistry, 286*(27), 24231–24241. http://dx.doi.org/10.1074/jbc.M111.253393.

Heo, J. O., Roszak, P., Furuta, K. M., & Helariutta, Y. (2014). Phloem development: current knowledge and future perspectives. *American Journal of Botany, 101*(9), 1393–1402. http://dx.doi.org/10.3732/ajb.1400197.

Herrick, J. W. (1977). *Iroqouis medical botany.* Albany: State University of New York.

Huang, Y. H., Colgrave, M. L., Clark, R. J., Kotze, A. C., & Craik, D. J. (2010). Lysine-scanning mutagenesis reveals an amendable face of the cyclotide kalata B1 for the optimization of nematocidal activity. *Journal of Biological Chemistry, 285*(14), 10797–10805. M109.089854.

Iwu, M. M. (1993). *Handbook of African medicinal plants.* Boca Raton, Florida: CRC Press, Inc.

Jagadish, K., Borra, R., Lacey, V., Majumder, S., Shekhtman, A., Wang, L., et al. (2013). Expression of fluorescent cyclotides using protein trans-splicing for easy monitoring of cyclotide-protein interactions. *Angewandte Chemie, International Edition in English, 52*(11), 3126–3131. http://dx.doi.org/10.1002/anie.201209219.

Jagadish, K., Gould, A., Borra, R., Majumder, S., Mushtaq, Z., Shekhtman, A., et al. (2015). Recombinant expression and phenotypic screening of a bioactive cyclotide against alpha-synuclein-induced cytotoxicity in Baker's yeast. *Angewandte Chemie, International Edition in English, 54*(29), 8390–8394. http://dx.doi.org/10.1002/anie.201501186.

Jennings, C., West, J., Waine, C., Craik, D., & Anderson, M. (2001). Biosynthesis and insecticidal properties of plant cyclotides: the cyclic knotted proteins from *Oldenlandia affinis. Proceedings of the National Academy of Sciences of the United States of America, 98*(19), 10614–10619. http://dx.doi.org/10.1073/pnas.191366898.

Jia, X., Kwon, S., Wang, C. I., Huang, Y. H., Chan, L. Y., Tan, C. C., et al. (2014). Semi-enzymatic cyclization of disulfide-rich peptides using Sortase A. *Journal of Biological Chemistry, 289*(10), 6627–6638. http://dx.doi.org/10.1074/jbc.M113.539262.

Kaas, Q., & Craik, D. J. (2010). Analysis and classification of circular proteins in CyBase. *Biopolymers.* http://dx.doi.org/10.1002/bip.21424.

Koehbach, J., O'Brien, M., Muttenthaler, M., Miazzo, M., Akcan, M., Elliott, A. G., et al. (2013). Oxytocic plant cyclotides as templates for peptide G protein-coupled receptor ligand design. *Proceedings of the National Academy of Sciences of the United States of America.* http://dx.doi.org/10.1073/pnas.1311183110.

Koehnke, J., Bent, A., Houssen, W. E., Zollman, D., Morawitz, F., Shirran, S., et al. (2012). The mechanism of patellamide macrocyclization revealed by the characterization of the PatG macrocyclase domain. *Nature Structural & Molecular Biology, 19*(8), 767–772. http://dx.doi.org/10.1038/nsmb.2340.

Leta Aboye, T., Clark, R. J., Craik, D. J., & Göransson, U. (2008). Ultra-stable peptide scaffolds for protein engineering-synthesis and folding of the circular cystine knotted cyclotide cycloviolacin O2. *ChemBioChem, 9*(1), 103–113. http://dx.doi.org/10.1002/cbic.200700357.

Lindgren, J., (1927). Läkemedelsnamn (Facsimile, Svensk Farmaceutisk Tidskrifts Förlags AB ed.) Lund: Berlingska Boktryckeriet.

Lindholm, P., Göransson, U., Johansson, S., Claeson, P., Gullbo, J., Larsson, R., et al. (2002). Cyclotides: a novel type of cytotoxic agents. *Molecular Cancer Therapeutics, 1*(6), 365–369.

Mahatmanto, T., Poth, A. G., Mylne, J. S., & Craik, D. J. (2014). A comparative study of extraction methods reveals preferred solvents for cystine knot peptide isolation from *Momordica cochinchinensis* seeds. *Fitoterapia, 95*, 22–33. http://dx.doi.org/10.1016/j.fitote.2014.02.016.

Mannich, C., & Krösche, W. (1912). Über ein Kondensationsprodukt aus Formaldehyd, Ammoniak und Antipyrin. *Archiv der Pharmazie, 250*(1), 647–667. http://dx.doi.org/10.1002/ardp.19122500151.

Maroti, G., Kereszt, A., Kondorosi, E., & Mergaert, P. (2011). Natural roles of antimicrobial peptides in microbes, plants and animals. *Research in Microbiology, 162*(4), 363–374. http://dx.doi.org/10.1016/j.resmic.2011.02.005.

Martinez-Crovetto, R. (1981). Fertility regulating plants used in popular medicine in Northeastern Argentina. *Parodiana, 1*(1), 97–117.

Melchior, H. (1925). Violaceae. In A. Engler, & K. Prantl (Eds.), *Die natürlichen Pflanzenfamilien* (2nd ed., Vol. 21, p. 21). Leipzig: Wilhelm Engelmann.

Mohimani, H., Liu, W. T., Mylne, J. S., Poth, A. G., Colgrave, M. L., Tran, D., et al. (2011). Cycloquest: Identification of cyclopeptides via database search of their mass spectra against genome databases. *Journal of Proteome Research*. http://dx.doi.org/10.1021/pr200323a.

Mulvenna, J. P., Wang, C., & Craik, D. J. (2006). CyBase: a database of cyclic protein sequence and structure. *Nucleic Acids Research, 34*(Database issue), D192–D194, 34/suppl_1/D192.

Mylne, J. S., Chan, L. Y., Chanson, A. H., Daly, N. L., Schaefer, H., Bailey, T. L., et al. (2012). Cyclic peptides arising by evolutionary parallelism via asparaginyl-endopeptidase-mediated biosynthesis. *Plant Cell, 24*(7), 2765–2778. http://dx.doi.org/10.1105/tpc.112.099085.

Nguyen, G. K., Wang, S., Qiu, Y., Hemu, X., Lian, Y., & Tam, J. P. (2014). Butelase 1 is an Asx-specific ligase enabling peptide macrocyclization and synthesis. *Nature Chemical Biology, 10*(9), 732–738. http://dx.doi.org/10.1038/nchembio.1586.

Northfield, S. E., Wang, C. K., Schroeder, C. I., Durek, T., Kan, M. W., Swedberg, J. E., et al. (2014). Disulfide-rich macrocyclic peptides as templates in drug design. *European Journal of Medicinal Chemistry*, 77, 248–257. http://dx.doi.org/10.1016/j.ejmech.2014.03.011.

Nuringtyas, T. R., Choi, Y. H., Verpoorte, R., Klinkhamer, P. G., & Leiss, K. A. (2012). Differential tissue distribution of metabolites in *Jacobaea vulgaris, Jacobaea aquatica* and their crosses. *Phytochemistry, 78*, 89–97. http://dx.doi.org/10.1016/j.phytochem.2012.03.011.

Ovesen, R. G., Brandt, K. K., Göransson, U., Nielsen, J., Hansen, H. C., & Cedergreen, N. (2011). Biomedicine in the environment: cyclotides constitute potent natural toxins in plants and soil bacteria. *Environmental Toxicology and Chemistry, 30*(5), 1190–1196. http://dx.doi.org/10.1002/etc.496.

Ovesen, R. G., Nielsen, J., & Hansen, H. C. (2011). Biomedicine in the environment: sorption of the cyclotide kalata B2 to montmorillonite, goethite, and humic acid. *Environmental Toxicology and Chemistry, 30*(8), 1785–1792. http://dx.doi.org/10.1002/etc.579.

Padovan, L., Scocchi, M., & Tossi, A. (2010). Structural aspects of plant antimicrobial peptides. *Current Protein and Peptide Science, 11*(3), 210–219.

Pallaghy, P. K., Nielsen, K. J., Craik, D. J., & Norton, R. S. (1994). A common structural motif incorporating a cystine knot and a triple-stranded beta-sheet in toxic and inhibitory polypeptides. *Protein Science, 3*(10), 1833–1839. http://dx.doi.org/10.1002/pro.5560031022.

Park, S., Strömstedt, A. A., & Göransson, U. (2014). Cyclotide structure-activity relationships: qualitative and quantitative approaches linking cytotoxic and anthelmintic activity to the clustering of physicochemical forces. *PLoS One, 9*(3), e91430. http://dx.doi.org/10.1371/journal.pone.0091430.

de Paula-Souza, J., & Ballard, H. E. (2014). Re-establishment of the name Pombalia, and new combinations from the polyphyletic *Hybanthus* (Violaceae). *Phytotaxa, 183*(1), 1–15.

de Paula-Souza, J., & Pirani, J. R. (2014). Reestablishment of *Calyptrion* (Violaceae). *Taxon, 63*(6), 1335–1339.

Perry, L. M. (1980). *Medicinal plants of East and Southeast Asia: Attributed properties and uses.* Cambridge, Massachusetts, and London, England: The MIT Press.

Plan, M. R., Saska, I., Cagauan, A. G., & Craik, D. J. (2008). Backbone cyclised peptides from plants show molluscicidal activity against the rice pest *Pomacea canaliculata* (golden apple snail). *Journal of Agricultural and Food Chemistry, 56*(13), 5237–5241. http://dx.doi.org/10.1021/jf800302f.

Poth, A. G., Mylne, J. S., Grassl, J., Lyons, R. E., Millar, A. H., Colgrave, M. L., et al. (2012). Cyclotides associate with leaf vasculature and are the products of a novel precursor in petunia (Solanaceae). *Journal of Biological Chemistry, 287*(32), 27033–27046. http://dx.doi.org/10.1074/jbc.M112.370841.

Pränting, M., Lööv, C., Burman, R., Göransson, U., & Andersson, D. I. (2010). The cyclotide cycloviolacin O2 from *Viola odorata* has potent bactericidal activity against Gram-negative bacteria. *Journal of Antimicrobial Chemotherapy, 65*(9), 1964–1971. dkq220.

Quisumbing, E. A. (1951). *Medicinal plants of the Philippines.* Dept. of Agriculture and Natural Resources.

Richards, F. M., & Knowles, J. R. (1968). Glutaraldehyde as a protein cross-linkage reagent. *Journal of Molecular Biology, 37*(1), 231–233.

Rosengren, K. J., Daly, N. L., Plan, M. R., Waine, C., & Craik, D. J. (2003). Twists, knots, and rings in proteins. Structural definition of the cyclotide framework. *Journal of Biological Chemistry, 278*(10), 8606–8616. http://dx.doi.org/10.1074/jbc.M211147200.

Saether, O., Craik, D. J., Campbell, I. D., Sletten, K., Juul, J., & Norman, D. G. (1995). Elucidation of the primary and three-dimensional structure of the uterotonic polypeptide kalata B1. *Biochemistry, 34*(13), 4147–4158.

Sandberg, F. (1965). *Étude sur les plantes médicinales et toxiques d'Afrique équatoriale, Tome 3, Fascicule 1* p. 27). Paris: Cahièrs de la Maboké, 12 Rue de Buffon.

Sando, L., Henriques, S. T., Foley, F., Simonsen, S. M., Daly, N. L., Hall, K. N., et al. (2011). A synthetic mirror image of kalata B1 reveals that cyclotide activity is independent of a protein receptor. *ChemBioChem, 12*(16), 2456–2462. http://dx.doi.org/10.1002/cbic.201100450.

Schöpke, T., Hasan, A. M. I., Kraft, R., Otto, A., & Hiller, K. (1993). Haemolytisch aktive Komponenten aus *Viola tricolor* L. und *Viola arvensis* murray. *Scientia Pharmaceutica, 61*, 145–153.

Schroeder, C. I., Swedberg, J. E., & Craik, D. J. (2013). Recent progress towards pharmaceutical applications of disulfide-rich cyclic peptides. *Current Protein and Peptide Science, 14*(6), 532–552.

Schultes, R. E., & Raffauf, R. F. (1990). *The healing forest: Medicinal and toxic plants of the Northwest Amazonia.* Portland, Oregon: Dioscorides Press.

Segura, A., Moreno, M., Madueno, F., Molina, A., & Garcia-Olmedo, F. (1999). Snakin-1, a peptide from potato that is active against plant pathogens. *Molecular Plant Microbe Interactions, 12*(1), 16–23. http://dx.doi.org/10.1094/MPMI.1999.12.1.16.

Seydel, P., & Dörnenburg, H. (2006). Establishment of in vitro plants, cell and tissue cultures from *Oldenlandia affinis* for the production of cyclic peptides. *Plant Cell, Tissue and Organ Culture, 85*(3), 247–255. http://dx.doi.org/10.1007/s11240-005-9056-0.

Simonsen, S. M., Sando, L., Ireland, D. C., Colgrave, M. L., Bharathi, R., Göransson, U., et al. (2005). A continent of plant defense peptide diversity: cyclotides in Australian *Hybanthus* (Violaceae). *Plant Cell, 17*(11), 3176–3189. tpc.105.034678.

Simonsen, S. M., Sando, L., Rosengren, K. J., Wang, C. K., Colgrave, M. L., Daly, N. L., et al. (2008). Alanine scanning mutagenesis of the prototypic cyclotide reveals a cluster of residues essential for bioactivity. *Journal of Biological Chemistry, 283*(15), 9805–9813. M709303200.

Slazak, B., Jacobsson, E., Kuta, E., & Göransson, U. (2015). Exogenous plant hormones and cyclotide expression in *Viola uliginosa* (Violaceae). *Phytochemistry, 117*, 527–536. http://dx.doi.org/10.1016/j.phytochem.2015.07.016.

Slazak, B., Sliwinska, E., Saluga, M., Ronikier, M., Bujak, J., Slomka, A., et al. (2015). Micropropagation of *Viola uliginosa* (Violaceae) for endangered species conservation and for somaclonal variation-enhanced cyclotide biosynthesis. *Plant Cell, Tissue and Organ Culture, 120*(1), 179–190. http://dx.doi.org/10.1007/s11240-014-0592-3.

Sletten, K., & Gran, L. (1973). Some molecular properties of kalatapeptide B-1. *Meddelelser fra Norsk Farmaceutisk Selskap, 35*, 69–82.

Stanger, K., Maurer, T., Kaluarachchi, H., Coons, M., Franke, Y., & Hannoush, R. N. (2014). Backbone cyclization of a recombinant cystine-knot peptide by engineered Sortase A. *FEBS Letters, 588*(23), 4487–4496. http://dx.doi.org/10.1016/j.febslet.2014.10.020.

Stopp, K. (1962). The medicinal plants used by the Mt. Hagen people (Mbowamb) in New Guinea. *Economic Botany, 17*, 16–22.

Svangård, E., Burman, R., Gunasekera, S., Lövborg, H., Gullbo, J., & Göransson, U. (2007). Mechanism of action of cytotoxic cyclotides: cycloviolacin O2 disrupts lipid membranes. *Journal of Natural Products, 70*(4), 643–647. http://dx.doi.org/10.1021/np070007v.

Tam, J. P., & Lu, Y. A. (1997). Synthesis of large cyclic cystine-knot peptide by orthogonal coupling strategy using unprotected peptide precursor. *Tetrahedron Letters, 38*(32), 5599–5602.

Tam, J. P., & Lu, Y. A. (1998). A biomimetic strategy in the synthesis and fragmentation of cyclic protein. *Protein Science, 7*(7), 1583–1592. http://dx.doi.org/10.1002/pro.5560070712.

Tam, J. P., Lu, Y. A., Yang, J. L., & Chiu, K. W. (1999). An unusual structural motif of antimicrobial peptides containing end-to-end macrocycle and cystine-knot disulfides. *Proceedings of the National Academy of Sciences of the United States of America, 96*(16), 8913–8918.

Thongyoo, P., Jaulent, A. M., Tate, E. W., & Leatherbarrow, R. J. (2007). Immobilized protease-assisted synthesis of engineered cysteine-knot microproteins. *ChemBioChem, 8*(10), 1107–1109. http://dx.doi.org/10.1002/cbic.200700187.

Thongyoo, P., Roque-Rosell, N., Leatherbarrow, R. J., & Tate, E. W. (2008). Chemical and biomimetic total syntheses of natural and engineered MCoTI cyclotides. *Organic & Biomolecular Chemistry, 6*(8), 1462–1470. http://dx.doi.org/10.1039/b801667d.

Tokuoka, T. (2008). Molecular phylogenetic analysis of Violaceae (Malpighiales) based on plastid and nuclear DNA sequences. *Journal of Plant Research, 121*(3), 253–260. http://dx.doi.org/10.1007/s10265-008-0153-0.

Trabi, M., & Craik, D. J. (2004). Tissue-specific expression of head-to-tail cyclized miniproteins in Violaceae and structure determination of the root cyclotide *Viola hederacea* root cyclotide1. *Plant Cell, 16*(8), 2204–2216. http://dx.doi.org/10.1105/tpc.104.021790.

Wahlert, G., Ballard, H., Jr., & de Paula-Souza, J. (2015). *Ixchelia*, a new genus of Violaceae from Mexico and Mesoamerica. *Brittonia*, 1–11. http://dx.doi.org/10.1007/s12228-015-9375-3.

Wahlert, G. A., Marcussen, T., de Paula-Souza, J., Feng, M., & Ballard, H. E. (2014). A phylogeny of the violaceae (Malpighiales) inferred from plastid DNA Sequences: Implications for generic diversity and intrafamilial classification. *Systematic Botany, 39*(1), 239–252. http://dx.doi.org/10.1600/036364414X678008.

Walch, A., Rauser, S., Deininger, S. O., & Hofler, H. (2008). MALDI imaging mass spectrometry for direct tissue analysis: a new frontier for molecular histology. *Histochemistry and Cell Biology, 130*(3), 421–434. http://dx.doi.org/10.1007/s00418-008-0469-9.

Wang, C. K., Colgrave, M. L., Gustafson, K. R., Ireland, D. C., Göransson, U., & Craik, D. J. (2008). Anti-HIV cyclotides from the Chinese medicinal herb *Viola yedoensis*. *Journal of Natural Products, 71*(1), 47–52. http://dx.doi.org/10.1021/np070393g.

Wang, C. K., Hu, S. H., Martin, J. L., Sjögren, T., Hajdu, J., Bohlin, L., et al. (2009). Combined X-ray and NMR analysis of the stability of the cyclotide cystine knot fold that underpins its insecticidal activity and potential use as a drug scaffold. *Journal of Biological Chemistry, 284*(16), 10672–10683. M900021200.

Witherup, K. M., Bogusky, M. J., Anderson, P. S., Ramjit, H., Ransom, R. W., Wood, T., et al. (1994). Cyclopsychotride A, a biologically active, 31-residue cyclic peptide isolated from *Psychotria longipes*. *Journal of Natural Products, 57*(12), 1619–1625.

Witkowska-Banaszczak, E., Bylka, W., Matlawska, I., Goslinska, O., & Muszynski, Z. (2005). Antimicrobial activity of *Viola tricolor* herb. *Fitoterapia, 76*(5), 458–461. http://dx.doi.org/10.1016/j.fitote.2005.03.005.

Wong, C. T., Taichi, M., Nishio, H., Nishiuchi, Y., & Tam, J. P. (2011). Optimal oxidative folding of the novel antimicrobial cyclotide from *Hedyotis biflora* requires high alcohol concentrations. *Biochemistry, 50*(33), 7275–7283. http://dx.doi.org/10.1021/bi2007004.

Yeshak, M. Y., Burman, R., Eriksson, C., & Göransson, U. (2012). Optimization of cyclotide extraction parameters. *Phytochemistry Letters, 5*(4), 776–781. http://dx.doi.org/10.1016/j.phytol.2012.09.001.

Yeshak, M. Y., Göransson, U., Burman, R., & Hellman, B. (2012). Genotoxicity and cellular uptake of cyclotides: evidence for multiple modes of action. *Mutation Research, 747*(2), 176–181. http://dx.doi.org/10.1016/j.mrgentox.2012.05.006.

Zarrabi, M., Dalirfardouei, R., Sepehrizade, Z., & Kermanshahi, R. K. (2013). Comparison of the antimicrobial effects of semipurified cyclotides from Iranian *Viola odorata* against some of plant and human pathogenic bacteria. *Journal of Applied Microbiology, 115*(2), 367–375. http://dx.doi.org/10.1111/jam.12251.

Zhang, J., Hua, Z., Huang, Z., Chen, Q., Long, Q., Craik, D. J., et al. (2015). Two blast-independent tools, CyPerl and CyExcel, for harvesting hundreds of novel cyclotides and analogues from plant genomes and protein databases. *Planta, 241*(4), 929–940. http://dx.doi.org/10.1007/s00425-014-2229-5.

Zhang, J., Li, J., Huang, Z., Yang, B., Zhang, X., Li, D., et al. (2015). Transcriptomic screening for cyclotides and other cysteine-rich proteins in the metallophyte *Viola baoshanensis*. *Journal of Plant Physiology, 178*, 17–26. http://dx.doi.org/10.1016/j.jplph.2015.01.017.

CHAPTER THREE

Cyclotides in the Rubiaceae

Johannes Koehbach*, Christian W. Gruber§,1
*School of Biomedical Sciences, The University of Queensland, St. Lucia, QLD, Australia
§Center for Physiology and Pharmacology, Medical University of Vienna, Vienna, Austria
[1]Corresponding author: E-mail: christian.w.gruber@meduniwien.ac.at

Contents

Abstract

Cyclotides are small disulfide-rich plant peptides. Their unique structural features, intrinsic bioactivities, and enormous sequence diversity make them interesting molecules for pharmaceutical applications. So far the amino acid sequences of nearly 400 individual cyclotides have been reported and they have been discovered in plant species of the Violaceae, Cucurbitaceae, Fabaceae, Solanaceae, Poaceae, and Rubiaceae families. Rubiaceae is the fourth largest angiosperm family and comprises about 650 genera and 13,000 species. The family includes coffee, which is a major export commodity and it was the top agricultural export for 12 countries in 2004, as well as the world's seventh largest legal agricultural export by value in 2005. Additionally, the family comprises timber species such as bilinga, many ornamental plants, and important plants of medicinal purpose such as quinine and ipecac. In part due to the wide geographical distribution, plants of the coffee family are estimated to constitute about 8% of the world's flora. In this book chapter, we will (1) summarize our efforts to identify and characterize novel cyclotide-expressing plants within the Rubiaceae family, (2)

Advances in Botanical Research, Volume 76
ISSN 0065-2296
http://dx.doi.org/10.1016/bs.abr.2015.09.002

analyze the sequence diversity of cyclotides from Rubiaceae, (3) discuss their evolution in flowering plants based on the distribution of cyclotide-bearing Rubiaceae species, and (4) provide a brief overview of bioactive properties of cyclotides that have been isolated from Rubiaceae plant species. Based on studies of cyclotides in Rubiaceae their number has been predicted to be greater than 50,000 and we believe cyclotides are one of the largest protein families within the plant kingdom.

1. INTRODUCTION

Cyclotides are a unique class of ribosomal-synthesized cysteine-rich peptides characterized by a head-to-tail cyclized backbone and three conserved disulfide-bonds in a knotted arrangement. So far, approximately 400 cyclotides have been discovered (Gerlach et al., 2013) and have been deposited via *CyBase*—the Web-based circular peptide database (www.cybase.org.au); the majority of cyclotides has been isolated from Rubiaceae (this chapter) or Violaceae plant species (Chapter 2). Indeed, the first cyclotide has been discovered from a Rubiaceae plant: *Oldenlandia affinis* DC. is a common weed that grows in tropical regions of Africa. In 1970, Lorents Gran reported that an extract of this plant is used as a medicinal tea by women in the Democratic Republic of Congo for its uterotonic properties (Gran, 1970). A few years later in 1973, the amino acid composition of the first cyclotide—kalata B1—has been characterized (Gran, 1973) and its three-dimensional structure has been elucidated in 1995 using NMR spectroscopy (Saether et al., 1995).

Now there is increasing evidence that cyclotides are expressed and distributed more widespread in various orders of the plant kingdom. In the context of understanding the distribution of cyclotides within the Rubiaceae, it is important to keep in mind the global distribution of cyclotides among flowering plants. Besides Rubiaceae species, cyclotide-containing plants have been identified in the violet family (Violaceae, order Malpighiales, Rosids) (see Chapter 2), the potato family (Solanaceae, order Solanales within Asterids) (Poth et al., 2012), the legume family (Fabaceae, order Fabales within Rosids) (Poth, Colgrave, Philip, et al., 2011), and the grasses (Poaceae, order Commelinids within monocots) (Nguyen, Lian, et al., 2012). In addition, cyclotide-like peptides have been described in the squash family (Cucurbitaceae, order Cucurbitales, Rosids) (Hernandez et al., 2000). The comparison of all cyclotides reported to date suggests that they contain common molecular, structural, and biophysical properties, which can be summarized as follows:

Cyclotides:

1. are between 27 and 38 amino acid residues in size,
2. contain six conserved cysteine residues,
3. have a cystine-knot motif,
4. are characterized by a cyclized peptide backbone,[1] and
5. are synthesized *in planta* via expression and processing of gene-encoded precursor molecules.

These common features may be utilized for discovering novel cyclotides in plants by means of analytical chemistry, peptidomics, and molecular biology. In the following, we will (1) describe methods for discovering of cyclotides, (2) report about the distribution of cyclotides in the Rubiaceae including brief summaries about their biosynthesis and molecular characteristics as well as give an outlook on the discovery of cyclotides in other Gentianales. In addition, we will (3) review the current understanding of cyclotide evolution based on their distribution in the Rubiaceae family and (4) highlight recent findings about the bioactivity of cyclotides isolated from Rubiaceae plants.

2. DISCOVERY OF CYCLOTIDES

2.1 Cyclotide Analytical Screening Workflow

Based on the above-mentioned properties, a number of criteria for the identification of cyclotides in plants have been established (Gruber et al., 2008; Hashempour, Koehbach, Daly, Ghassempour, & Gruber, 2013; Koehbach, Attah, et al., 2013; Poth, Colgrave, Philip, et al., 2011) as follows:

1. characteristic mass range,
2. mass shift upon chemical derivatization of cysteine residues,
3. late-eluting peaks in reversed-phase (RP) high-performance liquid chromatography (HPLC) and stability of native peptides during collision-induced dissociation tandem mass spectrometry (MS),
4. enzymatic ring-opening of the peptide backbone by the addition of water, and
5. transcriptome mining for annotation of cyclotide precursor sequences.

[1] There has been evidence of infrequent occurrence of acyclic cyclotides.

A typical cyclotide screening process starts with the collection of appropriate plant material (for review see Gruber, 2010) and chemical extraction. After solvent extraction of homogenized or pulverized plant material (e.g., a small piece of plant tissue), a single prepurification step is being carried out by solid phase extraction over prepacked C_{18} columns. The resulting samples, which are referred to as crude extracts will be initially analyzed by matrix-assisted laser desorption ionization time-of-flight (MALDI-TOF) MS, or alternative MS methods. All extracts that yield mass peaks within the expected range of 2500–4000 Da will be subjected to chemical modification to verify the presence of three disulfide bonds and six cysteines. Therefore, aliquots of the crude extracts will be reduced and carbamidomethylated using iodoacetamide, which in the presence of six cysteines leads to a distinct mass shift of 348 Da. In parallel, crude extracts will be analyzed by analytical RP-HPLC to confirm the presence of late-eluting, hydrophobic peptide peaks.

Often plants contain other disulfide-rich peptides besides cyclotides, such as knottins, defensins, or other antimicrobial peptides that may fulfill the above-described criteria of mass range, hydrophobicity, and cysteine content. Therefore, it has been necessary to define additional and more stringent analysis criteria that reflect distinct cyclotide features, for instance the circular backbone. This can be achieved the following way: Many cyclotides contain a single conserved glutamic acid residue in loop 1 and therefore enzymatic digestion with endoproteinase GluC yields single-site cleaved peptides (Craik, Daly, Bond, & Waine, 1999). In the case of backbone-cyclized peptides, such a single-site cleavage can be identified by the addition of a water molecule and thus in a resulting mass shift of +18 Da. It is important to note that there is evidence of the occasional occurrence of acyclic cyclotide homologs (Nguyen, Lian, et al., 2012; Nguyen, Lim, Nguyen, & Tam, 2012; Nguyen, Zhang, Wang, et al., 2011; Poth et al., 2012), as well as cyclotides that contain none, or more than one glutamic acid residues. These peptides may be identified as "false-negatives" using this criterion. Another opportunity to utilize the cyclic cystine-knot motif, in particular the circular backbone, for the purpose of cyclotide discovery is by observation of fragmentation patterns in MS. The typical knotted and cyclic backbone structure of cyclotides prohibits direct fragmentation analysis due to the lack of free N- and C-termini that are necessary for retaining charges. Spectra of head-to-tail circular peptides lack fragmentation, whereas spectra of linear peptides display multiple fragment ions (Poth, Colgrave, Philip, et al., 2011). Therefore, tandem MS fragmentation patterns of cyclotides are distinct and it is possible to distinguish stable cyclotides from commonly unstable linear peptides. Nevertheless, other peptides that contain three disulfide bonds in a non-cystine-knotted arrangement or other circular peptides may lead to "false-positives" using this criterion (Koehbach, Attah, et al., 2013). Ultimately for the accurate discovery of cyclotide plant species, it is essential to characterize at least partial peptide sequences that match known cyclotide sequences. These partial peptide sequences should consist of at least two adjacent intercysteine loops. Needless to say that no single criterion exists, which would suffice unambiguous identification of cyclotides in plant extracts, but only the positive evaluation of a combination of different criteria helps to determine cyclotide expression during high-throughput screening efforts. Hence, if one or more peptides in the crude plant extracts satisfy all of the above set criteria (summarized in Figure 1(A)), it is safe to assume the presence of cyclotides in the respective plant species. This screening procedure has been applied and validated in recent studies of cyclotide expression and distribution in Rubiaceae plants and offers a rapid, accurate, and reliable methodology for a high sample numbers and limited individual sample amount (Gruber et al., 2008; Koehbach, Attah, et al., 2013).

As an alternative, in studies with less focus on throughput, but detailed molecular structural analysis, cyclotide de novo sequencing is the method-of-choice. Appropriate protocols will be discussed in detail in Chapter 5. However, cyclotide identification is not restricted to the analysis of extracts or (semi-)pure peptides but can also be performed at nucleic acid level.

2.2 Cyclotides Sequence Analysis Using Transcriptomics and Peptidomics

As previously mentioned, cyclotides are true gene products and hence their identification can also be facilitated by isolating precursor protein sequences. Nowadays, due to advances in next generation sequencing, an increasing number of genome and transcriptome datasets are publicly available. Mining of in silico data and in particular transcriptome mining has great potential for the discovery of novel peptides (Gruber & Muttenthaler, 2012; Koehbach & Jackson, 2015). One advantage over MS-based peptide discovery is that transcriptome sequencing provides the correct primary sequence including the correct identification of isobaric residues leucine and isoleucine. Furthermore, this approach does not only yield mature peptides, but enables the characterization of partial or full length precursor sequences, which allows the analysis of processing sites involved in cyclotide biosynthesis. On the

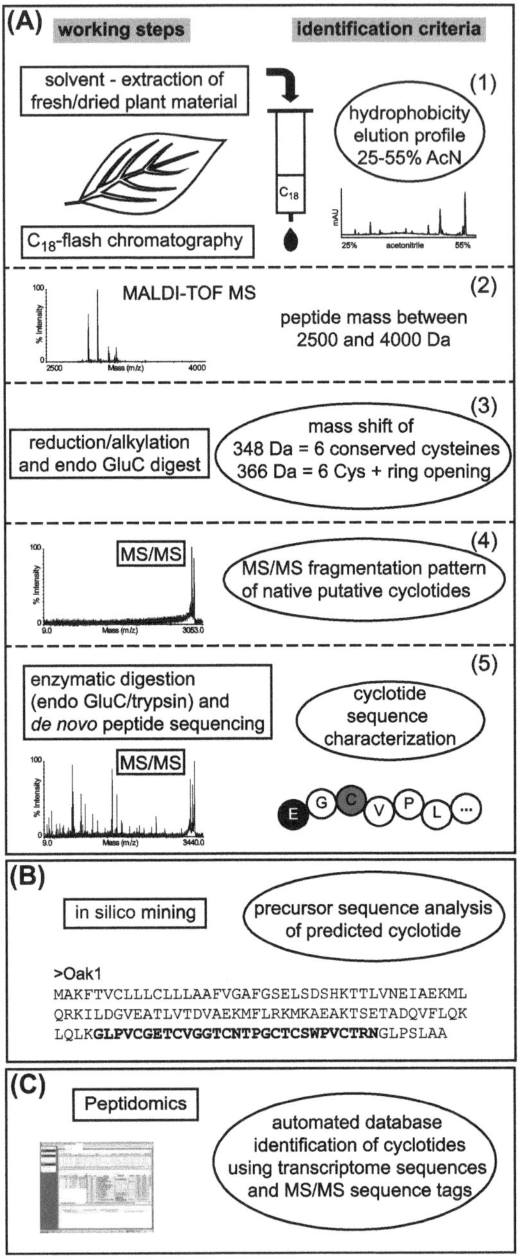

Figure 1 ***Cyclotide screening methodology.*** Schematic illustration of the experimental workflow and identification criteria that are being used for the discovery of novel cyclotide-containing species and peptide sequences (A) Plant extracts or processed samples are analyzed stepwise to assess their (1) hydrophobicity, (2) mass range, (3) disulfide

other hand, transcriptome sequencing has disadvantages in providing accurate information about posttranslational modifications such as cyclization. Hence, the identification of putative cyclotides by transcriptome or genome sequencing remains to be verified at peptide level. To avoid false hits, a combination of chemical and analytical identification using MS with high-throughput transcriptome sequencing provides the most accurate results for rapid cyclotide discovery (Koehbach, Attah, et al., 2013). The original idea of for this peptidomics workflow has been previously reported by Colgrave, Poth, Kaas, and Craik (2010); the automated ERA database tool aims to aid a rapid identification of putative cyclotide sequences in plant samples. Nevertheless, the discovery of novel sequences using such an automated peptidomics approach relies on a high similarity to those deposited in the database. It is therefore crucial to continuously update the database with sequences and sequence tags obtained from manual de novo sequencing as well as high-throughput transcriptome sequencing and in silico mining approaches. In summary, given the steadily growing number of published cyclotide sequences and availability of plant genome and transcriptome sequences, refined peptidomics workflows are likely to become a useful alternative for de novo peptide characterization in Rubiaceae and other plant families.

3. DISTRIBUTION OF CYCLOTIDES IN RUBIACEAE

3.1 Rubiaceae—An Overview

Rubiaceae is the fourth largest family within angiosperm (= Magnoliophyta) plants. With an estimated total number of approximately 13,000 species within 650 genera (Bremer & Eriksson, 2009; Robbrecht & Manen, 2006) they comprise about 4% of all flowering plants, making them one of the largest plant families among the eudicotyledons. Simplified, according to the angiosperm phylogeny system (The Angiosperm

bonds/cysteine content and circular backbone, and (4) MS/MS fragmentation pattern of native peptides. Only samples that allow the determination of partial sequence tags or full cyclotide sequences (5) can be assigned unambiguously as novel cyclotide-containing species. As alternative or complementary approach, in silico mining of transcriptome or genome datasets (B) allows the discovery of novel cyclotide sequences from their precursor proteins. For rapid and accurate identification of cyclotides automated peptidomics workflows have been developed (C). *Figure has been adapted and modified from Koehbach, Attah, et al. (2013).*

Phylogeny, 2009) flowering plants can be classified into basal angiosperms, magnoliids, monocotyledons, and dicotyledons, which can be further subclassified. The two major classes within dicotyl plants are rosids and asterids. Within asterid plants the family of Rubiaceae are placed in the order Gentianales.

Rubiaceae have adapted to a range of environmental conditions. Thus, Rubiaceae plants are found in almost any type of habitat around the world. This includes arid to desert environments, but equally humid rainforests as well as subarctic cold climate or tropical hot climatic regions (Delprete, 2004). In temperate regions, Rubiaceae represent mostly herbaceous species whereas the bulk of species occurs in tropical regions and is predominantly woody. Rubiaceae plants fall into two major lineages, i.e., Cinchonoideae/Ixoroideae and Rubioideae based on distinguishing features such as morphology, anatomy, flower biology, and chemical compounds (Robbrecht & Manen, 2006). Further subclassification into clades, tribes, and subfamilies is trying to categorize the vast array of different but related species and underpins the high variety of this plant family. Similar names of tribes or clades and genera can be confusing and the correct use of taxonomy is not trivial. Moreover, their classification is subject to ongoing rearrangements based on molecular phylogenetic studies. In line with the often difficult phylogenetic classification, Rubiaceae plants produce a different set of chemical compounds. For example, *Psychotria* and *Palicourea* spp. are known as rich source of alkaloids whereas the genus *Ronabea* and the tribe Lasiantheae are characterized by the presence of iridoid glucosids, but lack alkaloids (Berger et al., 2011). Although chemosystematics applies to secondary metabolites, it is intriguing if true gene products such as cyclotides may serve in the future as chemosystematic parameters to distinguish Rubiaceae species and delineate phylogenetic relationships.

The presence or absence of chemical constituents has raised particular interest of certain Rubiaceae species for being used as timber, ornamental, or medicinal plants. Several species of the Rubiaceae family are commonly well known, with the genus *Coffea* being the most prominent representative. For this reason, the family of Rubiaceae plants is often referred to as the "coffee family." Nevertheless, several other species are of commercial or pharmaceutical relevance. This includes plants with reported medicinal properties, such as *Cinchona officinalis* (quinine) (Achan et al., 2011) or *Carapichea ipecacuanha* (ipecac, formerly known as *Psychotria ipecacuanha*) (WHO, 2007). Another representative member of Rubiaceae is the species *Rubia tinctorum* (madder), which has been used for its red coloring properties.

Other species are used for their high-quality timber (e.g., *Nauclea diderichii*) or as ornamentals (*Ixora, Gardenia*). Moreover, for a variety of Rubiaceae species aphrodisiac or psychoactive properties have been reported (Adkins, Boyer, & McCurdy, 2011). This brief summary highlights the unique variety of the Rubiaceae family and therefore it is not surprising that identification of Rubiaceae species is not trivial and requires a profound knowledge of their morphology.

3.2 Morphology and Taxonomy of Rubiaceae

The Rubiaceae family experiences an ongoing reclassification and renaming of species between tribes and genera, respectively. These novel rearrangements are based on molecular phylogenetic studies (Bremer & Eriksson, 2009; Robbrecht & Manen, 2006), whereas older classifications rely on morphological features. Although being a very large family there are certain features that allow the identification and discrimination of Rubiaceae plants from other families. Previously, we have reported the following guide to Rubiaceae identification (Gruber, 2010):

"*Most Rubiaceae are shrubs or small to large trees. They are easily recognized at family level by decussate and entire leaves, presence of stipules, 4- or 5-merous actinomorphic flowers, and inferior ovary.*

Besides these characteristics, the following list of features is used to classify Rubiaceae:

- *Floral features: Flowers are actinomorphic (rarely zygomorphic) and bisexual (rarely unisexual); corolla connate (i.e., sympetalous); inflorescences are various; flowers often show heterostyly and their ovaries are inferior and often with axile placentation*
- *Fruit and seed features: Dicotyledon; fruits are berries, capsules, drupe, or schizocarp; seeds are single or many and with or without endosperm*
- *Vegetative features: Occurrence as herbs, shrubs, trees, lianas, or epiphytes; leaves are simple, entire (rarely lobed), opposite or whorled; stipules are often fused at the node and they may be leaf-like*"

Furthermore, we have previously described a five-point guide to Rubiaceae identification for the nonexpert (Gruber, 2010). This guide may be "*useful during field collections to identify plants to the family level and in a few cases to the genus level. With the help of adequate literature, it may be possible to identify plants to the species level. Please note: This guide should be used as a first 'hands-on guide' only and not as a reliable tool for the correct identification of Rubiaceae plants to the species level. However, if a plant meets all the five features described below, the chance of it belonging to the Rubiaceae family is very high!*

1. *Red color: The word Rubiaceae is thought to originate etymologically from the Latin word 'rubeus', meaning red. As an example, the most famous member of the Rubiaceae family, i.e., coffee (Coffea arabica) has bright red fruits when ripe. However, the color red as a feature of Rubiaceae seems rather anecdotal, as there are many Rubiaceae plants that cannot be characterized by their red color and there are many examples of non-Rubiaceae plants which also have red-colored plant parts. In the context of a field guide to identify Rubiaceae, the color red is a useful indicator. When walking through a forest, red colored fruits, flowers or leaves can easily be recognized and it is worth investigating that particular plant and by using secondary indicators (such as leaf and fruit morphology) to determine whether the species belongs to the Rubiaceae family or not.*
2. *Opposite leaves: Rubiaceae leaves are opposite (i.e., two leaves directly opposite—180°—attached at the same node) or in rare cases whorled (i.e., three or more leaves attached at the same node). This feature is easily recognized when seeking Rubiaceae in the field, although other plant families have opposite leaves as well.*
3. *Stipules: Stipules can often be recognized. They are an outgrowth of the lower zone of a young leaf and part of the leaf base. Sometimes they can be very small and the use of a pocket magnifying glass is necessary to recognize them.*
4. *Leaf structure: The leaves are entire or whole (i.e., they have a smooth margin) and they have abaxial veins (i.e., the veins exit the axis of the leaf in alternate fashion and bend towards the apex of the leaves).*
5. *Flowers/fruits: Flowers and fruits are mostly actinomorphic, which means that the flowers and reproductive organs can be divided into symmetrical planes."*

The above-described knowledge about the morphology and taxonomy of Rubiaceae is indispensable prior to attempting studying the distribution of cyclotides in species of Rubiaceae. We would therefore like to point out that cyclotide discovery relies on the accurate species determination with the help of experienced plant taxonomists.

3.3 Distribution of Cyclotides within Rubiaceae

Contrary to the family of violet plants (see Chapter 2), for which to date every analyzed species has been confirmed to contain cyclotides, the occurrence within Rubiaceae is not that clear and by far not every species sampled has been identified to contain cyclotides. At this point, it has to be mentioned that the family of violet plants only comprise around 800 species compared to ~13,000 of Rubiaceae, and for example, the genus *Psychotria* alone comprises around 2000 different species and is larger than most of the other angiosperm plant families. Therefore a correct placement of a given species is crucial to assess the systematic value of cyclotide distribution. To date, cyclotides have been identified within several tribes of the subfamily Rubioideae, i.e., Lasiantheae, Psychotrieae, Palicoureeae, and Spermacoceae s.l. (Figure 2). The current distribution of cyclotides is in line with recent reclassifications of Rubiaceae species (Bremer & Eriksson, 2009; Robbrecht & Manen, 2006) and their occurrence within the large genus *Psychotria* (Nepokroeff, Bremer, & Sytsma, 1999; Taylor, Lorence, & Gereau, 2010) is of particular interest. For example, the Micronesian species

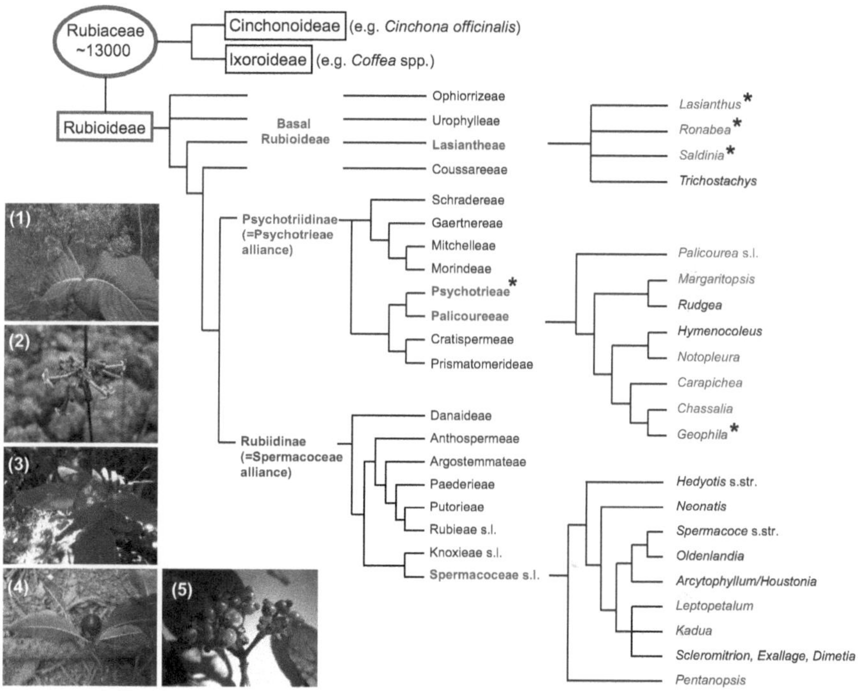

Figure 2 ***Phylogenetic distribution of cyclotide-containing plants in the Rubiaceae.*** The presence or absence of cyclotide-containing species within the genera listed (right column, italic font) and tribe (middle column, bold font) is indicated with green colored font. Cyclotides have been discovered within the tribes Spermacoceae (supertribe Rubiidinae), Psychotrieae and Palicoureeae (Psychotriidinae), as well as Lasiantheae (Basal Rubioideae). Genera and tribes wherein cyclotide-containing species have been identified according to Gruber et al. (2008), but for which no cyclotide sequence has been reported hitherto are marked with an asterisk. Phylogeny of Rubiaceae follows the literature (Bremer & Eriksson, 2009; Neupane et al., 2015; Robbrecht & Manen, 2006). s.l.—sensu lato, s.str.—sensu. stricto. Illustrations of the floral diversity and colorful fruits of cyclotide-containing species are shown; (1) *Palicourea rigida* Kunth, (2) *Kadua centranthoides* Hook. & Arn., (3) *Psychotria poeppigiana* Müll. Arg., (4) *Psychotria suterella* Müll. Arg., (5) *Chassalia discolor* K. Schum. (See color plate)

Psychotria leptothyrsa has recently been transferred to the genus *Magaritopsis* based on DNA sequence data (Barrabé et al., 2012) and thus within the tribe Palicoureeae, all but two genera, i.e., *Rudgea* and *Hymenocoleus* have been reported to contain cyclotides. Interestingly *Psychotria punctata* remains the only species from the tribe Psychotrieae that has been identified as cyclotide-containing species. However, no sequence has been reported hitherto and hence it remains unclear whether this species actually expresses cyclotides. To date, more than 300 different species have been sampled (Gruber et al., 2008; Koehbach, Attah, et al., 2013) and a total number of 88 different cyclotide sequences have been characterized. Many of these are a result of MS analysis, but due to advances in next generation sequencing, it is obvious that future cyclotide discovery will rely on transcriptome and genome mining of cyclotide precursor proteins.

3.4 Cyclotide Precursors of Rubiaceae Plants

Cyclotides are synthesized at the ribosome and are processed from larger precursor proteins. A detailed description of the processes involved in cyclotide biosynthesis can be found in Chapter 8. Briefly, a cyclotide precursor protein contains an endoplasmic reticulum signal sequence, a pro-domain, a region corresponding to the mature cyclotide domain, and a small C-terminal tail of commonly hydrophobic residues (Jennings, West, Waine, Craik, & Anderson, 2001) (Figure 3(A)). Some precursors contain multiple copies of the mature domain ($N = 1-3$) and in all such cases, a small section of the pro-domain just upstream of the mature peptide region is also repeated and is referred to as the N-terminal repeat. The processing events that result in excision and cyclization of the mature peptide sequence from the precursor protein are not fully understood yet. However, the enzyme asparaginyl endopeptidase has been implicated in the processing and cyclization of cyclotides (Conlan et al., 2012; Gillon et al., 2008; Saska et al., 2007). Much less is known about the biological folding process that results in cyclotides forming a cystine-knot within a cyclic peptide backbone. Likely, the oxidative folding of the cysteine residues occurs before cyclization, and may be facilitated by biological chaperones (Gruber, Cemazar, Clark, et al., 2007; Jennings et al., 2001). Indeed, we previously reported the discovery and biochemical characterization of protein disulfide isomerase from several cyclotide-expressing plant species (Burman et al., 2010; Gruber, Cemazar, Clark, et al., 2007; Gruber, Cemazar, Mechler, Martin, & Craik, 2009) and showed that the enzyme accelerates folding of cyclotides in vitro (Gruber, Cemazar, Clark, et al., 2007).

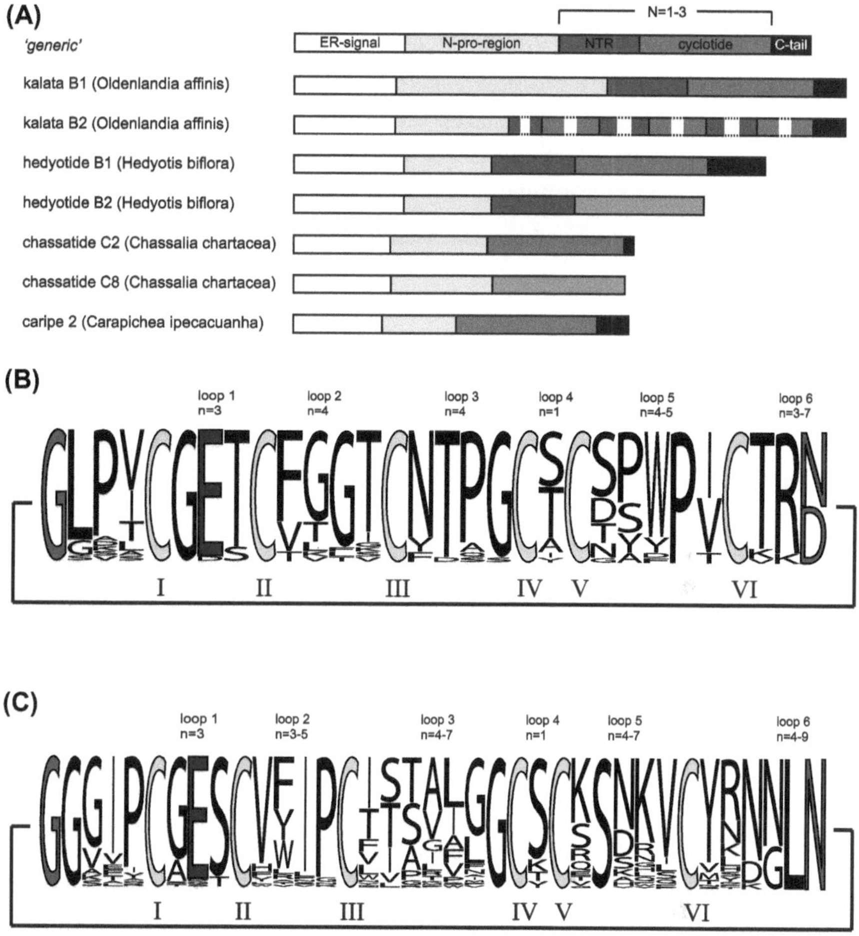

Figure 3 ***Rubiaceae cyclotide precursors and mature peptides.*** (A) An overview of the architecture of cyclotide precursor proteins among different species is shown. The typical cyclotide precursor architecture is depicted as "generic" and its domains are color-coded, i.e., ER-signal (white), N-pro-region (light gray), N-terminal repeats (NTR, dark gray), mature cyclotide domain (green), and C-terminal tail sequence (black). Orange indicates cyclotide domains encoding for linear cyclotides. Rubiaceae precursor proteins may contain up to three cyclotide domains encoding the same or different cyclotide sequences (illustrated as N = 1–3). Sequence logos of all Rubiaceae cyclotide sequences as presented in Table 1 are shown for Moebius-type (B) and bracelet-type cyclotides (C). For illustration purposes "hybrid cyclotides," i.e., kalata B8, B9, and Oak7 cyclotide have been excluded. Conserved cysteines (yellow), glutamic acid (red), asparagine/aspartic acid, and glycine residues (green) at the predicted cyclization site are highlighted and loop sizes are given. (See color plate)

As cyclotides are true gene products, discovery of cyclotide-containing species is not limited to peptide analysis by MS-based approaches, but may also be performed at nucleic acid level—using genome- or transcriptome-mining approaches—as mentioned earlier. The cyclotide precursor Oak1 (O. *affinis* kalata B1) that has been isolated from the Rubiaceae plant *O. affinis* by molecular cloning of cDNA libraries (Jennings et al., 2001), has been considered as prototypical precursor protein that aided isolation of cyclotide precursor proteins from other plant families. To date, respective precursor proteins have been reported from all plant families that are currently known to contain cyclotide expressing species (Koehbach, Attah, et al., 2013). The discovery of cyclotides precursor proteins in Fabaceae plants was interesting: Unlike all previously reported cyclotides, the domain corresponding to the mature cyclotide from Fabaceae is embedded within an albumin precursor protein. Hence, these precursor proteins are arranged as chimeras and this atypical genetic arrangement implies a different mechanism of biosynthetic processing of cyclotides in the Fabaceae (Nguyen, Zhang, Nguyen, et al., 2011; Poth, Colgrave, Lyons, Daly, & Craik, 2011). Following the description of cyclotide distribution and the characteristics of cyclotide precursors, we would like to analyze and compare all currently known cyclotide sequences of Rubiaceae species.

3.5 Cyclotide Sequence Variations within Rubiaceae

A total number of 88 different cyclotide sequences have been reported from Rubiaceae species so far (Table 1). This includes both partial and full-length sequences that have been identified at peptide level or within genome or transcriptome datasets. Recent work by Gruber and coworkers predicted the number of individual cyclotide sequences within the Rubiaceae alone to surpass 50,000 and single species have been reported to contain more than 70 different cyclotides (Gruber et al., 2008; Gründemann, Koehbach, Huber, & Gruber, 2012; Seydel, Gruber, Craik, & Dörnenburg, 2007). Based on the presence or absence of a proline residue in loop 5, cyclotides can be divided into two subfamilies, i.e. Moebius (Figure 3(B)) and bracelet (Figure 3(C)) cyclotides. Other sequence motifs specific for either subfamily have been identified (Table 1 and Figure 3(B) and (C)); these are located in loops 2, 3, 5, and 6, with loops 3 and 5 being overall the most diverse intercysteine loops. These molecular differences have important implications for their physicochemical properties, which ultimately influence their folding and cyclization behavior during biosynthesis as well as their bioactivity. Interestingly, there are peptides such as kalata B8 that share sequence motifs

Table 1 Overview of Cyclotide Sequences from Rubiaceae Species

Cyclotide	Sequence[g]												Species	Tribe	Reference(s)
Loop		1		2		3		4		5		6			
Caripe 1	*cyclo*-**C**	GES	**C**	VFIP	**C**	ISTVIG	**C**	S	**C**	KDKV	**C**	YRNGVIP	*Carapichea ipecacuanha*	Palicoureeae	Koehbach, Attah, et al. (2013)
Caripe 2	**C**	GES	**C**	VFIR	**C**	TITALLG	**C**	S	**C**	SNNV	**C**	YKNGIP			
Caripe 3	**C**	GES	**C**	VFIP	**C**	ISAVVG	**C**	S	**C**	SNKV	**C**	YNN			
Caripe 4	**C**	SST	**C**	LRIP	**C**	LSPR	**C**	T	**C**	RHHI	**C**	YLNLI			
Caripe 5	**C**	GES	**C**	VFIP	**C**	FTSVFG	**C**	S	**C**	KDKV	**C**	YRN X			
Caripe 6	**C**	TGT	**C**	FRNP	**C**	LSRR	**C**	T	**C**	RHYI	**C**	YLNGAI			
Caripe 7	**C**	GES	**C**	VFIP	**C**	TVTALLG	**C**	S	**C**	KNKV	**C**	YRNGLN			
Caripe 8	**C**	GES	**C**	VFIP	**C**	ITAAIG	**C**	S	**C**	KKKV	**C**	YRNGVIP			
Caripe 9	**C**	X	**C**	VFIP	**C**	TITALLG	**C**	S	**C**	SNNV	**C**	YKN X			
Circulin A[a]	**C**	GES	**C**	VWIP	**C**	ISAALG	**C**	S	**C**	KNKV	**C**	YRNGIP	*Chassalia parviflora*	Palicoureeae	Gustafson et al. (1994)
Circulin B	**C**	GES	**C**	VFIP	**C**	ISTLLG	**C**	S	**C**	KNKV	**C**	YRNGVIP			
Circulin C	**C**	GES	**C**	VFIP	**C**	ITSVAG	**C**	S	**C**	KSKV	**C**	YRNGIP			Gustafson et al. (2000)
Circulin D	**C**	GES	**C**	VWIP	**C**	VTSIFN	**C**	K	**C**	ENKV	**C**	YHDKIP			
Circulin E	**C**	GES	**C**	VWIP	**C**	LTSVFN	**C**	K	**C**	ENKV	**C**	YHDKIP			
Circulin F	**C**	GES	**C**	VWIP	**C**	ISAAIG	**C**	S	**C**	KNKV	**C**	YRAIP			
CD-1	**C**	GES	**C**	YVIP	**C**	ISYLVG	**C**	S	**C**	DTIEKV	**C**	KRNGADGF	*Chassalia discolor*	Palicoureeae	Gruber et al. (2008)
Chacur 1	**C**	GET	**C**	VGGT	**C**	NTPG	**C**	T	**C**	SWPI	**C**	TRNGLPV	*Chassalia curviflora*	Palicoureeae	Koehbach, Attah, et al. (2013)
Chassatide 1	**C**	GET	**C**	FTGI	**C**	FTAG	**C**	S	**C**	NPWPT	**C**	TRNGDA	*Chassalia chartacea*	Palicoureeae	Nguyen, Lim, et al. (2012)
Chassatide 2	**C**	AES	**C**	VWIP	**C**	TITALMG	**C**	S	**C**	KNNV	**C**	YNNGIP			
Chassatide 3	**C**	GES	**C**	WIP	**C**	ISSALG	**C**	S	**C**	KNKV	**C**	YRNGIP			
Chassatide 4	**C**	GET	**C**	FTGI	**C**	FTAG	**C**	S	**C**	NPWPT	**C**	TRNGAS			
Chassatide 5	**C**	GES	**C**	VFIP	**C**	ISSVVG	**C**	S	**C**	KNKV	**C**	YRNGVIP			
Chassatide 6	**C**	GES	**C**	VFIP	**C**	ISSVIG	**C**	S	**C**	KNKV	**C**	YRNGVIP			
Chassatide 7	**C**	GES	**C**	VWIP	**C**	LTAIAG	**C**	S	**C**	KNKV	**C**	YTGIP			
Chassatide 8	**C**	GES	**C**	VWIP	**C**	ISTVIG	**C**	S	**C**	SNKV	**C**	YRGAIP			
Chassatide 9	**C**	GES	**C**	VFIP	**C**	VTTVIG	**C**	S	**C**	KDKV	**C**	YNNGIP			
Chassatide 10	**C**	GES	**C**	YLIP	**C**	FTPG	**C**	Y	**C**	VSRQ	**C**	VNKNGEY			

(Continued)

Table 1 Overview of Cyclotide Sequences from Rubiaceae Species—cont'd

Cyclotide	Sequence[g]												Species	Tribe	Reference(s)
Chassatide 11	C	GES	C	VWIP	C	ISGMFG	C	S	C	KDKV	C	YSIP			
Chassatide 12	C	GES	C	YLIP	C	FTPG	C	Y	C	VSRQ	C	VNKNEY			
Chassatide 13	C	AES	C	VYIP	C	TVTALLG	C	S	C	RNRV	C	YRNGFP			
Chassatide 14	C	AES	C	VYIP	C	TITALFG	C	S	C	KDKV	C	YNNGIP			
Chassatide 15	C	AES	C	VYIP	C	TITALLG	C	S	C	KDKV	C	YKNGIP	*C. chartacea*	Palicoureeae	
Chassatide 16	C	AES	C	VYIP	C	TITALFG	C	S	C	KDKV	C	YNNGVP			
Chassatide 17	C	GES	C	VYIP	C	ISAVLG	C	S	C	QNKV	C	YRGIP			
Chassatide 18	C	GES	C	VFIP	C	ISALLG	C	S	C	SNKV	C	YNNGIP			
Paltet 1	C	GET	C	FTGT	C	NTPG	C	T	C	SYPV	C	TRNGLPI	*Palicourea tetragona*	Palicoureeae	Koehbach, Attah, et al. (2013)
Vibi B[b]	C	GET	C	FGGT	C	NTPG	C	T	C	SYPI	C	TRNGPLV			
Cycloviolacin O22[c]	C	GET	C	VGGT	C	NTPG	C	T	C	SWPV	C	TRNGLPI			
PS-1	C	GET	C	IWDKT	C	HAAG	C	S	C	SVANI	C	VRNGFIP	*Psychotria suterella*	Palicoureeae	Gruber et al. (2008)
Psyle A	C	GES	C	VFLG	C	FIPG	C	S	C	KSKV	C	YFNGIA	*Psychotria leptothyrsa*	Palicoureeae	Gerlach, Burman, Bohlin, Mondal, and Göransson (2010)
Psyle B	C	GET	C	VAFG	C	WIPG	C	S	C	KDKL	C	YYDGIP			
Psyle C	C	GET	C	FKFK	C	YTPG	C	S	C	SYFP	C	KKL			
Psyle D	C	GES	C	VFIP	C	TVTALLG	C	S	C	QNKV	C	YRDGIP			
Psyle E	C	GES	C	VFIP	C	ISSVLG	C	S	C	KNKV	C	YRDGVIP			
Psyle F	C	GES	C	VFIP	C	ITAAVG	C	S	C	KNKV	C	YRDGVIP			
Psybra 1	C	GET	C	TLGT	C	NTPG	C	T	C	SWPI	C	TKNGLPI	*Psychotria brachiata*	Palicoureeae	Koehbach, Attah, et al. (2013)
Notcap 1	C	X S	C	VW X	C	ITSPSAG	C	K	C	X	C	X	*Notopleura capacifolia*	Palicoureeae	Koehbach, Attah, et al. (2013)
Psydef 1	C	X	C	X	C	NTSG	C	T	C	KW X	C	TR X	*Psychotria deflexa*	Palicoureeae	Koehbach, Attah, et al. (2013)
Psydef 2	C	XES	C	WTSN	C	FTSP	C	X	C	X HP	C	TR X			
Psysue 1	C	X	C	X	C	X IAG	C	S	C	SSALL	C	V X	*Psychotria suerensis*	Palicoureeae	Koehbach, Attah, et al. (2013)
Psysue 2	C	X	C	X	C	X IAG	C	S	C	SSALL	C	V X			
Psysol 1	C	X	C	X	C	YTPG	C	T	C	GSYFV	C	N X	*Psychotria solitudinum*	Palicoureeae	Koehbach, Attah, et al. (2013)

Psysol 2	C	GES	C	VGGT	C	NTPG	C	T	C	TWPV	C	TRNGLPI			Hellinger et al. (2015)
Psypoe 1	C	GET	C	FTTV	C	NTPG	C	Y	C	GAY X	C	TRNGSVI	*Psychotria poeppigiana*	Palicoureeae	Koehbach, Attah, et al. (2013)
Parigidin-br1	C	GES	C	VFIP	C	ITSLAG	C	S	C	KNKV	C	YYDGGSVP	*Palicourea rigida*	Palicoureeae	Pinto et al. (2012)
Palicourein	C	GET	C	RVIPV	C	TYSAALG	C	T	C	DDRSDGL	C	KRNGDPTF	*Palicourea condensata*	Palicoureeae	Bokesch et al. (2001)
Cyclopsychotride A	C	GES	C	VFIP	C	TVTALLG	C	S	C	KSKV	C	YKNSIP	*Palicourea jungiana*	Palicoureeae	Witherup et al. (1994)
HCF-1	C	GES	C	HYIP	C	VTSAIG	C	S	C	RNRS	C	MRNGIP	*Kadua centranthoides*	Spermacoceae	CyBase[d]
HCF-1 variant	C	GES	C	HIP	C	VTSAIG	C	S	C	RNRS	C	MRNGIP			CyBase
HTF-1	C	GDS	C	HYIP	C	VTSTIG	C	S	C	TNGS	C	MRNGIP	*Kadua affinis*	Spermacoceae	CyBase
Hedyotide B1	C	GET	C	FVLP	C	WSAKFG	C	Y	C	QKGF	C	YRNGTR	*Leptopetalum biflorum*	Spermacoceae	Wong, Taichi, Nishio, Nishiuchi, and Tam (2011)
Hedyotide B2	C	GES	C	VWIP	C	ISSAWG	C	S	C	KNKI	C	SSGIQ			
Kalata B1[e]	C	GET	C	VGGT	C	NTPG	C	T	C	SWPV	C	TRNGLPV	*Oldenlandia affinis*	Spermacoceae	Saether et al. (1995)
Kalata B2	C	GET	C	FGGT	C	NTPG	C	S	C	TWPI	C	TRDGLPV			Craik et al. (1999)
Kalata B3	C	GET	C	FGGT	C	NTPG	C	T	C	DPWPI	C	TRDGLPT			
Kalata B4	C	GET	C	VGGT	C	NTPG	C	T	C	SWPV	C	TRDGLPV			
Kalata B5	C	GES	C	VYIP	C	ISGVIG	C	S	C	TDKV	C	YLNGTP			
Kalata B6	C	GET	C	FGGT	C	NTPG	C	S	C	SSWPI	C	TRNGLPT			Jennings et al. (2001)
Kalata B7	C	GET	C	TLGT	C	YTQG	C	T	C	SWPI	C	KRNGLPV			
Kalata B8	C	GET	C	LLGT	C	YTTG	C	T	C	NKYRV	C	TKDGSVLN			Daly, Clark, Plan, and Craik (2006)
Kalata B9	C	GET	C	VLGT	C	YTPG	C	T	C	NTYRV	C	TKDGSVFN			Plan et al. (2007)
Kalata B9 linear	C	GET	C	VLGT	C	YTPG	C	T	C	NTYRV	C	TKDGSVFN			
Kalata B10	C	GET	C	FGGT	C	NTPG	C	S	C	SSWPI	C	TRDGLPT			
Kalata B10 linear	C	GET	C	FGGT	C	NTPG	C	S	C	SSWPI	C	TRDGLPT			
Kalata B11	C	GET	C	FGGT	C	NTPG	C	S	C	TDPI	C	TRDGLPV			
Kalata B12	C	GDT	C	FVLG	C	NDSS	C	S	C	NYPI	C	VKDGSL			

(*Continued*)

Table 1 Overview of Cyclotide Sequences from Rubiaceae Species—cont'd

Cyclotide	Sequence[g]												Species	Tribe	Reference(s)
Kalata B13	C	GET	C	FGGT	C	NTPG	C	A	C	DPWPV	C	TRDGLPV			
Kalata B14	C	GES	C	FGGT	C	NTPG	C	A	C	DPWPV	C	TRDGLPV			
Kalata B15	C	GES	C	FGGS	C	YTPG	C	S	C	TWPI	C	TRDGLPV			
Kalata B16	C	AES	C	VYIP	C	TITALLG	C	K	C	QDKV	C	YDGIP			
Kalata B17	C	AES	C	VYIP	C	TITALLG	C	K	C	KDQV	C	YNGIP			
Kalata B18	C	AES	C	VYIP	C	ISTVLG	C	S	C	SNQV	C	YRNGVP			Seydel et al. (2007)
Kalata B19	C	GES	C	VYVP	C	LTAAIG	C	S	C	SNKV	C	YKNGFP			Qin et al. (2010)
Oak6 cyclotide 1	C	GET	C	FGGT	C	NTPG	C	A	C	DPWPV	C	TRNGLPV			Mylne, Wang, van der Weerden, and Craik (2010)
Oak6 cyclotide 2	C	GET	C	FGGT	C	NTPG	C	I	C	DPWPV	C	TRDGLPI			
Oak7 cyclotide	C	GET	C	FFFG	C	YKPG	C	S	C	DELRQ	C	YKNGSH			
Oak8 cyclotide	C	GES	C	VFIP	C	LTAVVG	C	S	C	SNKV	C	YLNGVP			
Kalata S (= varv A)[f]	C	GET	C	VGGT	C	NTPG	C	S	C	SWPV	C	TRNGLPV			Craik et al. (1999)

[a]Peptide also identified in *Chassalia chartacea* (Nguyen, Lim, et al., 2012).
[b]Originally identified from *Viola biflora* (Herrmann et al., 2008).
[c]Originally identified from *Viola odorata* (Ireland et al., 2006).
[d]Available at www.cybase.org.au.
[e]Peptide also present in *V. odorata* (Ireland et al., 2006), *Viola baoshanensis* (Zhang et al., 2009), *Viola yedoensis* (Wang et al., 2008), and *Viola philippica* (He et al., 2011).
[f]Originally identified as varv peptide A from *Viola arvensis* (Claeson, Goransson, Johansson, Luijendijk, & Bohlin, 1998), also found in *Viola tricolor* (Svangård et al., 2003), *V. odorata* (Ireland et al., 2006), *V. biflora* (Herrmann et al., 2008), *V. baoshanensis* (Zhang et al., 2009), *V. yedoensis* (Wang et al., 2008), and *V. philippica* (He et al., 2011).
[g]X indicates one or more unidentified amino acid residues.

found within both subfamilies (Table 1). Besides this high variability, in particular within the intercysteine spanning sequence loops, several positions in the cyclotide backbone are conserved throughout all sequences. This includes in particular a glutamic acid in loop 1, which has been shown to be important for the three-dimensional stability of cyclotides (Göransson, Herrmann, Burman, Haugaard-Jonsson, & Rosengren, 2009). The only exceptions that do not contain this conserved residue are kalata B12 (Asp instead of Glu) (Wang et al., 2011) as well as caripe 4 (Ser) and caripe 6 (Gly) (Koehbach, Attah, et al., 2013). However, the latter examples have not yet been confirmed at peptide level. Nearly all identified Rubiaceae cyclotides contain glycine and asparagine/aspartic acid residues within loop 6, underpinning their importance for cyclization and processing during biosynthesis. The only cyclotides that have different residues at these positions are cyclopsychotride A (Asn—Ser), circulin D, E (Asp—Lys), and F (Arg—Ala) allowing speculations regarding alternative cyclization mechanisms. The analysis of the molecular structure of cyclotides from Rubiaceae and the characteristic distribution of cyclotides in various species of Rubiaceae led to studies of cyclotides in other plant families of the Gentianales.

3.6 Distribution of Cyclotides in Other Gentianales

Detection of cyclotides in the basal grade Rubioideae (i.e., genera Lasianthus, Ronabea, and Saldinia) led to the hypothesis that cyclotide occurrence had its origin early in the evolution of the Rubioideae and hence these findings indicated that other families within the order Gentianales are likely to contain cyclotides, based on the evolution and relationships of the families within the order Gentianales (Backlund, Oxelman, & Bremer, 2000; Struwe, Albert, & Bremer, 1994). Therefore several species from four plant families that are phylogenetically closely related to Rubiaceae, i.e. Apocynaceae, Gentianaceae, Loganiaceae, and Potaliaceae have been screened and 12 potential cyclotide containing species within the Apocynaceae s.l. (including Asclepiadaceae) have been identified. Although it is likely that Apocynaceae s.l. do contain cyclotides, it is important that they are still regarded as putative hits until actual sequence data have been provided.

4. EVOLUTION OF CYCLOTIDES

Following several studies in which we and others have screened and identified cyclotides from Rubiaceae plants (for overview, see Table 1), it became clear that they are much more abundant and diverse than earlier anticipated. Taking into account the occurrence of cyclotides in the violet family (Violaceae) (Burman et al., 2010; Ireland, Colgrave, & Craik, 2006; Simonsen et al., 2005) (see Chapter 2), the legume family (Fabaceae) (Nguyen, Zhang, Nguyen, et al., 2011; Poth, Colgrave, Philip, et al., 2011), the nightshade family (Solanaceae) (Poth et al., 2012), the grasses (Poaceae) (Mulvenna et al., 2006; Nguyen, Lian, et al., 2012), and cyclotide-like molecules from the squash family (Cucurbitaceae) (Felizmenio-Quimio, Daly, & Craik, 2001; Hernandez et al., 2000), respectively, it is evident that the number of existing and unique cyclotides may be by far greater than 50,000. Hence, the family of cyclotides could form one of the largest distinct protein families in the plant kingdom and a very large group of naturally occurring plant compounds.

While trying to understand their number and diversity, we and others have generated data allowing to discuss the origin and evolution of cyclotides in plants. In particular, the knowledge from cyclotide screening in Rubiaceae and identification of cyclotide precursor proteins has helped to develop and build a theory of their evolutionary mechanism. Considering the presently known distribution of cyclotides in Rubiaceae, the most favorable explanation for their occurrence is to assume at least four independent origins within this family. This is supported by the distribution of cyclotide- and cyclotide-like-containing plant families as pointed out above. There is no direct phylogenetic link between these families, which is another indicator that cyclotides likely have originated independently more than once, and in distantly related families (Gruber et al., 2008). Although this theory is based on a process of elimination of possible other evolutionary mechanisms such as lateral gene transfer or divergent evolution by loss of cyclotide genes, it is likely that cyclotides evolved as multiple independent gains of function, i.e., by convergent evolution (Gruber, 2010; Gruber et al., 2008).

As mentioned above, the enzyme asparaginyl endopeptidase is important for the processing and cyclization of cyclotides during their biosynthesis. The discovery of noncircular (syn. acyclic) cyclotides in *Chassalia chartaceae* (Nguyen, Lim, et al., 2012), *Leptopetalum biflorum* (formerly known as *Hedyotis biflora*) (Nguyen, Zhang, Wang, et al., 2011) (Rubiaceae), *Petunia* × *hybrida* (Poth et al., 2012) (Solanaceae), *Panicum laxum* (Nguyen, Lian, et al., 2012) (Poaceae), and other grasses (Mulvenna et al., 2006) suggested that a mutation which introduces Asn/Asp residues at a crucial cyclization point near the terminus of the mature cyclotide domain is the driving force of cyclotide evolution (Gruber et al., 2008). In addition, those acyclic cyclotides may also be the evolutionary link between cyclic cystine-knot and other (noncyclic) cystine-knot peptides, such as peptides of the inhibitor and growth factor cystine-knot families that are distributed not only in plants, but also in other kingdoms of life (McDonald & Hendrickson, 1993; Smith et al., 2011; Zhu, Darbon, Dyason, Verdonck, & Tytgat, 2003).

In the plant kingdom there are at least three classes of gene-encoded backbone-cyclized peptides. Each of these classes uses asparaginyl endopeptidase-mediated processing for biosynthesis, but express different type of peptides such as the sunflower trypsin inhibitor SFTI-I, the Momordica trypsin inhibitor MCoTI-I, or the cyclotide kalata B1. Not only do the peptides differ structurally, but they are also embedded within precursor proteins of very different architectures and in unrelated plant families. Despite these differences in the peptide and precursor structure, they seem to share common features, for instance, the proto-N-terminal Gly and the proto-C-terminal Asp/Asn (Mahatmanto et al., 2015; Mylne et al., 2012) and those residues that are important for processing and cyclization (Gillon et al., 2008; Saska et al., 2007). These three classes of circular peptides are produced by at least six plant lineages that are phylogenetically quite distantly related; it appears they have converged to use the same processing mechanism, and this provides strong evidence of evolutionary parallelism (Mylne et al., 2012). The biological purpose of this convergent occurrence of circular peptides has not been elucidated, yet, but obviously backbone cyclization confers greater stability to these molecules, and likely the restricted degree of flexibility due to the lack of free termini could have led to different and distinct bioactive properties.

5. BIOACTIVITIES OF RUBIACEAE CYCLOTIDES

The biological function of cyclotides will be discussed in detail elsewhere (in particular Chapter 7). Briefly, it appears that cyclotides are defense peptides against herbivores, based on observations that they modulate the growth and development of several invertebrates, such as insects (Barbeta, Marshall, Gillon, Craik, & Anderson, 2008; Gruber, Cemazar, Anderson, & Craik, 2007), molluscs (Plan, Saska, Cagauan, & Craik, 2008), and nematodes (Colgrave, Kotze, Ireland, Wang, & Craik, 2008). Besides cyclotides exhibit a range of bioactivities that may be relevant for pharmaceutical applications: for example, they inhibit the growth and proliferation of viruses (Gustafson, McKee, & Bokesch, 2004), bacteria

(Pranting, Loov, Burman, Goransson, & Andersson, 2010), and cancer cells (Gerlach, Rathinakumar, et al., 2010).

The first cyclotide that has been studied for its bioactivity has been isolated from plant extracts of the Rubiaceae species *O. affinis*. Indeed as mentioned previously, the initial identification of cyclotides is based on the observed uterotonic activity of a preparation from this herbal medicine called "kalata–kalata" (Gran, 1973). Recently we were able to provide pharmacological evidence for its traditional use by the identification of the oxytocin and vasopressin receptors as molecular targets for the cyclotide kalata B7 (Figure 4) (Koehbach, O'Brien, et al., 2013). In addition, the presence of cyclotides has been reported in other medicinal plants, for instance *C. ipecacuanha* (Koehbach, Attah, et al., 2013). This plant is used to prepare the traditional "syrup of ipecac"—a medicine known for its strong and powerful emetic activity.

Rubiaceae cyclotides can still be considered as prototypes to discover and study novel activities and applications. For example, kalata cyclotides were characterized for their immunosuppressive activity to inhibit the proliferation of T-lymphocytes (Gründemann et al., 2012, 2013), and a cyclotide isolated from *Psychotria solitudinum* was found to inactivate the enzymatic activity of prolyl-oligopeptidase (Hellinger et al., 2015).

6. CONCLUSION AND OUTLOOK

Nearly half a century after the discovery of the first cyclotide from the Rubiaceae plant *O. affinis*, there has been recent progress about the molecular diversity and distribution of those fascinating molecules within the plant kingdom. Although, unlike within the Violaceae family, not every member has been shown to express cyclotides, the analysis of cyclotides in the Rubiaceae has been very fruitful. Indeed their scattered distribution in the subfamily Rubioideae has led to developing a theory of their evolution. Importantly, based on studies of cyclotides in Rubiaceae their number has been predicted to be greater than 50,000 and we believe cyclotides are one of the largest protein families within the plant kingdom. With a modest number of approximately 400 published cyclotide sequences, we are convinced that the combination of high-throughput next generation sequencing and state-of-the-art peptidomics workflows will lead to the characterization of many more cyclotide sequences in the future and will advance our understanding of cyclotide diversity, their distribution and

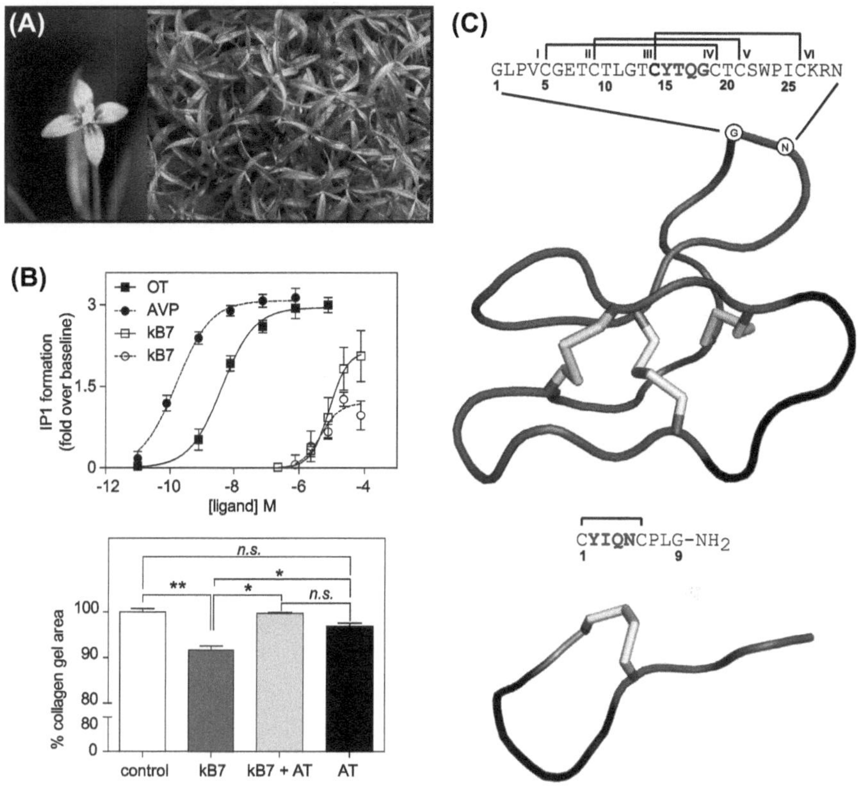

Figure 4 ***Bioactivity of Rubiaceae cyclotides.*** The African plant *Oldenlandia affinis* (photograph by David Wilson, James Cook University, Australia) (A) contains a number of different cyclotides, known as kalata peptides (Table 1) and is used in traditional medicine for its uterotonic properties. Recent pharmacological characterization of this activity revealed the oxytocin (OTR) and vasopressin V_{1a} ($V_{1a}R$) receptors as molecular targets of the cyclotide kalata B7 (kB7). Functional receptor activation showed kB7 (blank symbols) to be a partial agonist at both receptors (OTR—solid lines, $V_{1a}R$—dashed lines) (B, upper panel). Oxytocin (OT, filled squares) and vasopressin (AVP, filled circles) were used as control peptides. The contractile effects of kB7 were studied using a collagen-gel assay (B, lower panel). *(Figure has been adapted and modified from Koehbach, O'Brien, et al. (2013).)* Sequence alignment and NMR structural studies revealed a high similarity of kalata B7 (PDB ID: 2M9O) with human oxytocin (PDB ID: 2MGO) (C). Structures are shown as ribbon diagrams with disulphide bonds highlighted in yellow. The "active" loop of kalata B7 (-YTQG-) showing sequence similarity to oxytocin (-YIQN-) is highlighted in red. (See color plate)

occurrence not only in Rubiaceae but all plants. Lastly, detailed bioactivity studies of prototypical kalata cyclotides may soon help to elucidate their biological role in plants, and potentially provide lead molecules for pharmaceutical applications.

ACKNOWLEDGMENTS

The authors would like to thank all colleagues and friends whose work has been referred to in the reference list. In particular we are grateful for comments about this manuscript from Andreas Berger (University of Vienna). Work on cyclotides in the laboratory of C.W.G. has been supported by the Austrian Science Fund FWF through projects P22889 and P24743 and the Austria Wirtschaftsservice GmbH (Prize, P1308423).

REFERENCES

Achan, J., Talisuna, A. O., Erhart, A., Yeka, A., Tibenderana, J. K., Baliraine, F. N., et al. (2011). Quinine, an old anti-malarial drug in a modern world: role in the treatment of malaria. *Malaria Journal, 10*, 144.

Adkins, J. E., Boyer, E. W., & McCurdy, C. R. (2011). *Mitragyna speciosa*, a psychoactive tree from Southeast Asia with opioid activity. *Current Topics in Medicinal Chemistry, 11*, 1165–1175.

Backlund, M., Oxelman, B., & Bremer, B. (2000). Phylogenetic relationships within the Gentianales based on NDHF and RBCL sequences, with particular reference to the Loganiaceae. *American Journal of Botany, 87*, 1029–1043.

Barbeta, B. L., Marshall, A. T., Gillon, A. D., Craik, D. J., & Anderson, M. A. (2008). Plant cyclotides disrupt epithelial cells in the midgut of lepidopteran larvae. *Proceedings of the National Academy of Sciences of the United States of America, 105*, 1221–1225.

Barrabé, L., Buerki, S., Mouly, A., Davis, A. P., Munzinger, J., & Maggia, L. (2012). Delimitation of the genus *Margaritopsis* (Rubiaceae) in the Asian, Australasian and Pacific region, based on molecular phylogenetic inference and morphology. *Taxon, 61*, 1251–1268.

Berger, A., Fasshuber, H., Schinnerl, J., Robien, W., Brecker, L., & Valant-Vetschera, K. (2011). Iridoids as chemical markers of false ipecac (*Ronabea emetica*), a previously confused medicinal plant. *Journal of Ethnopharmacology, 138*, 756–761.

Bokesch, H. R., Pannell, L. K., Cochran, P. K., Sowder, R. C., McKee, T. C., & Boyd, M. R. (2001). A novel anti-HIV macrocyclic peptide from *Palicourea condensata*. *Journal of Natural Products, 64*, 249–250.

Bremer, B., & Eriksson, T. (2009). Time tree of Rubiaceae: phylogeny and dating the family, subfamilies, and tribes. *International Journal of Plant Sciences, 170*, 766–793.

Burman, R., Gruber, C. W., Rizzardi, K., Herrmann, A., Craik, D. J., Gupta, M. P., et al. (2010). Cyclotide proteins and precursors from the genus *Gloeospermum*: filling a blank spot in the cyclotide map of Violaceae. *Phytochemistry, 71*, 13–20.

Claeson, P., Goransson, U., Johansson, S., Luijendijk, T., & Bohlin, L. (1998). Fractionation protocol for the isolation of polypeptides from plant biomass. *Journal of Natural Products, 61*, 77–81.

Colgrave, M. L., Kotze, A. C., Ireland, D. C., Wang, C. K., & Craik, D. J. (2008). The anthelmintic activity of the cyclotides: natural variants with enhanced activity. *Chembiochem, 9*, 1939–1945.

Colgrave, M. L., Poth, A. G., Kaas, Q., & Craik, D. J. (2010). A new "era" for cyclotide sequencing. *Biopolymers, 94*, 592–601.

Conlan, B. F., Colgrave, M. L., Gillon, A. D., Guarino, R., Craik, D. J., & Anderson, M. A. (2012). Insights into processing and cyclization events associated with biosynthesis of the cyclic Peptide kalata B1. *Journal of Biological Chemistry, 287*, 28037–28046.

Craik, D. J., Daly, N. L., Bond, T., & Waine, C. (1999). Plant cyclotides: a unique family of cyclic and knotted proteins that defines the cyclic cystine knot structural motif. *Journal of Molecular Biology, 294*, 1327–1336.

Daly, N. L., Clark, R. J., Plan, M. R., & Craik, D. J. (2006). Kalata B8, a novel antiviral circular protein, exhibits conformational flexibility in the cystine knot motif. *Biochemical Journal, 393*, 619–626.

Delprete, P. G. (2004). Rubiaceae. In N. P. Smith, S. V. Heald, A. Henderson, S. A. Mori, & D. W. Stevenson (Eds.), *Flowering plant families of the American tropics* (pp. 328–333). New York: Princeton University Press/New York Botanical Garden Press.

Felizmenio-Quimio, M. E., Daly, N. L., & Craik, D. J. (2001). Circular proteins in plants – solution structure of a novel macrocyclic trypsin inhibitor from *Momordica cochinchinensis*. *Journal of Biological Chemistry, 276*, 22875–22882.

Gerlach, S. L., Burman, R., Bohlin, L., Mondal, D., & Göransson, U. (2010). Isolation, characterization, and bioactivity of cyclotides from the Micronesian plant *Psychotria leptothyrsa*. *Journal of Natural Products, 73*, 1207–1213.

Gerlach, S. L., Goransson, U., Kaas, Q., Craik, D. J., Mondal, D., & Gruber, C. W. (2013). A systematic approach to document cyclotide distribution in plant species from genomic, transcriptomic, and peptidomic analysis. *Biopolymers, 100*, 433–437.

Gerlach, S. L., Rathinakumar, R., Chakravarty, G., Göransson, U., Wimley, W. C., Darwin, S. P., et al. (2010). Anticancer and chemosensitizing abilities of cycloviolacin O_2 from *Viola odorata* and psyle cyclotides from *Psychotria leptothyrsa*. *Biopolymers, 94*, 617–625.

Gillon, A. D., Saska, I., Jennings, C. V., Guarino, R. F., Craik, D. J., & Anderson, M. A. (2008). Biosynthesis of circular proteins in plants. *Plant Journal, 53*, 505–515.

Göransson, U., Herrmann, A., Burman, R., Haugaard-Jonsson, L. M., & Rosengren, K. J. (2009). The conserved glu in the cyclotide cycloviolacin O_2 has a key structural role. *Chembiochem, 10*, 2354–2360.

Gran, L. (1970). An oxytocic principle found in *Oldenlandia affinis* DC. *Meddelelser fra Norsk Farmaceutisk Selskap, 12*, 173–180.

Gran, L. (1973). On the effect of a polypeptide isolated from "Kalata-Kalata" (*Oldenlandia affinis* DC) on the oestrogen dominated uterus. *Acta Pharmacologica et Toxicologica (Copenh), 33*, 400–408.

Gruber, C. W. (2010). Global cyclotide adventure: a journey dedicated to the discovery of circular peptides from flowering plants. *Biopolymers, 94*, 565–572.

Gruber, C. W., Cemazar, M., Anderson, M. A., & Craik, D. J. (2007). Insecticidal plant cyclotides and related cystine knot toxins. *Toxicon, 49*, 561–575.

Gruber, C. W., Cemazar, M., Clark, R. J., Horibe, T., Renda, R. F., Anderson, M. A., et al. (2007). A novel plant protein disulfide isomerase involved in the oxidative folding of cystine knot defense proteins. *Journal of Biological Chemistry, 282*, 20435–20446.

Gruber, C. W., Cemazar, M., Mechler, A., Martin, L. L., & Craik, D. J. (2009). Biochemical and biophysical characterization of a novel plant protein disulfide isomerase. *Biopolymers, 92*, 35–43.

Gruber, C. W., Elliott, A. G., Ireland, D. C., Delprete, P. G., Dessein, S., Goransson, U., et al. (2008). Distribution and evolution of circular miniproteins in flowering plants. *Plant Cell, 20*, 2471–2483.

Gruber, C. W., & Muttenthaler, M. (2012). Discovery of defense- and neuropeptides in social ants by genome-mining. *PLoS One, 7*, e32559.

Gründemann, C., Koehbach, J., Huber, R., & Gruber, C. W. (2012). Do plant cyclotides have potential as immunosuppressant peptides? *Journal of Natural Products, 75*, 167–174.

Gründemann, C., Thell, K., Lengen, K., Garcia-Kaufer, M., Huang, Y. H., Huber, R., et al. (2013). Cyclotides suppress human T-lymphocyte proliferation by an interleukin 2-dependent mechanism. *PLoS One, 8*, e68016.

Gustafson, K. R., McKee, T. C., & Bokesch, H. R. (2004). Anti-HIV cyclotides. *Current Protein and Peptide Science, 5*, 331–340.

Gustafson, K. R., Sowder, R. C., Henderson, L. E. I., Parsons, I. C., Kashman, Y., Cardellina, J. H., et al. (1994). Circulins A and B. Novel human immunodeficiency virus (HIV)-inhibitory macrocyclic peptides from the tropical tree *Chassalia parvifolia*. *Journal of the American Chemical Society, 113*, 9337–9338.

Gustafson, K. R., Walton, L. K., Sowder, R. C., Jr., Johnson, D. G., Pannell, L. K., Cardellina, J. H., Jr., et al. (2000). New circulin macrocyclic polypeptides from *Chassalia parvifolia*. *Journal of Natural Products, 63*, 176–178.

Hashempour, H., Koehbach, J., Daly, N. L., Ghassempour, A., & Gruber, C. W. (2013). Characterizing circular peptides in mixtures: sequence fragment assembly of cyclotides from a violet plant by MALDI-TOF/TOF mass spectrometry. *Amino Acids, 44*, 581–595.

He, W., Chan, L. Y., Zeng, G., Daly, N. L., Craik, D. J., & Tan, N. (2011). Isolation and characterization of cytotoxic cyclotides from *Viola philippica*. *Peptides, 32*, 1719–1723.

Hellinger, R., Koehbach, J., Puigpinós, A., Clark, R. J., Tarragó, T., Giralt, E., et al. (2015). Inhibition of human prolyl oligopeptidase activity by the cyclotide psysol 2 isolated from *Psychotria solitudinum*. *Journal of Natural Products, 78*, 1073–1082.

Hernandez, J. F., Gagnon, J., Chiche, L., Nguyen, T. M., Andrieu, J. P., Heitz, A., et al. (2000). Squash trypsin inhibitors from *Momordica cochinchinensis* exhibit an atypical macrocyclic structure. *Biochemistry, 39*, 5722–5730.

Herrmann, A., Burman, R., Mylne, J. S., Karlsson, G., Gullbo, J., Craik, D. J., et al. (2008). The alpine violet, *Viola biflora*, is a rich source of cyclotides with potent cytotoxicity. *Phytochemistry, 69*, 939–952.

Ireland, D. C., Colgrave, M. L., & Craik, D. J. (2006). A novel suite of cyclotides from *Viola odorata*: sequence variation and the implications for structure, function and stability. *Biochemical Journal, 400*, 1–12.

Jennings, C., West, J., Waine, C., Craik, D., & Anderson, M. (2001). Biosynthesis and insecticidal properties of plant cyclotides: the cyclic knotted proteins from *Oldenlandia affinis*. *Proceedings of the National Academy of Sciences of the United States of America, 98*, 10614–10619.

Koehbach, J., Attah, A. F., Berger, A., Hellinger, R., Kutchan, T. M., Carpenter, E. J., et al. (2013). Cyclotide discovery in Gentianales revisited-identification and characterization of cyclic cystine-knot peptides and their phylogenetic distribution in Rubiaceae plants. *Biopolymers, 100*, 438–452.

Koehbach, J., & Jackson, K. A. V. (2015). Unravelling peptidomes by in silico mining. *Peptidomics, 2*, 17–25.

Koehbach, J., O'Brien, M., Muttenthaler, M., Miazzo, M., Akcan, M., Elliott, A. G., et al. (2013). Oxytocic plant cyclotides as templates for peptide G protein-coupled receptor ligand design. *Proceedings of the National Academy of Sciences of the United States of America, 110*, 21183–21188.

Mahatmanto, T., Mylne, J. S., Poth, A. G., Swedberg, J. E., Kaas, Q., Schaefer, H., et al. (2015). The evolution of Momordica cyclic peptides. *Molecular Biology and Evolution, 32*, 392–405.

McDonald, N. Q., & Hendrickson, W. A. (1993). A structural superfamily of growth factors containing a cystine knot motif. *Cell, 73*, 421–424.

Mulvenna, J. P., Mylne, J. S., Bharathi, R., Burton, R. A., Shirley, N. J., Fincher, G. B., et al. (2006). Discovery of cyclotide-like protein sequences in graminaceous crop plants: ancestral precursors of circular proteins? *Plant Cell, 18*, 2134–2144.

Mylne, J. S., Chan, L. Y., Chanson, A. H., Daly, N. L., Schaefer, H., Bailey, T. L., et al. (2012). Cyclic peptides arising by evolutionary parallelism via asparaginyl-endopeptidase-mediated biosynthesis. *Plant Cell, 24*, 2765–2778.

Mylne, J. S., Wang, C. K., van der Weerden, N. L., & Craik, D. J. (2010). Cyclotides are a component of the innate defense of *Oldenlandia affinis*. *Peptide Science, 94*, 635–646.

Nepokroeff, M., Bremer, B., & Sytsma, K. J. (1999). Reorganization of the genus *Psychotria* and tribe *Psychotrieae* (Rubiaceae) inferred from ITS and rbcL sequence data. *Systematic Botany, 24*, 5–27.

Neupane, S., Dessein, S., Wikström, N., Lewis, P. O., Long, C., Bremer, B., et al. (2015). The Hedyotis-Oldenlandia complex (Rubiaceae: Spermacoceae) in Asia and the Pacific: phylogeny revisited with new generic delimitations. *Taxon, 64*, 299–322.

Nguyen, G. K., Lian, Y., Pang, E. W., Nguyen, P. Q., Tran, T. D., & Tam, J. P. (2012). Discovery of linear cyclotides in monocot plant Panicum laxum of poaceae family provides new insights into evolution and distribution of cyclotides in plants. *Journal of Biological Chemistry, 288*, 3370–3380.

Nguyen, G. K., Lim, W. H., Nguyen, P. Q., & Tam, J. P. (2012). Novel cyclotides and uncyclotides with highly shortened precursors from *Chassalia chartacea* and effects of methionine oxidation on bioactivities. *Journal of Biological Chemistry, 287*, 17598–17607.

Nguyen, G. K., Zhang, S., Nguyen, N. T. K., Nguyen, P. Q., Chiu, M., Hardjojo, A., et al. (2011). Discovery and characterization of novel cyclotides originated from chimeric precursors consisting of albumin-1 chain a and cyclotide domains in the Fabaceae family. *Journal of Biological Chemistry, 286*, 4275–4287.

Nguyen, G. K. T., Zhang, S., Wang, W., Wong, C. T. T., Nguyen, N. T. K., & Tam, J. P. (2011). Discovery of a linear cyclotide from the bracelet subfamily and its disulfide mapping by top-down mass spectrometry. *Journal of Biological Chemistry, 286*, 44833–44844.

Pinto, M. F., Fensterseifer, I. C., Migliolo, L., Sousa, D. A., de Capdville, G., Arboleda-Valencia, J. W., et al. (2012). Identification and structural characterization of novel cyclotide with activity against an insect pest of sugar cane. *Journal of Biological Chemistry, 287*, 134–147.

Plan, M. R., Goransson, U., Clark, R. J., Daly, N. L., Colgrave, M. L., & Craik, D. J. (2007). The cyclotide fingerprint in *Oldenlandia affinis*: elucidation of chemically modified, linear and novel macrocyclic peptides. *Chembiochem, 8*, 1001–1011.

Plan, M. R., Saska, I., Cagauan, A. G., & Craik, D. J. (2008). Backbone cyclised peptides from plants show molluscicidal activity against the rice pest *Pomacea canaliculata* (golden apple snail). *Journal of Agricultural and Food Chemistry, 56*, 5237–5241.

Poth, A. G., Colgrave, M. L., Lyons, R. E., Daly, N. L., & Craik, D. J. (2011). Discovery of an unusual biosynthetic origin for circular proteins in legumes. *Proceedings of the National Academy of Sciences of the United States of America, 108*, 10127–10132.

Poth, A. G., Colgrave, M. L., Philip, R., Kerenga, B., Daly, N. L., Anderson, M. A., et al. (2011). Discovery of cyclotides in the fabaceae plant family provides new insights into the cyclization, evolution, and distribution of circular proteins. *ACS Chemical Biology, 20*, 20.

Poth, A. G., Mylne, J. S., Grassl, J., Lyons, R. E., Millar, A. H., Colgrave, M. L., et al. (2012). Cyclotides associate with leaf vasculature and are the products of a novel precursor in *Petunia* (Solanaceae). *Journal of Biological Chemistry, 287*, 27033–27046.

Pranting, M., Loov, C., Burman, R., Goransson, U., & Andersson, D. I. (2010). The cyclotide cycloviolacin O_2 from *Viola odorata* has potent bactericidal activity against Gram-negative bacteria. *Journal of Antimicrobial Chemotherapy, 65*, 1964–1971.

Qin, Q., McCallum, E. J., Kaas, Q., Suda, J., Saska, I., Craik, D. J., et al. (2010). Identification of candidates for cyclotide biosynthesis and cyclisation by expressed sequence tag analysis of *Oldenlandia affinis*. *BMC Genomics, 11*, 111.

Robbrecht, E., & Manen, J.-F. (2006). The major evolutionary lineages of the coffee family (Rubiaceae, angiosperms). Combined analysis (nDNA and cpDNA) to infer the position of Coptosapelta and Luculia, and supertree construction based on rbcL, rps16, trnL-trnF and atpB-rbcL data. A new classification in two subfamilies, Cinchonoideae and Rubioideae. *Systematics and Geography of Plants, 76*, 85–145.

Saether, O., Craik, D. J., Campbell, I. D., Sletten, K., Juul, J., & Norman, D. G. (1995). Elucidation of the primary and three-dimensional structure of the uterotonic polypeptide kalata B1. *Biochemistry, 34*, 4147–4158.

Saska, I., Gillon, A. D., Hatsugai, N., Dietzgen, R. G., Hara-Nishimura, I., Anderson, M. A., et al. (2007). An asparaginyl endopeptidase mediates in vivo protein backbone cyclization. *Journal of Biological Chemistry, 282*, 29721–29728.

Seydel, P., Gruber, C. W., Craik, D. J., & Dörnenburg, H. (2007). Formation of cyclotides and variations in cyclotide expression in *Oldenlandia affinis* suspension cultures. *Applied Microbiology and Biotechnology, 77*, 275–284.

Simonsen, S. M., Sando, L., Ireland, D. C., Colgrave, M. L., Bharathi, R., Goransson, U., et al. (2005). A continent of plant defense peptide diversity: cyclotides in Australian *Hybanthus* (Violaceae). *Plant Cell, 17*, 3176–3189.

Smith, J. J., Hill, J. M., Little, M. J., Nicholson, G. M., King, G. F., & Alewood, P. F. (2011). Unique scorpion toxin with a putative ancestral fold provides insight into evolution of the inhibitor cystine knot motif. *Proceedings of the National Academy of Sciences of the United States of America, 108*, 10478–10483.

Struwe, L., Albert, V. A., & Bremer, B. (1994). Cladistics and family level classification of the Gentianales. *Cladistics, 10*, 175–205.

Svangård, E., Göransson, U., Hocaoglu, Z., Gullbo, J., Larsson, R., Claeson, P., et al. (2003). Cytotoxic cyclotides from *Viola tricolor* L. *Journal of Natural Products, 67*, 144–147.

Taylor, C. M., Lorence, D. H., & Gereau, R. E. (2010). Rubiacearum americanarum magna hama pars XXV: the nocturnally flowering *Psychotria domingensis*–Coussarea hondensis group plus three other Mesoamerican psychotria species transfer to Palicourea. *Novon, 20*, 481–492.

The Angiosperm Phylogeny, G. (2009). An update of the Angiosperm Phylogeny Group classification for the orders and families of flowering plants: APG III. *Botanical Journal of the Linnean Society, 161*, 105–121.

Wang, C. K., Clark, R. J., Harvey, P. J., Rosengren, K. J., Cemazar, M., & Craik, D. J. (2011). The role of conserved Glu residue on cyclotide stability and activity: a structural and functional study of kalata B12, a naturally occurring Glu to Asp mutant. *Biochemistry, 50*, 4077–4086.

Wang, C. K., Colgrave, M. L., Gustafson, K. R., Ireland, D. C., Goransson, U., & Craik, D. J. (2008). Anti-HIV cyclotides from the Chinese medicinal herb *Viola yedoensis*. *Journal of Natural Products, 71*, 47–52.

WHO. (2007). *WHO Monographs on selected medicinal plants* (Vol. 3). Available at http://apps.who.int/medicinedocs/en/m/abstract/Js14213e/. Retrieved 17.12.12 from http://apps.who.int/medicinedocs/documents/s14213e/s14213e.pdf#page=212.

Witherup, K. M., Bogusky, M. J., Anderson, P. S., Ramjit, H., Ransom, R. W., Wood, T., et al. (1994). Cyclopsychotride A, A biologically active, 31-residue cyclic peptide isolated from *Psychotria longipes*. *Journal of Natural Products, 57*, 1619–1625.

Wong, C. T., Taichi, M., Nishio, H., Nishiuchi, Y., & Tam, J. P. (2011). Optimal oxidative folding of the novel antimicrobial cyclotide from *Hedyotis biflora* requires high alcohol concentrations. *Biochemistry, 50*, 7275–7283.

Zhang, J., Liao, B., Craik, D. J., Li, J. T., Hu, M., & Shu, W. S. (2009). Identification of two suites of cyclotide precursor genes from metallophyte *Viola baoshanensis*: cDNA sequence variation, alternative RNA splicing and potential cyclotide diversity. *Gene, 431*, 23–32.

Zhu, S., Darbon, H., Dyason, K., Verdonck, F., & Tytgat, J. (2003). Evolutionary origin of inhibitor cystine knot peptides. *FASEB Journal, 17*, 1765–1767.

CHAPTER FOUR

Cyclotides from Chinese Plants

Ning-Hua Tan[1], Wen-Jun He
State Key Laboratory of Phytochemistry and Plant Resources in West China, Kunming Institute of Botany, Chinese Academy of Sciences, Kunming, Yunnan, People's Republic of China
[1]Corresponding author: E-mail: nhtan@mail.kib.ac.cn

Contents

Abstract

Intense global research interest has been directed toward the study of cyclotides in plants. However, it is only in the last decade that Chinese scientists have become involved in this field. Many cyclotide-containing plants are used in traditional Chinese medicines (TCMs), and these plants have therefore become hot resources in the cyclotide area. In this chapter, we systematically describe progress in understanding the chemistry and biology of cyclotides discovered from Chinese plants, including methods of chemical detection, distribution, chemotaxonomy, extraction, isolation, structural elucidation, and biological activity. We focus on our newly developed thin layer chromatography chemical method for the detection of cyclotides, which is a cost-effective alternative to mass spectroscopy. We have also optimized the isolation procedures to develop an approach for the cost- and time-effective extraction of cyclotides from plants. Using these chemical detection and isolation methods, we have characterized new cyclotides in Chinese plants with cytotoxic, hemolytic, and trypsin inhibitory activities. This work expands the number of characterized cyclotides in the plant kingdom and provides interesting results about the active ingredients of TCMs.

1. INTRODUCTION

Cyclotides are disulfide-rich macrocyclic plant proteins comprising approximately 30 amino acids, which not only contain a unique head-to-tail cyclized peptide backbone, but also incorporate a cystine knot motif.

Advances in Botanical Research, Volume 76
ISSN 0065-2296
http://dx.doi.org/10.1016/bs.abr.2015.09.009

Cyclotides display a range of interesting bioactivities, including uterotonic, hemolytic, anti-HIV, antimicrobial, cytotoxic, trypsin inhibitory, and insecticidal activities (Craik, 2013; Craik, Daly, Bond, & Waine, 1999; Tan & Zhou, 2006). The six backbone loops between the conserved cysteine residues display a variety of sequences and provide cyclotide-containing plants with a natural combinatorial library of bioactive cyclic peptides for host defense. The sequence variation occurs mainly in loops 2, 3, 5, and 6, with loops 1 and 4 being more highly conserved (Kaas & Craik, 2010). The first cyclotide discovered, kalata B1, a 29-residue cyclopeptide, was structurally characterized from the tropical African plant *Oldenlandia affinis* after coming to notice because of its uterotonic activity (Saether et al., 1995; Sletten & Gran, 1973). It was first discovered as the active agent in a native medicine used by local women in Africa to accelerate labor (Gran, 1973; Gran, Sandberg, & Sletten, 2000). It is estimated that the total number of cyclotides in the Violaceae plant family alone exceeds 9000 (Simonsen et al., 2005), but may extend to 50,000 in all plant species. So far, about 400 cyclotides have been isolated and characterized from 38 plants in the Violaceae (Schöepke, Hasan Agha, Kraft, & Otto, 1993), Rubiaceae (Saether et al., 1995; Sletten & Gran, 1973), Cucurbitaceae (Chan et al., 2013; Hernandez et al., 2000; Mylne et al., 2012), Fabaceae (Nguyen et al., 2011; Poth, Colgrave, Lyons, Daly, & Craik, 2011; Poth et al., 2010), Solanaceae (Poth et al., 2012), and Poaceae (Nguyen et al., 2013) families.

It is only in the past decade that Chinese scientists have become involved in the cyclotide field of research. Many cyclotide-containing plants are used in traditional Chinese medicines (TCMs), and these plants have therefore become hot resources in the cyclotide area. In this chapter, we systematically describe the progress in understanding the chemistry and biology of cyclotides discovered from Chinese plants, including methods of chemical detection, distribution, chemotaxonomy, extraction, isolation, structural elucidation, and biological activity.

2. CHEMICAL DETECTION METHOD FOR CYCLOTIDES

Coomassie brilliant blue G-250 and ninhydrin reagents were used to detect and distinguish cyclotides from amino acids, cyclopeptides or linear peptides using thin layer chromatography (TLC), as previously described (Xu, Tang, Ji, He, & Tan, 2008; Zhou & Tan, 2000). The protocol is

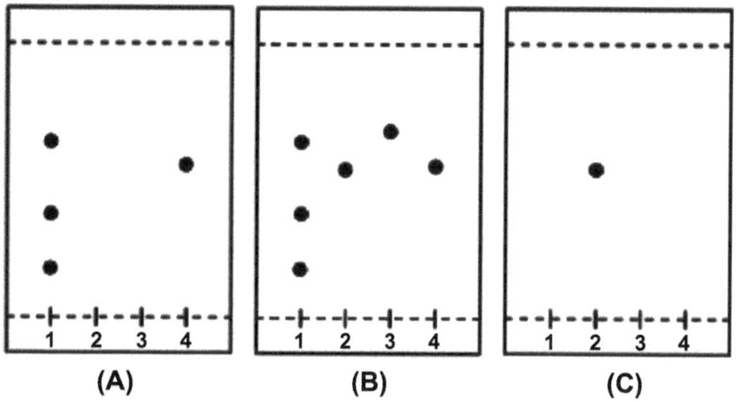

Figure 1 TLC results of amino acids, cyclotides, cyclopeptides, and linear peptides. (A) Nonhydrolyzed plate spraying with ninhydrin, (B) hydrolyzed plate spraying with ninhydrin, and (C) nonhydrolyzed plate spraying with G-250. Samples 1–4 on plates A–C: amino acids (mixture of proline, lysine, glutamic acid, methionine, and tryptophane), cyclotides (cycloviolacin O2), cyclopeptides (heterophyllin B), and linear peptides (linear-heterophyllin B). TLC conditions: solid phase, silica gel G; mobile phase, *n*-butanol:acetic acid:water (3:1:1, v:v:v). Spot colors were purplish-red or yellow, while the background was white in plates (A) and (B), and spot color was cyan, while the background was red in plate (C). *Reproduced with permission from Xu et al. (2008).*

summarized in Figure 1. Specifically, plant extracts are dotted on three TLC plates A–C, and the plates are then developed with *n*-butanol:acetic acid: water (3:1:1, v:v:v). Plates A and C are detected with 0.2% ninhydrin reagent and Coomassie brilliant blue G-250 reagent, respectively. Plate B is hung in a sealed glass vessel with about 1 ml concentrated HCl and hydrolyzed in an incubator at 110 °C for 2 h, then detected with 0.2% ninhydrin reagent. Spots with color in plates B and C, but not plate A are presumed to be cyclotides, which are then confirmed by HPLC-MS. Using this method 31 cyclotides were detected, isolated, and characterized from *Momordica cochinchinensis* (**77–78**), *Viola labradorica* (**97**, **134–137**), *Viola philippica* (**46**, **55**, **76**, **82**, **97**, **116**, **120**, **122**, **139–146**), and *Viola tricolor* (**46**, **55**, **97**, **100–101**, **103**, **147–152**) (Table 1).

3. DISTRIBUTION AND CHEMOTAXONOMY OF CYCLOTIDES IN CHINESE PLANTS

Among 38 plants from 6 families and 13 genera containing about 400 cyclotides, 13 of the plants (from 4 families and 4 genera) contain 152

Table 1 Cyclotides from Chinese Plants

Cyclotides	Sequences I loop 1 II loop 2 III lopp3 IV (loop 4) V loop 5 VI loop 6	Species (Parts)#	References
Cliotide T1 (**1**)	CGES—CVFIP——CITGAIG-CSCKSKV–CYRNGIP	*C. ternatea* (whole plant)	Nguyen et al. (2011)
Cliotide T2 (**2**)	CGES—CVQGE——CYTPG—CSCDWPI–CKKNGEFLK	*C. ternatea* (whole plant)	Nguyen et al. (2011)
Cliotide T3 (**3**)	CGET—CTLGT——CYVPD—CSCSWPI–CMKNGGLPT	*C. ternatea* (whole plant)	Nguyen et al. (2011)
Cliotide T6 (**4**)	CGES—CVYIP——CITTIVG-CSCKNSV–CYSNSIP	*C. ternatea* (whole plant)	Nguyen et al. (2011)
Cliotide T8 (**5**)	CGES—CVFIP——CISSVVG-CSCKSKV–CYNNGIP	*C. ternatea* (whole plant)	Nguyen et al. (2011)
Cliotide T9 (**6**)	CGES—CVFIP——CITTVVG-CSCKNKV–CYNNGIP	*C. ternatea* (whole plant)	Nguyen et al. (2011)
Cliotide T10 (**7**)	CGES—CVYIP——CTVTALLGCSCKDKV–CYKNGIP	*C. ternatea* (whole plant)	Nguyen et al. (2011)
Cliotide T11 (**8**)	CGES—CVFIP——CTITALLGCSCKDKV–CYKNGIP	*C. ternatea* (whole plant)	Nguyen et al. (2011)
Cliotide T12 (**9**)	CGES—CVFIP——CITGAIG-CSCKSKV–CYRDGIP	*C. ternatea* (whole plant)	Nguyen et al. (2011)
cO26 (**10**)	CGES—CFRGK——CYTPG—CSCSKYPL-CAKDGSIPA	*V. odorata* (aerial parts)	Colgrave et al. (2010)
cO27 (**11**)	CGES—CFKGW——CYTPG—CSCSKYPL-CAKDSIPAG	*V. odorata* (aerial parts)	Colgrave et al. (2010)
cO28 (**12**)	CGET—CVGGT——CNTPG—CSCSWPV–CFRDGLPV	*V. odorata* (aerial parts)	Colgrave et al. (2010)
cO29 (**13**)	CGES—CVWIP——CISGAIG-CSCKSKV–CYKNGIP	*V. odorata* (aerial parts)	Colgrave et al. (2010)

cO30 (**14**)	**C**GES—**C**VWIP——**C**ISSAIG-**C**S**C**KNKV–**C**FKNGIP	*V. odorata* (aerial parts)	Colgrave et al. (2010)
cO31 (**15**)	**C**GET—**C**VGGT——**C**NTPG—**C**S**C**SIPV–**C**TRNGLPV	*V. odorata* (aerial parts)	Colgrave et al. (2010)
cO32 (**16**)	**C**GET—**C**FGGT——**C**NTPG—**C**T**C**DPWPV-**C**TNDGAPV	*V. odorata* (aerial parts)	Colgrave et al. (2010)
cO33 (**17**)	**C**GET—**C**VGGT——**C**NTPYCT-**C**SWPV——**C**TRDGLPV	*V. odorata* (aerial parts)	Colgrave et al. (2010)
cO34 (**18**)	**C**GET—**C**VGGT——**C**NTEY—**C**T**C**SWPV–**C**TRDGLPV	*V. odorata* (aerial parts)	Colgrave et al. (2010)
cO35 (**19**)	**C**GET—**C**VGGT——**C**NTPY—**C**F**C**SWPV–**C**TRDGLPV	*V. odorata* (aerial parts)	Colgrave et al. (2010)
cO36 (**20**)	**C**GET—**C**FGGT——**C**NTPG—**C**T**C**DPFPV-**C**THDGLPT	*V. odorata* (aerial parts)	Colgrave et al. (2010)
Cter A (**21**)	**C**GES—**C**VFIP——**C**ISTVIG-**C**S**C**KNKV–**C**YRNGVIP	*C. ternatea* (whole plant, seeds)	Nguyen et al. (2011) and Poth et al. (2010)
Cter B (**22**)	**C**AES—**C**VWIP——**C**TVTALLG**C**S**C**KDKV–**C**YLNGVP	*C. ternatea* (whole plant, seeds)	Nguyen et al. (2011) and Poth et al. (2010)
Cter C (**23**)	**C**AES—**C**VWIP——**C**TVTALLG**C**S**C**KDKV–**C**YLDGVP	*C. ternatea* (seeds)	Poth et al. (2010)
Cter D (**24**)	**C**AES—**C**VWIP——**C**TVTALLG**C**S**C**KDKV–**C**YLNGIP	*C. ternatea* (whole plant, seeds)	Nguyen et al. (2011) and Poth et al. (2010)
Cter E (**25**)	**C**AES—**C**VWIP——**C**TVTALLG**C**S**C**KDKV–**C**YLDGIP	*C. ternatea* (seeds)	Poth et al. (2010)
Cter F (**26**)	**C**GES—**C**VFIP——**C**ISSVVG-**C**S**C**KSKV–**C**YLDGIP	*C. ternatea* (seeds)	Poth et al. (2010)
Cter G (**27**)	**C**GES—**C**VFIP——**C**ITTVVG-**C**S**C**KNKV–**C**YNNGLP	*C. ternatea* (seeds)	Poth et al. (2010)

(Continued)

Table 1 Cyclotides from Chinese Plants—cont'd

Cyclotides	Sequences loop 4 ↓ **I** loop 1 **II** loop 2 **III** lopp3 **IV V** loop 5 **VI** loop 6	Species (Parts)#	References
Cter H (**28**)	**C**GES—**C**VFIP——**C**ITTVVG-**C**S**C**KNKV–**C**YNDGLP	*C. ternatea* (seeds)	Poth et al. (2010)
Cter I (**29**)	**C**GES—**C**VFIP——**C**ITGIAG-**C**S**C**KNKV–**C**YINGTVP	*C. ternatea* (seeds)	Poth et al. (2010)
Cter J (**30**)	**C**GES—**C**VFIP——**C**ITGIAG-**C**S**C**KNKV–**C**YIDGTVP	*C. ternatea* (seeds)	Poth et al. (2010)
Cter K (**31**)	**C**GES—**C**VFIP——**C**ITTVVG-**C**S**C**KNKV–**C**YNHEP	*C. ternatea* (seeds)	Poth et al. (2010)
Cter L (**32**)	**C**GES—**C**VFIP——**C**ITTVVG-**C**S**C**KNKV–**C**YDHEP	*C. ternatea* (seeds)	Poth et al. (2010)
Cter M (**33**)	**C**GET—**C**TLGT——**C**YVPD—**C**S**C**SWPI–**C**MKNGLPT	*C. ternatea* (leaves)	Poth et al. (2011)
Cter N (**34**)	**C**GET—**C**VLGT——**C**YTPD—**C**S**C**TALV–**C**LKNGSAF	*C. ternatea* (leaves)	Poth et al. (2011)
Cter O (**35**)	**C**GES—**C**VFIP——**C**ITGIAG-**C**S**C**KSKV–**C**YRNGIP	*C. ternatea* (leaves)	Poth et al. (2011)
Cter P (Cliotide T4, **36**)	**C**GES—**C**VFIP——**C**ITAAIG-**C**S**C**KSKV–**C**YRNGIP	*C. ternatea* (leaves)	Poth et al. (2011)
Cter Q (Cliotide T5, **37**)	**C**GES—**C**VFIP——**C**ISTVIG-**C**S**C**KNKV–**C**YRNGIP	*C. ternatea* (leaves)	Poth et al. (2011)
Cter R (Cliotide T7, **38**)	**C**GES—**C**VFIP——**C**TVTALLG**C**S**C**KDKV–**C**YKNGIP	*C. ternatea* (leaves)	Poth et al. (2011)
Cycloviolacin H1 (**39**)	**C**GES—**C**VYIP——**C**LTSAIG-**C**S**C**KSKV–**C**YRNGIP	*V. hederaceae* (whole plant)	Chen, Colurave, Wang, and Craik (2006)

Cycloviolacin H2 (**40**)	**C**GES—**C**VYIP——**C**FIPG—**C**S**C**RNRV–**C**YLNSAIA	*V. hederaceae* (whole plant)	Chen et al. (2006)
Cycloviolacin H3 (**41**)	**C**GET—**C**FGGT——**C**NTPG—**C**I**C**DPWPV–**C**TRNGLPV	*V. hederaceae* (whole plant)	Chen et al. (2006)
Cycloviolacin H4 (**42**)	**C**AES—**C**VWIP——**C**TVTALLG**C**S**C**SNNV–**C**YNGIP	*V. hederaceae* (whole plant)	Chen et al. (2006)
Cycloviolacin O2 (**43**)	**C**GES—**C**VWIP——**C**ISSAI-G**C**S**C**KSKV–**C**YRNGIP	*V. biflora* (aerial parts) *V. odorata* (whole plant) *V. philippica* (whole plant) *V. tricolor* (whole plant)	He et al. (2011), Herrmann et al. (2008), Ireland, Colgrave, and Craik (2006), and Tang et al. (2010a, 2010b)
Cycloviolacin O3 (**44**)	**C**GES—**C**VWIP——**C**LTSAIG-**C**S**C**KSKV–**C**YRNGIP	*V. odorata* (whole plant)	Ireland et al. (2006)
Cycloviolacin O4 (**45**)	**C**GES—**C**VWIP——**C**ISSAIG-**C**S**C**KNKV–**C**YRNGIP	*V. odorata* (whole plant)	Ireland et al. (2006)
Cycloviolacin O5 (**46**)	**C**GES—**C**VWIP——**C**ISSAVG-**C**S**C**KNKV–**C**YKNGTP	*V. odorata* (whole plans)	Ireland et al. (2006)
Cycloviolacin O6 (**47**)	**C**GES—**C**VWIP——**C**ISAAVG-**C**S**C**KSKV–**C**YKNGTLP	*V. odorata* (whole plant)	Ireland et al. (2006)
Cycloviolacin O7 (**48**)	**C**GES—**C**VWIP——**C**TITALAG**C**K**C**KSKV–**C**YNSIP	*V. odorata* (whole plant)	Ireland et al. (2006)
Cycloviolacin O8 (**49**)	**C**GES—**C**VWIP——**C**ISSVVG-**C**S**C**KSKV–**C**YKNGTLP	*V. odorata* (whole plant)	Ireland et al. (2006)
Cycloviolacin O9 (**50**)	**C**GES—**C**VWIP——**C**LTSAVG-**C**S**C**KSKV–**C**YRNGIP	*V. biflora* (aerial parts) *V. odorata* (whole plant)	Herrmann et al. (2008) and Ireland et al. (2006)
Cycloviolacin O11 (**51**)	**C**GES—**C**VWIP——**C**ISAVVG-**C**S**C**KSKV–**C**YKNGTLP	*V. odorata* (whole plant)	Ireland et al. (2006)

(*Continued*)

Table 1 Cyclotides from Chinese Plants—cont'd

Cyclotides	Sequences (loop 4 ↓) **I** loop 1 **II** loop 2 **III** lopp3 **IV** **V** loop 5 **VI** loop 6	Species (Parts)#	References
Cycloviolacin O12 (Varv peptide E, **52**)	**C**GET—**C**VGGT——**C**NTPG—**C**S**C**SWPV–**C**TRNGLPI	*Viola baoshanensis* (leaves and roots) *V. odorata* (whole plant) *V. philippica* (whole plant) *V. tianshanica* (whole plant) *V. tricolor* (aerial parts) *V. yedoensis* (whole plant)	He et al. (2011), Ireland et al. (2006), Liu et al. (2014), Tang et al. (2010b), Wang et al. (2007), Xiang et al. (2010), and Zhang et al. (2009)
Cycloviolacin O13 (**53**)	**C**GES—**C**VWIP——**C**ISAAIG-**C**S**C**KSKV–**C**YRNGIP	*V. odorata* (whole plant)	Ireland et al. (2006)
Cycloviolacin O14 (**54**)	**C**GES—**C**FKGK——**C**YTPG—**C**S**C**SKYPL-**C**AKNGSIPA	*V. odorata* (whole plant)	Ireland et al. (2006)
Cycloviolacin O15 (**55**)	**C**GET—**C**FTGK——**C**YTPG—**C**S**C**SYPI–**C**KKNGLVP	*V. odorata* (whole plant)	Ireland et al. (2006)
Cycloviolacin O16 (**56**)	**C**GET—**C**FTGK——**C**YTPG—**C**S**C**SYPI–**C**KKINGLP	*V. odorata* (whole plant)	Ireland et al. (2006)
Cycloviolacin O17 (**57**)	**C**GES—**C**VWIP——**C**ISAAIG-**C**S**C**KNKV–**C**YRNGIP	*V. odorata* (whole plant)	Ireland et al. (2006)
Cycloviolacin O18 (**58**)	**C**GES—**C**VYIP——**C**TVTALAG**C**K**C**KSKV–**C**YNGIP	*V. odorata* (whole plant)	Ireland et al. (2006)
Cycloviolacin O19 (**59**)	**C**GES—**C**VWIP——**C**ISSVVG-**C**S**C**KSKV–**C**YKDGTLP	*V. odorata* (whole plant)	Ireland et al. (2006)
Cycloviolacin O20 (**60**)	**C**GES—**C**VWIP——**C**LTSAIG-**C**S**C**KSKV–**C**YRDGIP	*V. odorata* (whole plant)	Ireland et al. (2006)
Cycloviolacin O21 (**61**)	**C**GET—**C**VTGS——**C**YTPG—**C**T**C**SWPV–**C**TRNGLPV	*V. odorata* (whole plant)	Ireland et al. (2006)

Cycloviolacin O22 (**62**)	CGET—CVGGT——CNTPG—CTCSWPV–CTRNGLPI	*V. odorata* (whole plant)	Ireland et al. (2006)
Cycloviolacin O23 (**63**)	CGET—CFGGT——CNTPG—CTCDSSWPICTHNGLPT	*V. odorata* (whole plant)	Ireland et al. (2006)
Cycloviolacin O24 (**64**)	CGET—CFGGT——CNTPG—CTCDPWPV-CTHNGLPT	*V. odorata* (whole plant)	Ireland et al. (2006)
Cycloviolacin O25 (**65**)	CGET—CAFIP——CITHVPGTCSCKSKV–CYFNDIF	*V. odorata* (whole plant)	Ireland et al. (2006)
Cycloviolacin T1 (**66**)	CGET—CVGGT——CNTPG—CSCSWPV–CTRNGIPV	*V. tianshanica* (whole plant)	Xiang et al. (2010)
Cycloviolacin VY1 (**67**)	CGES—CVFIP——CITTVLG-CSCSIKV–CYKNGSIP	*V. yedoensis* (whole plant)	Liu et al. (2014)
Cycloviolacin Y1 (**68**)	CGET—CFLGT——CYTPG—CSCGNYGF-CYGTNGGTIFD	*V. yedoensis* (whole plant)	Wang et al. (2007)
Cycloviolacin Y2 (**69**)	CGES—CFLGT——CYTAG—CSCGNWGL-CYGTNGGTIFD	*V. yedoensis* (whole plant)	Wang et al. (2007)
Cycloviolacin Y3 (**70**)	CGET—CFLGT——CYTAG—CSCGNWGL-CYGTNGGTIFD	*V. yedoensis* (whole plant)	Wang et al. (2007)
Cycloviolacin Y4 (**71**)	CGES—CVFIP——CITGVIG-CSCSSNV–CYLNGVP	*V. yedoensis* (whole plant)	Wang et al. (2007)
Cycloviolacin Y5 (**72**)	CAES—CVWIP——CTVTALVGCSCSDKV–CYNGIP	*V. baoshanensis* (leaves and roots) *V. yedoensis* (whole plant)	Liu et al. (2014), Wang et al. (2007), and Zhang et al. (2009)
Kalata B1 (**73**)	CGET—CVGGT——CNTPG—CTCSWPV–CTRNGLPV	*V. baoshanensis* (leaves and roots) *V. odorata* (whole plant) *V. philippica* (whole plant) *V. yedoensis* (whole plant)	He et al. (2011), Tang et al. (2010a), Wang et al. (2007), and Zhang et al. (2009)

(Continued)

Table 1 Cyclotides from Chinese Plants—cont'd

Cyclotides	Sequences loop 4 ↓ **I** loop 1 **II** loop 2 **III** lopp3 **IV V** loop 5 **VI** loop 6	Species (Parts)#	References
MCoTI-I (**74**)	**C**PKILQR**C**RRDSD—**C**PGA——**C**I**C**RGNGY-**C**GSGSDGGV	*M. cochinchinensis* (seeds)	Chan et al. (2013) and Hernandez et al. (2000)
MCoTI-II (**75**)	**C**PKILKK**C**RRDSD—**C**PGA——**C**I**C**RGNGY-**C**GSGSDGGV	*M. cochinchinensis* (seeds)	Chan et al. (2013) and Hernandez et al. (2000)
MCoTI-IV (**76**)	**C**PRILKK**C**RRDSD—**C**PGA——**C**V**C**QGNGY-**C**GSGSDGGA	*M. cochinchinensis* (seeds)	Mylne et al. (2012)
MCoTI-VII (**77**)	**C**PRILKK**C**RRDSD—**C**PGA——**C**V**C**KGNGY-**C**GSGSDGGA	*M. cochinchinensis* (seeds)	Mylne et al. (2012)
MCoTI-VIII (**78**)	**C**PKILQR**C**RRDSD—**C**PGA——**C**I**C**LGNGY-**C**GSGSDGGV	*M. cochinchinensis* (seeds)	Mylne et al. (2012)
Mram 8 (**79**)	**C**GES—**C**VFIP——**C**LTSAIG-**C**S**C**KSKV–**C**YRNGIP	*V. philippica* (whole plant)	He et al. (2011)
Phyb A (**80**)	**C**GES—**C**VWIP——**C**VSAAIG-**C**S**C**SNKI–**C**YRNGIG	*P.* × *hybrida* (whole plant)	Poth et al. (2012)
Phyb B (**81**)	**C**GES—**C**VWMY——**C**ISAAMG-**C**S**C**RNKV–**C**YRNGVP	*P.* × *hybrida* (whole plant)	Poth et al. (2012)
Phyb C (**82**)	**C**GES—**C**VWMY——**C**ITATMG-**C**S**C**RNKV–**C**YKNGIP	*P.* × *hybrida* (whole plant)	Poth et al. (2012)
Phyb D (**83**)	**C**GES—**C**MWIP——**C**ISAAIG-**C**S**C**TNHV–**C**YKNGIP	*P.* × *hybrida* (whole plant)	Poth et al. (2012)
Phyb E (**84**)	**C**GES—**C**VWIP——**C**ISGVQG-**C**S**C**SNKI–**C**YRNGIP	*P.* × *hybrida* (whole plant)	Poth et al. (2012)

Phyb F (**85**)	**C**GGS—**C**VWIP——**C**ISGVQG-**C**S**C**SNKI–**C**YRNGIP	*P.* × *hybrida* (whole plant)	Poth et al. (2012)
Phyb G (**86**)	**C**GES—**C**AWIP——**C**ISAVQG-**C**S**C**RNKI–**C**YRNGIP	*P.* × *hybrida* (whole plant)	Poth et al. (2012)
Phyb H (**87**)	**C**GES—**C**IWIE——**C**ISGAIG-**C**S**C**RNKV–**C**YRNGLP	*P.* × *hybrida* (whole plant)	Poth et al. (2012)
Phyb I (**88**)	**C**GES—**C**IWIP——**C**TTTALLG**C**S**C**SNKV–**C**YKNGIP	*P.* × *hybrida* (whole plant)	Poth et al. (2012)
Phyb J (**89**)	**C**GES—**C**LWIP——**C**TVTAAFG**C**Y**C**SNKV–**C**VKDSYT	*P.* × *hybrida* (whole plant)	Poth et al. (2012)
Phyb K (**90**)	**C**GEP—**C**VYIP——**C**TITALLG**C**S**C**LNKV–**C**VRPSTD	*P.* × *hybrida* (whole plant)	Poth et al. (2012)
Phyb L (**91**)	**C**AET—**C**VWIP——**C**ATSLIG-**C**S**C**VNSI–**C**TYTNQSIS	*P.* × *hybrida* (whole plant)	Poth et al. (2012)
Tricyclon A (**92**)	**C**GES—**C**FLGT——**C**YTKG—**C**S**C**GEWKL–**C**YGTNGGTIFD	*V. tricolor* (flowers)	Mulvenna, Sando, and Craik (2005)
Tricyclon B (**93**)	**C**GES—**C**FLGT——**C**YTKG—**C**S**C**GEWKL–**C**YGENGGTIFD	*V. tricolor* (flowers)	Mulvenna et al. (2005)
Varv peptide A (**94**)	**C**GET—**C**VGGT——**C**NTPG—**C**S**C**SWPV–**C**TRNGLPV	*V. baoshanensis* (leaves and roots) *V. biflora* (aerial parts) *V. labradorica* (whole plant) *V. odorata* (whole plant) *V. philippica* (whole plant) *V. tricolor* (aerial parts) *V. yedoensis* (whole plant)	He et al. (2011), Herrmann et al. (2008), Ireland et al. (2006), Tang et al. (2010a, 2010b), Wang et al. (2007), and Zhang et al. (2009)

(*Continued*)

Table 1 Cyclotides from Chinese Plants—cont'd

Cyclotides	Sequences loop 4 ↓ **I** loop 1 **II** loop 2 **III** lopp3 **IV V** loop 5 **VI** loop 6	Species (Parts)#	References
Varv peptide B (**95**)	**C**GET—**C**FGGT——**C**NTPG—**C**S**C**DPWPM-**C**SRNGLPV	*V. arvensis* (aerial parts)	Göransson, Luijendijk, Johansson, Bohlin, and Claeson (1999)
Varv peptide C (**96**)	**C**GET—**C**VGGT——**C**NTPG—**C**S**C**SWPV–**C**TRNGVPI	*V. arvensis* (aerial parts)	Göransson et al. (1999)
Varv peptide D (**97**)	**C**GET—**C**VGGS——**C**NTPG—**C**S**C**SWPV–**C**TRNGLPI	*V. arvensis* (aerial parts) *V. tricolor* (aerial parts) *V. yedoensis* (whole plant)	Göransson et al. (1999), Tang et al. (2010b), and Wang et al. (2007)
Varv peptide F (**98**)	**C**GET—**C**TLGT——**C**YTAG—**C**S**C**SWPV–**C**TRNGVPI	*V. arvensis* (aerial parts) *V. tricolor* (aerial parts)	Göransson et al. (1999) and Tang et al. (2010b)
Varv peptide G (**99**)	**C**GET—**C**FGGT——**C**NTPG—**C**S**C**DPWPV-**C**SRNGVPV	*V. arvensis* (aerial parts)	Göransson et al. (1999)
Varv peptide H (**100**)	**C**GET—**C**FGGT——**C**NTPG—**C**S**C**ETWPV-**C**SRNGLPV	*V. arvensis* (aerial parts)	Göransson et al. (1999) and Tang et al. (2010b)
Vhl-1 (**101**)	**C**GES—**C**AMISF—**C**FTEVIG-**C**S**C**KNKV–**C**YLNSIS	*V. hederaceae* (whole plant)	Chen et al. (2005)
Vhl-2 (**102**)	**C**GET—**C**FTGT——**C**YTNG—**C**T**C**DPWPV-**C**TRNGLPV	*V. hederaceae* (whole plant)	Chen et al. (2005)
Viba 1 (**103**)	**C**GEG—**C**VYLP——**C**FTAPLG-**C**S**C**SSKV–**C**YRNGIP	*V. baoshanensis* (leaves and roots)	Zhang et al. (2009)
Viba 2 (**104**)	**C**GES—**C**VYLP——**C**FTAPLG-**C**S**C**SSKV–**C**YRNGIP	*V. baoshanensis* (leaves and roots)	Zhang et al. (2009)
Viba 3 (**105**)	**C**GES—**C**VWIP——**C**LTAAIG-**C**S**C**SSKV–**C**YRNGIP	*V. baoshanensis* (leaves and roots)	Zhang et al. (2009)

Viba 4 (**106**)	**C**GES—**C**VWIP——**C**LTSAIG-**C**S**C**KSSV–**C**YRNGVP	*V. baoshanensis* (leaves and roots)	Zhang et al. (2009)
Viba 5 (**107**)	**C**GES—**C**VWIP——**C**LTATIG-**C**S**C**KSKV–**C**YRNGIP	*V. baoshanensis* (leaves and roots)	Zhang et al. (2009)
Viba 6 (**108**)	**C**GES—**C**VLIP——**C**ISSVIG-**C**S**C**KSKV–**C**YRNGIP	*V. baoshanensis* (leaves and roots)	Zhang et al. (2009)
Viba 7 (**109**)	**C**GES—**C**VFIP——**C**ISSVIG-**C**S**C**KSKV–**C**YRNGVIP	*V. baoshanensis* (leaves and roots)	Zhang et al. (2009)
Viba 8 (**110**)	**C**IET—**C**YTFP——**C**ISEMIN-**C**S**C**KNSR–**C**QKNGAG	*V. baoshanensis* (leaves and roots)	Zhang et al. (2009)
Viba 9 (**111**)	**C**GES—**C**VWIP——**C**ISSAIG-**C**S**C**KNKV–**C**YRKGIP	*V. baoshanensis* (leaves and roots)	Zhang et al. (2009)
Viba 10 (**112**)	**C**AES—**C**VYLP——**C**VTIVIG-**C**S**C**KDKV–**C**YNGIP	*V. baoshanensis* (leaves and roots)	Zhang et al. (2009)
Viba 11 (**113**)	**C**GES—**C**VWIP——**C**ISGAIG-**C**S**C**KSKV–**C**YRNG-IP	*V. baoshanensis* (leaves and roots) *V. philippica* (whole plant)	He et al. (2011)
Viba 12 (**114**)	**C**AES—**C**VWIP——**C**TVTALLG**C**S**C**KDKV–**C**YNGIP	*V. baoshanensis* (leaves and roots)	Zhang et al. (2009)
Viba 13 (**115**)	**C**AES—**C**VWIP——**C**TVTALLG**C**S**C**KDKV–**C**YNTIP	*V. baoshanensis* (leaves and roots)	Zhang et al. (2009)
Viba 14 (**116**)	**C**GER—**C**VIERTRAW**C**RTVG—**C**I**C**SLHTLE**C**VRNGRL	*V. baoshanensis* (leaves and roots)	Zhang et al. (2009)
Viba 15 (**117**)	**C**GET—**C**VGGT——**C**NTPG—**C**A**C**SWPV–**C**TRNGLPV	*V. philippica* (whole plant)	He et al. (2011) and Zhang et al. (2009)
Viba 16 (**118**)	**C**GET—**C**TLGT——**C**YTVG—**C**T**C**SWPI–**C**TRNGLPI	*V. baoshanensis* (leaves and roots)	Zhang et al. (2009)

(*Continued*)

Table 1 Cyclotides from Chinese Plants—cont'd

Cyclotides	Sequences loop 4 ↓ **I** loop 1 **II** loop 2 **III** lopp3 **IV V** loop 5 **VI** loop 6	Species (Parts)[#]	References
Viba 17 (**119**)	**C**GET—**C**VGGT——**C**NTPG—**C**G**C**SWPV–**C**TRNGLPV	*V. philippica* (whole plant)	He et al. (2011) and Zhang et al. (2009)
Vibi A (**120**)	**C**GET—**C**FGGT——**C**NTPG—**C**S**C**SYPI–**C**TRNGLPV	*V. biflora* (aerial parts)	Herrmann et al. (2008)
Vibi B (**121**)	**C**GET—**C**FGGT——**C**NTPG—**C**T**C**SYPI–**C**TRNGLPV	*V. biflora* (aerial parts)	Herrmann et al. (2008)
Vibi C (**122**)	**C**GET—**C**AFGS——**C**YTPG—**C**S**C**SWPV–**C**TRNGLPV	*V. biflora* (aerial parts)	Herrmann et al. (2008)
Vibi D (**123**)	**C**GET—**C**FGGR——**C**NTPG—**C**T**C**SYPI–**C**TRNGLPV	*V. biflora* (aerial parts)	Herrmann et al. (2008)
Vibi E (**124**)	**C**AES—**C**VWIP——**C**TVTALIG**C**G**C**SNKV–**C**YNGIP	*V. biflora* (aerial parts)	Herrmann et al. (2008)
Vibi F (**125**)	**C**GES—**C**VFIP——**C**LTSALG-**C**S**C**KSKV–**C**YKNGTIP	*V. biflora* (aerial parts)	Herrmann et al. (2008)
Vibi G (**126**)	**C**GES—**C**VFIP——**C**LTSAIG-**C**S**C**KSKV–**C**YKNGTFP	*V. biflora* (aerial parts)	Herrmann et al. (2008)
Vibi H (**127**)	**C**AES—**C**VYIP——**C**LTTVIG-**C**S**C**KSKV–**C**YKNGLLP	*V. biflora* (aerial parts)	Herrmann et al. (2008)
Vibi I (**128**)	**C**GES—**C**VWIP——**C**LTSTVG-**C**S**C**KSKV–**C**YRNGIP	*V. biflora* (aerial parts)	Herrmann et al. (2008)
Vibi J (**129**)	**C**GES—**C**VWIP——**C**ISKVIG-**C**A**C**KSKV–**C**YKNGTFP	*V. biflora* (aerial parts)	Herrmann et al. (2008)
Vibi K (**130**)	**C**GES—**C**VWIP——**C**LTSAVG-**C**P**C**KSKV–**C**YRNGIP	*V. biflora* (aerial parts)	Herrmann et al. (2008)

Vila A (**131**)	CGES—CVWIP——CISSAIG-CSCKDKV–CYRDGIP	*V. labradorica* (whole plant)	Tang et al. (2010a)
Vila B (**132**)	CGES—CVWIP——CISSAIG-CSCRSKV–CYRDGIP	*V. labradorica* (whole plant)	Tang et al. (2010a)
Vila C (**133**)	CGES—CVWIP——CISSVVG-CSCKDKV–CYKDGTLP	*V. labradorica* (whole plant)	Tang et al. (2010a)
Vila D (**134**)	CGET—CAFGT——CYTGG—CSCSWPV–CTRNGIPV	*V. labradorica* (whole plant)	Tang et al. (2010a)
Violapeptide 1 (**135**)	CGET—CVGGT——CNTPG—CSCSRPV–CTXNGLPV	*V. arvensis* (aerial parts) *V. tricolor* (whole plant)	Schöepke et al. (1993)
Viphi A (**136**)	CGES—CVFIP——CISSVIG-CACKSKV–CYKNGSIP	*V. philippica* (whole plant)	He et al. (2011)
Viphi B (**137**)	CGET—CTIGT——CYTAG—CTCSWPI–CTRNGLPV	*V. philippica* (whole plant)	He et al. (2011)
Viphi C (**138**)	CGES—CVYIP——CITSVI-GCSCSSKV–CYINGVP	*V. philippica* (whole plant)	He et al. (2011)
Viphi D (**139**)	CGES—CVFIP——CISSVI-GCSCSSKV–CYRNGIP	*V. philippica* (whole plant)	He et al. (2011)
Viphi E (**140**)	CGES—CVFIP——CISAVI-GCSCSNKV–CYKNGSIP	*V. philippica* (whole plant)	He et al. (2011)
Viphi F (**141**)	CGES—CVFIP——CISAII-GCSCSSKV–CYKNGSIP	*V. philippica* (whole plant)	He et al. (2011)
Viphi G (**142**)	CGES—CVFIP——CISAII-GCSCSNKV–CYKNGSIP	*V. philippica* (whole plant)	He et al. (2011)
Viphi H (**143**)	CAES—CVWIP——CTVTAIVGCSCSWGV–CYNGIP	*V. philippica* (whole plant)	He et al. (2011)

(Continued)

Table 1 Cyclotides from Chinese Plants—cont'd

Cyclotides	Sequences loop 4 ↓ **I** loop 1 **II** loop 2 **III** lopp3 **IV V** loop 5 **VI** loop 6	Species (Parts)#	References
Vitri A (**144**)	**C**GES—**C**VWIP——**C**ITSAIG**C**S**C**KSKV—**C**YRNGIP	*V. tricolor* (aerial parts)	Svangard et al. (2004) and Tang et al. (2010b)
Vitri B (**145**)	**C**GES—**C**VGGT——**C**NTPG–**C**S**C**SWPV—**C**TTNGVPI	*V. tricolor* (whole plant)	Tang et al. (2010b)
Vitri C (**146**)	**C**GET—**C**VGGT——**C**NTPG–**C**F**C**TWPV—**C**TRNGLPI	*V. tricolor* (whole plant)	Tang et al. (2010b)
Vitri D (**147**)	**C**GET—**C**FTGS——**C**YTPG–**C**S**C**NWPV—**C**NRNGLPV	*V. tricolor* (whole plant)	Tang et al. (2010b)
Vitri E (**148**)	**C**GET—**C**VGGT——**C**NTPG–**C**S**C**SWPV—**C**FRNGLPV	*V. tricolor* (whole plant)	Tang et al. (2010b)
Vitri F (**149**)	**C**GES—**C**VWIP——**C**ISSVVG**C**A**C**KSKV—**C**YKDGLTP	*V. tricolor* (whole plant)	Tang et al. (2010b)
VocC (**150**)	**C**GET—**C**VGGT——**C**NTPG–**C**S**C**SWPV—**C**IRNGLPV	*V. odorata* (leaves)	Zarrabi, Dalirfardouei, Sepehrizade, and Kermanshahi (2013)
Vodo M (**151**)	**C**GES—**C**FTGK——**C**YTVQ–**C**S**C**SWPV—**C**TRNGAPI	*V. odorata* (aerial parts)	Ireland et al. (2006) and Svangard et al. (2003)
Vodo N (**152**)	CGET---CTLGK----CYTAG--CSCSWPV---CYRNGLPV	*V. odorata* (aerial parts)	Ireland et al. (2006) and Svangard et al. (2003)

C.: Clitoria; V.: Viola; M.: Momordica; P.: Petunia.

cyclotides and are distributed throughout China (Table 1, Editorial Committee of Chinese Flora, 1978).

To gain further insight into the distribution and chemotaxonomy of cyclotides in the plant kingdom, 203 Chinese plant species (from 82 families and 153 genera) were collected and detected for cyclotides using the aforementioned chemical detection method (see Section 2) (Table 2). The results indicated that cyclotides were found only in seven plants: *M. cochinchinensis* (51), *Momordica subangulata* subsp. *renigera* (52), *Momordica subangulata* subsp. *subangulata* (53), *Viola hamiltoniana* (196), *V. labradorica* (197), *V. philippica* (198), and *V. tricolor* (199) (Figure 2). In line with previously published reports (Craik et al., 1999; Mahatmanto et al., 2015), we suggest that cyclotides are characteristic components in the genera *Viola* of Violaceae and *Momordica* of Cucurbitaceae and can be used as a marker of secondary metabolites for plants in these families.

4. EXTRACTION, ISOLATION AND STRUCTURAL ELUCIDATION OF CYCLOTIDES IN CHINESE PLANTS

We have previously reviewed protocols concerning the extraction, isolation, and structural elucidation of cyclotides in plants (Xu et al., 2008; Tan & Zhou, 2006); here, we give a brief introduction.

4.1 Extraction and Isolation of Cyclotides in Chinese Plants

Air-dried plant materials were powdered and extracted with ethanol at room temperature, followed by extraction with ethanol/water (1:1, v/v) three times (Tang et al., 2010a, 2010b). The combined extracts were concentrated, suspended in H_2O, and partitioned successively with petroleum ether, EtOAc, and *n*-BuOH. The chemical detection method described in Section 2 was used to guide the following separating steps with an ethanol—H_2O gradient. The *n*-BuOH fraction was subjected to macroporous resin (D 101) column chromatography. The cyclotide-containing fraction was loaded on polyamide column chromatography to remove tannins and then purified by reversed-phase C_{18} column chromatography and size-exclusion column chromatography on Sephadex LH-20. Final purification was achieved by HPLC (Figure 3; He et al., 2011). This isolation procedure was designed to develop a cost- and time-effective extraction from plants, which led to the isolation of cyclotides in gram amounts amenable to industrial scale-up (Tang et al., 2010a, 2010b).

Table 2 List of Chinese Plants Collected and Tested for Cyclotide Detection

	Family Name	Species Name	Parts Used	Place and Time Collected
1	Acanthaceae	*Pseuderanthemum polyanthum*	Branches and leaves	Yunnan, 2012
2	Aceraceae	*Acer davidii*	Leaves and stem	Yunnan, 2011
3	Anacardiaceae	*Spondias pinnata*	Fruit	Yunnan, 2011
4	Apiaceae	*Eryngium foetidum*	Whole plant	Yunnan, 2011
5	Apiaceae	*Foeniculum vulgare*	Whole plant	Yunnan, 2011
6	Apiaceae	*Ligusticum sinense*	Roots	Yunnan, 2011
7	Apiaceae	*Saposhnikovia divaricata*	Roots	Yunnan, 2011
8	Aquifoliaceae	*Ilex godajam*	Fruit and roots	Yunnan, 2011
9	Aquifoliaceae	*Ilex macrocarpa*	Leaves and stem	Yunnan, 2011
10	Aquifoliaceae	*Ilex szechuanensis*	Branches and leaves	Yunnan, 2011
11	Araceae	*Colocasia fallax*	Tubers	Yunnan, 2011
12	Araceae	*Lasia spinosa*	Tubers	Yunnan, 2011
13	Araliaceae	*Nothopanax delavayi*	Roots	Yunnan, 2010
14	Arecaceae	*Trachycarpus fortunei*	Flowers	Yunnan, 2011
15	Asclepiadaceae	*Cynanchum stauntonii*	Tubers	Yunnan, 2011
16	Asteraceae	*Arctium lappa*	Tubers	Yunnan, 2011
17	Asteraceae	*Aster tataricus*	Roots	Yunnan, 2010
18	Asteraceae	*Carthamus tinctorius*	Flowers	Yunnan, 2011
19	Asteraceae	*Cirsium japonicum*	Roots	Yunnan, 2010
20	Asteraceae	*Eupatorium japonicum*	Leaves and roots	Yunnan, 2010
21	Asteraceae	*Helianthus tuberosus*	Roots	Yunnan, 2011
22	Asteraceae	*Taraxacum mongolicum*	Flowers	Yunnan, 2011
23	Asteraceae	*Vernonia arborea*	Branches and leaves	Yunnan, 2011
24	Balsaminaceae	*Impatiens macrovexilla*	Whole plant	Yunnan, 2011
25	Brassicaceae	*Thlaspi arvense*	Whole plant	Yunnan, 2011

26	Burseraceae	*Canarium strictum*	Leaves	Yunnan, 2010
27	Burseraceae	*Garuga pinnata*	Fruit	Yunnan, 2011
28	Buxaceae	*Sarcococca ruscifolia*	Branches and leaves	Yunnan, 2011
29	Caesalpiniaceae	*Caesalpinia sappan*	Heartwoods	Yunnan, 2010
30	Campanulaceae	*Adenophora tetraphylla*	Roots	Yunnan, 2011
31	Campanulaceae	*Campanumoea javanica*	Roots	Yunnan, 2010
32	Campanulaceae	*Codonopsis bulleyana*	Roots	Yunnan, 2010
33	Campanulaceae	*Codonopsis pilosula*	Roots	Yunnan, 2011
34	Capparaceae	*Capparis masaikai*	Fruit and seeds	Yunnan, 2010
35	Capparaceae	*Capparis yunnanensis*	Fruit and seeds	Yunnan, 2010
36	Caprifoliaceae	*Lonicera japonica*	Leaves	Yunnan, 2011
37	Caprifoliaceae	*Lonicera maackii*	Leaves and fruit	Yunnan, 2012
38	Caprifoliaceae	*Viburnum foetidum*	Leaves	Yunnan, 2013
39	Caprifoliaceae	*Viburnum sargentii*	Fruit	Yunnan, 2010
40	Caricaceae	*Carica papaya*	Fruit	Yunnan, 2011
41	Caryophyllaceae	*Pseudostellaria heterophylla*	Tubers	Yunnan, 2010
42	Caryophyllaceae	*Stellaria monosperma*	Roots	Yunnan, 2010
43	Caryophyllaceae	*Stellaria saxatilis*	Whole plant	Yunnan, 2011
44	Convolvulaceae	*Cuscuta australis*	Whole plant	Yunnan, 2010
45	Coriariaceae	*Coriaria nepalensis*	Branches and leaves	Yunnan, 2011
46	Cornaceae	*Dendrobenthamia capitata*	Leaves, flowers, and stem	Yunnan, 2011
47	Cornaceae	*Swida oblonga*	Leaves, fruit, and stem	Yunnan, 2011
48	Cucurbitaceae	*Gynostemma longipes*	Whole plant	Yunnan, 2010
49	Cucurbitaceae	*Hemsleya longivillosa*	Tubers	Yunnan, 2010
50	Cucurbitaceae	*Momordica charantia*	Seeds, stem, fruit, and vines	Guangxi, 2010
51	Cucurbitaceae	*Momordica cochinchinensis*	Seeds, stem, fruit, and vines	Yunnan, 2010
52	Cucurbitaceae	*Momordica subangulata* subsp. *renigera*	Seeds, stem, fruit, and vines	Yunnan, 2013

(*Continued*)

Table 2 List of Chinese Plants Collected and Tested for Cyclotide Detection—cont'd

	Family Name	Species Name	Parts Used	Place and Time Collected
53	Cucurbitaceae	*Momordica subangulata* subsp. *subangulata*	Seeds, stem, fruit, and vines	Yunnan, 2014
54	Cucurbitaceae	*Siraitia grosvenorii*	Seeds, stem, fruit, and vines	Yunnan, 2011
55	Cucurbitaceae	*Trichosanthes cucumeroides*	Fruit	Yunnan, 2011
56	Cucurbitaceae	*Trichosanthes kirilowii*	Fruit	Yunnan, 2011
57	Cyperaceae	*Eleocharis dulcis*	Fruit	Yunnan, 2011
58	Daphniphyllaceae	*Daphniphyllum paxianum*	Branches and leaves	Yunnan, 2011
59	Dilleniaceae	*Dillenia indica*	Fruit	Yunnan, 2011
60	Dioscoreaceae	*Dioscorea bulbifera*	Tubers	Yunnan, 2010
61	Dioscoreaceae	*Dioscorea opposita*	Tubers	Yunnan, 2011
62	Ebenaceae	*Diospyros morrisiana*	Fruit	Yunnan, 2011
63	Elaeagnaceae	*Elaeagnus angustifolia*	Fruit	Yunnan, 2011
64	Elaeocarpaceae	*Elaeocarpus japonicus*	Seed husks	Yunnan, 2011
65	Elaeocarpaceae	*Elaeocarpus lanceifolius*	Seed husks	Yunnan, 2011
66	Ericaceae	*Rhododendron hancockii*	Flowers and leaves	Yunnan, 2011
67	Euphorbiaceae	*Phyllanthus emblica*	Fruit	Yunnan, 2010
68	Euphorbiaceae	*Sauropus androgynus*	Leaves	Yunnan, 2009
69	Euphorbiaceae	*Sauropus garrettii*	Leaves and stem	Yunnan, 2010
70	Fagaceae	*Cassia acutifolia*	Leaves	Yunnan, 2011
71	Fagaceae	*Castanopsis remotidenticulata*	Branches and leaves	Yunnan, 2011
72	Fagaceae	*Lithocarpus dealbatus*	Leaves and stem	Yunnan, 2011
73	Fagaceae	*Lithocarpus litseifolius*	Leaves and stem	Yunnan, 2011
74	Fagaceae	*Lithocarpus xylocarpus*	Branches and leaves	Yunnan, 2011
75	Fagaceae	*Pisum sativum*	Seeds	Yunnan, 2011
76	Fagaceae	*Quercus pseudosemecarpifolia*	Leaves and stem	Yunnan, 2011
77	Gramineae	*Saccharum spontaneum*	Roots	Yunnan, 2011

78	Guttiferae	*Garcinia xanthochymus*	Fruit	Yunnan, 2011
79	Hamamelidaceae	*Hamamelis mollis*	Fruit	Yunnan, 2010
80	Hypericaceae	*Hypericum henryi*	Branches and leaves	Yunnan, 2011
81	Icacinaceae	*Gomphandra mollis*	Branches and leaves	Yunnan, 2012
82	Juglandaceae	*Cyclocarya paliurus*	Leaves	Yunnan, 2011
83	Juglandaceae	*Engelhardia aceriflora*	Leaves	Yunnan, 2010
84	Juglandaceae	*Engelhardia spicata*	Leaves	Yunnan, 2010
85	Juglandaceae	*Platycarya strobilacea*	Branches and leaves	Yunnan, 2011
86	Juncaceae	*Juncus effusus*	Stem	Yunnan, 2011
87	Labiatae	*Glechoma longituba*	Leaves	Yunnan, 2011
88	Labiatae	*Salvia yunnanensis*	Roots	Yunnan, 2010
89	Lamiaceae	*Elsholtzia blanda*	Leaves and flowers	Yunnan, 2010
90	Lamiaceae	*Leonurus heterophyllus*	Whole plant	Yunnan, 2012
91	Lamiaceae	*Perilla frutescens*	Stem	Yunnan, 2009
92	Lardizabalaceae	*Akebia trifoliata*	Fruit	Yunnan, 2011
93	Lauraceae	*Cinnamomum cassia*	Barks	Yunnan, 2011
94	Leguminosae	*Astragalus membranaceus*	Roots	Yunnan, 2011
95	Leguminosae	*Canavalia gladiata*	Fruit	Yunnan, 2011
96	Leguminosae	*Derris eriocarpa*	Stem	Yunnan, 2011
97	Lemnaceae	*Marsilea quadrifolia*	Leaves and stem	Yunnan, 2011
98	Liliaceae	*Allium hookeri*	Tubers	Yunnan, 2011
99	Liliaceae	*Allium sativum*	Flowers	Yunnan, 2011
100	Liliaceae	*Anemarrhena asphodeloides*	Tubers	Yunnan, 2011
101	Liliaceae	*Asparagus filicinus*	Tubers	Yunnan, 2011
102	Liliaceae	*Lilium brownii*	Tubers	Yunnan, 2011
103	Liliaceae	*Lilium lancifolium*	Tubers	Yunnan, 2010
104	Liliaceae	*Lilium pumilum*	Tubers	Yunnan, 2012

(*Continued*)

Table 2 List of Chinese Plants Collected and Tested for Cyclotide Detection—cont'd

	Family Name	Species Name	Parts Used	Place and Time Collected
105	Liliaceae	*Polygonatum cyrtonema*	Roots	Anhui, 2011
106	Liliaceae	*Polygonatum kingianum*	Tubers	Yunnan, 2010
107	Liliaceae	*Polygonatum odoratum*	Roots	Zhejiang, 2011
108	Liliaceae	*Polygonatum verticillatum*	Tubers	Yunnan, 2012
109	Liliaceae	*Smilax glabra*	Tubers	Guangdong, 2011
110	Liliaceae	*Smilax scobinicaulis*	Leaves, fruit, and stem	Yunnan, 2011
111	Loranthaceae	*Scurrula ferruginea*	Branches and leaves	Yunnan, 2011
112	Loranthaceae	*Scurrula parasitica*	Branches and leaves	Yunnan, 2011
113	Loranthaceae	*Taxillus delavayi*	Leaves, flowers, and stem	Yunnan, 2011
114	Moraceae	*Broussonetia papyrifera*	Leaves	Yunnan, 2011
115	Moraceae	*Ficus carica*	Fruit	Xinjiang, 2011
116	Moraceae	*Ficus racemosa*	Fruit	Yunnan, 2011
117	Moraceae	*Morus alba*	Leaves and roots	Yunnan, 2011
118	Moraceae	*Morus australis*	Leaves, fruit, and stem	Yunnan, 2011
119	Moraceae	*Streblus macrophyllus*	Branches and leaves	Yunnan, 2012
120	Myrsinaceae	*Embelia longifolia*	Fruit	Yunnan, 2010
121	Myrsinaceae	*Myrsine africana*	Leaves and stem	Yunnan, 2011
122	Myrtaceae	*Melaleuca bracteata*	Leaves and stem	Yunnan, 2011
123	Myrtaceae	*Psidium guajava*	Fruit	Sichuan, 2012
124	Myrtaceae	*Rhodomyrtus tomentosa*	Fruit	Guangdong, 2011
125	Myrtaceae	*Syzygium buxifolium*	Leaves, fruit, and stem	Yunnan, 2012
126	Opiliaceae	*Yunnanopilia longistaminata*	Leaves	Yunnan, 2009
127	Orobanchaceae	*Cistanche deserticola*	Stem	Yunnan, 2011
128	Papilionaceae	*Caragana sinica*	Roots	Yunnan, 2011
129	Papilionaceae	*Pueraria lobata*	Roots	Yunnan, 2010

130	Papilionaceae	*Robinia pseudoacacia*	Flowers	Shanxi, 2012
131	Pedaliaceae	*Sesamum indicum*	Seeds	Yunnan, 2010
132	Poaceae	*Imperata cylindrica*	Roots	Yunnan, 2011
133	Poaceae	*Lophatherum gracile*	Leaves	Yunnan, 2011
134	Poaceae	*Phragmites australis*	Roots	Jiangsu, 2012
135	Podocarpaceae	*Podocarpus nagi*	Fruit and stem	Yunnan, 2010
136	Polygalaceae	*Polygala saxicola*	Stem and leaves	Yunnan, 2010
137	Polygonaceae	*Polygonum amphibium*	Whole plant	Yunnan, 2011
138	Polygonaceae	*Polygonum chinense*	Fruit	Yunnan, 2010
139	Polygonaceae	*Polygonum hydropiper*	Whole plant	Yunnan, 2011
140	Polygonaceae	*Polygonum orientale*	Whole plant	Yunnan, 2011
141	Polygonaceae	*Fagopyrum tataricum*	Leaves and stem	Yunnan, 2011
142	Primulaceae	*Lysimachia christinae*	Whole plant	Yunnan, 2011
143	Ranunculaceae	*Clematis rehderiana*	Leaves and stem	Yunnan, 2011
144	Rhamnaceae	*Hovenia acerba*	Fruit and leaves	Yunnan, 2010
145	Rhamnaceae	*Hovenia dulcis*	Fruit and leaves	Yunnan, 2010
146	Rhamnaceae	*Pallium ramosissimus*	Fruit	Yunnan, 2012
147	Rhamnaceae	*Ziziphus jujuba*	Fruit	Yunnan, 2011
148	Rhamnaceae	*Ziziphus mauritiana*	Seeds	Yunnan, 2007
149	Rhizopogonaceae	*Rhizopogon luteolus*	Fruit	Yunnan, 2011
150	Rosaceae	*Armeniaca vulgaris*	Fruit	Yunnan, 2012
151	Rosaceae	*Cotoneaster melanocarpus*	Fruit	Yunnan, 2011
152	Rosaceae	*Cotoneaster microphyllus*	Fruit	Yunnan, 2010
153	Rosaceae	*Crataegus pinnatifida*	Leaves	Yunnan, 2011
154	Rosaceae	*Dichotomanthus tristaniaecarpa*	Fruit	Yunnan, 2011
155	Rosaceae	*Docynia delavayi*	Leaves, fruit, and stem	Yunnan, 2011
156	Rosaceae	*Eriobotrya japonica*	Leaves	Yunnan, 2011

(*Continued*)

Table 2 List of Chinese Plants Collected and Tested for Cyclotide Detection—cont'd

	Family Name	Species Name	Parts Used	Place and Time Collected
157	Rosaceae	*Laurocerasus phaeosticta*	Branches and leaves	Yunnan, 2011
158	Rosaceae	*Laurocerasus undulata*	Branches and leaves	Yunnan, 2011
159	Rosaceae	*Potentilla anserina*	Fruit	Yunnan, 2011
160	Rosaceae	*Potentilla fruticosa*	Fruit and stem	Yunnan, 2010
161	Rosaceae	*Pyracantha fortuneana*	Fruit	Yunnan, 2010
162	Rosaceae	*Pyrus pashia*	Leaves, fruit, and stem	Yunnan, 2011
163	Rosaceae	*Pyrus pyrifolia*	Fruit	Yunnan, 2010
164	Rosaceae	*Rosa chinensis*	Flowers	Yunnan, 2011
165	Rosaceae	*Rosa laevigata*	Fruit	Yunnan, 2011
166	Rosaceae	*Rosa multiflora*	Fruit	Yunnan, 2010
167	Rosaceae	*Rosa roxburghii*	Fruit	Yunnan, 2011
168	Rosaceae	*Rosa rugosa*	Flowers	Yunnan, 2011
169	Rosaceae	*Sorbus ochracea*	Fruit	Yunnan, 2011
170	Rubiaceae	*Hedyotis hedyotidea*	Fruit	Yunnan, 2011
171	Rubiaceae	*Silvianthus tonkinensis*	Branches and leaves	Yunnan, 2012
172	Rutaceae	*Citrus sinensis*	Seed coats	Yunnan, 2012
173	Sapindaceae	*Sapindus mukorossi*	Fruit	Yunnan, 2010
174	Saurauiaceae	*Saurauia napaulensis*	Fruit	Yunnan, 2010
175	Saurauiaceae	*Saurauia tristyla*	Branches and leaves	Yunnan, 2012
176	Schisandraceae	*Kadsura coccinea*	Fruit	Yunnan, 2011
177	Schisandraceae	*Kadsura heteroclita*	Branches and leaves	Yunnan, 2011
178	Scrophulariaceae	*Scoparia dulcis*	Whole plant	Yunnan, 2011
179	Solanaceae	*Capsicum tetragonum*	Fruit	Yunnan, 2011
180	Solanaceae	*Lycium chinense*	Twigs, fruit, and roots	Yunnan, 2011
181	Solanaceae	*Saccharum sinense*	Stem	Yunnan, 2012

182	Solanaceae	*Solanum nigrum*	Fruit	Yunnan, 2012
183	Sonneratiaceae	*Duabanga grandiflora*	Branches and leaves	Yunnan, 2012
184	Symplocaceae	*Symplocos adenophylla*	Leaves and stem	Yunnan, 2010
185	Symplocaceae	*Symplocos adenopus*	Leaves and stem	Yunnan, 2010
186	Symplocaceae	*Symplocos chinensis*	Branches and leaves	Yunnan, 2011
187	Symplocaceae	*Symplocos dryophila*	Leaves and stem	Yunnan, 2011
188	Symplocaceae	*Symplocos paniculata*	Leaves and stem	Yunnan, 2011
189	Symplocaceae	*Symplocos pilosa*	Leaves and stem	Yunnan, 2010
190	Tetracentraceae	*Tetracentron sinense*	Branches and leaves	Yunnan, 2011
191	Theaceae	*Camellia longicaudata*	Branches and leaves	Yunnan, 2011
192	Ulmaceae	*Celtis tetrandra*	Seed husks	Yunnan, 2011
193	Ulmaceae	*Celtis timorensis*	Seed husks	Yunnan, 2011
194	Urticaceae	*Debregeasia orientalis*	Fruit	Yunnan, 2012
195	Vacciniaceae	*Gaultheria leucocarpa*	Fruit	Yunnan, 2010
196	Violaceae	*Viola hamiltoniana*	Whole plant	Yunnan, 2006
197	Violaceae	*Viola labradorica*	Whole plant	Yunnan, 2005
198	Violaceae	*Viola philippica*	Whole plant	Yunnan, 2010
199	Violaceae	*Viola tricolor*	Whole plant	Yunnan, 2006
200	Vitaceae	*Tetrastigma obovatum*	Fruit	Yunnan, 2011
201	Zingiberaceae	*Curcuma aromatica*	Tubers	Yunnan, 2011
202	Zingiberaceae	*Kaempferia galanga*	Roots	Yunnan, 2011
203	Zygophyllaceae	*Zingiber officinale*	Fruit	Yunnan, 2011

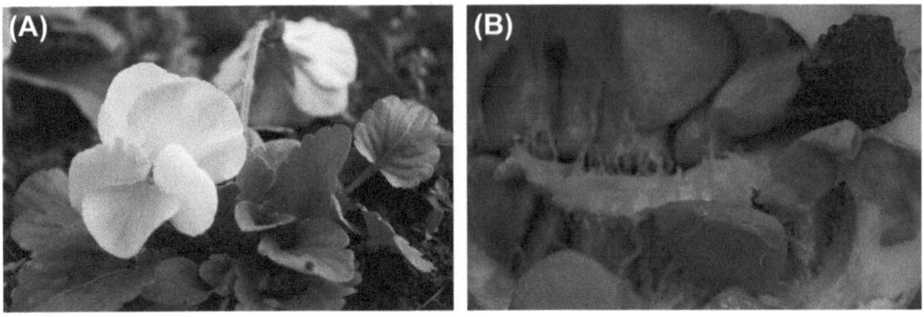

Figure 2 Two Chinese plants containing cyclotides, taken by Wen-Jun He. (A) The flowers of *Viola tricolor*; (B) the fruit of *Momordica cochinchinensis*. (See color plate)

4.2 Tandem MS Sequencing of Cyclotides in Chinese Plants

Tandem MS sequencing is a rapid and sensitive analytical method and in particular is efficient for the characterization of the primary structure of cyclotides (Colgrave, Poth, Kaas, & Craik, 2010; Craik et al., 1999; He et al., 2011). It is an optimum method for analyzing samples available only in small amounts. Furthermore, it is possible to unambiguously obtain amino acid sequences for a mixture of cyclotides (Colgrave et al., 2010; He et al., 2011). Chapter 5 of this book provides additional detail on sequencing approaches for cyclotides.

An example of the successful use of this method is the determination of four cyclotide sequences from *V. philippica* (He et al., 2011). For the mixture of the four cyclotides, Viba 17, Viba 15, Varv A, and kalata B1 were coeluted in a single HPLC peak (Figure 3(A)). Their molecular weights were determined by LCMS to be 2846.02, 2860.04, 2876.04, and 2890.06 Da, respectively. The cyclotides were reduced using TCEP tris(2-carboxyethyl) phosphine hydrochloride and each gained 6 mass units, indicating the presence of six Cys residues involved in three disulfide bonds. The reduced cyclotides were digested with trypsin and endoGluC. The monoisotopic masses of the major fragments that resulted from this digestion are 2101.68 Da (Viba 17), 2115.69 Da (Viba 15), 2131.69 Da (Varv A), and 2145.70 Da (kalata B1) (Figure 4(A)). These fragments were sequenced using nanospray MS/MS and led to the sequences TCVGGTCNTPGCGCSWPVCTR (Viba 17, Figure 4(C)), TCVGGTCNTPGCACSWPVCTR (Viba 15), TCVGGTCNTPGCSCSWPVCTR (Varv A), and TCVGGTCNTPGCT CSWPVCTR (kalata B1), respectively. Another fragment, NGLPVCGE, was sequenced using MS/MS (Figure 4(B)). The combination of these fragments yielded the complete sequences of Viba 17, Viba 15, Varv A, and kalata B1 (Figure 3(B)). All fragments were confirmed with chymotrypsin and

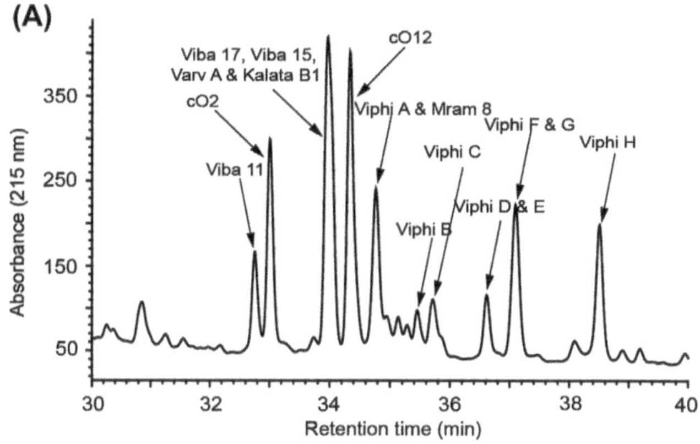

(B)

	loop 1 II loop 2 III loop 3 IV ↓ (loop 4) V loop 5 VI loop 6	Expt. *m/z*	Theory
Viphi A	CGESCVFIPCISSVI-GCACKSKVCYKNGSIP	3170.28	3170.47
Viphi B	CGETCTIGTCYTAG---CTCSWPICTRNGLPV	2985.18	2985.25
Viphi C	CGESCVYIPCITSVI-GCSCSSKVCYING-VP	3059.18	3059.35
Viphi D	CGESCVFIPCISSVI-GCSCSSKVCYRNG-IP	3086.30	3086.37
Viphi E	CGESCVFIPCISAVI-GCSCSNKVCYKNGSIP	3156.24	3156.41
Viphi F	CGESCVFIPCISAII-GCSCSSKVCYKNGSIP	3143.30	3143.42
Viphi G	CGESCVFIPCISAII-GCSCSNKVCYKNGSIP	3170.36	3170.43
Viphi H	CAESCVWIPCTVTAIVGCSCSWGVCY-NG-IP	3096.24	3096.32
Viba 17	CGETCVGGTCNTPG---CGCSWPVCTRNGLPV	2846.02	2846.16
Viba 15	CGETCVGGTCNTPG---CACSWPVCTRNGLPV	2860.04	2860.18
Varv A	CGETCVGGTCNTPG---CSCSWPVCTRNGLPV	2876.04	2876.17
Kalata B1	CGETCVGGTCNTPG---CTCSWPVCTRNGLPV	2890.06	2890.19
Mram 8	CGESCVFIPCLTSAI-GCSCKSKVCYRNG-IP	3113.26	3113.42
cO2	CGESCVWIPCISSAI-GCSCKSKVCYRNG-IP		
cO12	CGETCVGGTCNTPG---CSCSWPVCTRNGLPI		
Viba 11	CGESCVWIPCISGAI-GCSCKSKVCYRNG-IP		

Figure 3 (A) Separation profile of a crude cyclotide extract by RP-HPLC from *Viola philippica* showing the retention times and relative intensities of the cyclotides isolated, (B) sequence alignment of the isolated cyclotides. *Reproduced with permission from He et al. (2011).*

endoGluC digestions. These four cyclotides differ at only one position in their sequence, in loop 4 (marked with dashed circle in the upper-left corner of Figure 4(A)).

4.3 NMR Characterization of Cyclotides in Chinese Plants

NMR is another important tool for the structural investigation of cyclotides; however, its use is limited in some cases where it is difficult to unambiguously obtain all sequential connectivities due to the overlapping of some

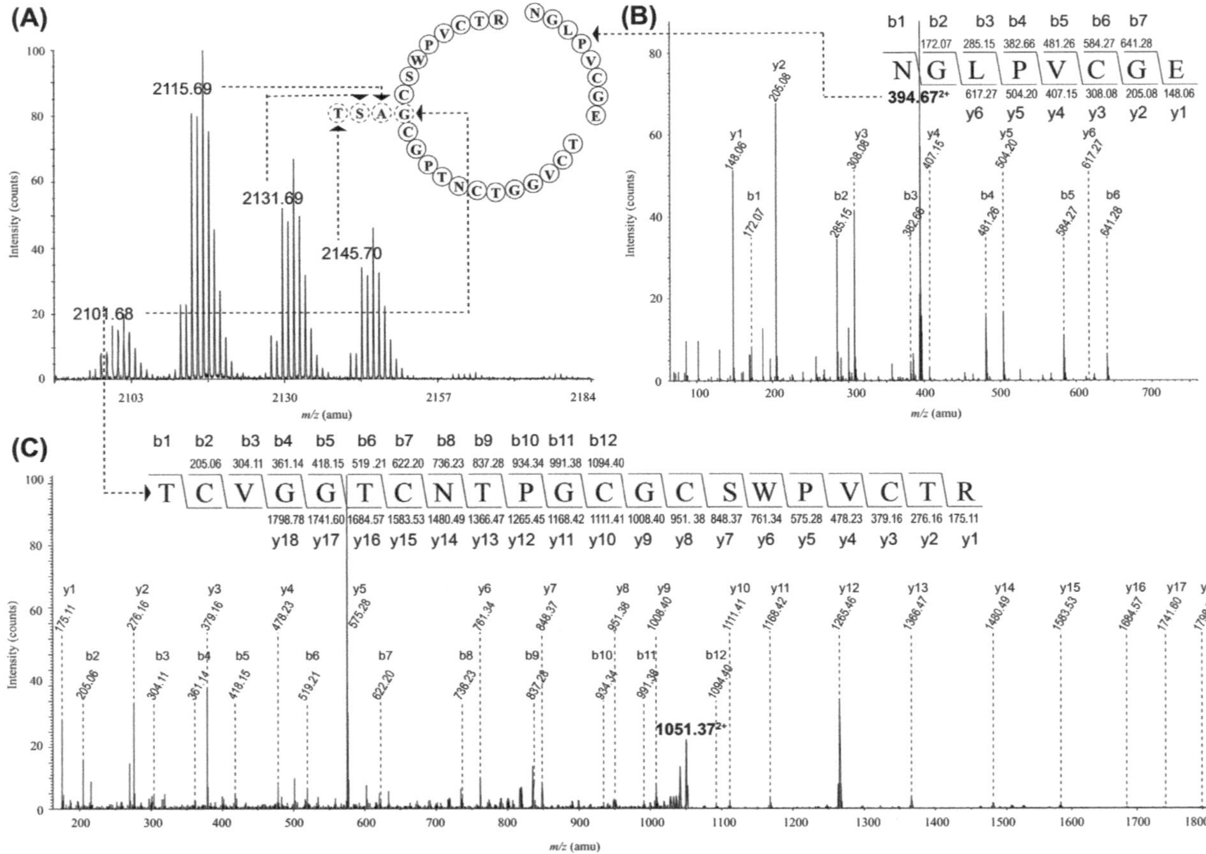

Figure 4 (A) MALDI spectrum of the coeluted mixture of Viba 17, Viba 15, Varv A and, kalata B1 in a single HPLC peak (Figure 3(A)), (B) fragment sequence and MS/MS spectrum of the doubly charged ion *m/z* 394.67 of Viba 17, (C) fragment sequence and MS/MS spectrum of the doubly charged ion m/z 1051.37 of Viba 17. *Reproduced with permission from He et al. (2011).*

amino acid residues. Although the prior MS/MS sequencing is important in the NMR structure determination process, NMR is very useful for distinguishing some amino acid residues that are indistinguishable with MS analyses, such as leucine and isoleucine. An example of the determination of leucine and isoleucine in the cyclotide Vitri D from *V. tricolor* serves to illustrate this point (Tang et al., 2010b). The complete sequence of Vitri D was determined as GL(or I)PVCGETCFTGSCYTPGCSCNWPVCNRN using the aforementioned MS/MS method (Section 4.2), in which L (leucine) or I (isoleucine) could not be distinguished because of their identical molecular weight of 113 Da. TOCSY and NOESY spectra were recorded (Figure 5) and all spin systems except Pro3, Pro17, Trp23, and Pro24 were observed in TOCSY spectrum, allowing the second residue to be identified as leucine. The sequence (except Pro3, Pro17, Trp23, and Pro24) was confirmed by the connectivities between the αH of one residue to the NH proton of the next residue in the NOESY spectrum.

5. BIOLOGICAL ACTIVITY OF CYCLOTIDES IN CHINESE PLANTS

Although cyclotides exhibit a range of interesting bioactivities (Tan & Zhou, 2006), bioactivity studies of cyclotides in Chinese plants have focused on cytotoxic, hemolytic, anti-HIV, anti-influenza A H1N1 virus, and trypsin inhibitory activities. All cyclotide-containing Chinese plants described here are TCMs. *Viola philippica*, an edible wild vegetable, is a TCM used for reduction of fever, detoxification, anti-inflammation, and pain relief. Sixteen cyclotides were isolated from this plant, and all were found to exhibit cytotoxic activity (He et al., 2011). From *V. tricolor*, a TCM for treating epilepsy, asthma, and skin diseases, 14 cyclotides were characterized and were shown to have hemolytic and cytotoxic activities (Tang et al., 2010b). From *Viola yedoensis*, a TCM for treating swelling, sores, boils, snakebites, bronchitis, and hepatitis, nine cyclotides were characterized and these shown to have anti-HIV and anti-influenza A H1N1 virus activities (Liu et al., 2014). Cyclotides were also isolated from *M. cochinchinensis*, a TCM for treating inflammation and pain, and were found to be trypsin inhibitors (Cemazar et al., 2006; Chan et al., 2013; Hernandez et al., 2000; Mylne et al., 2012). These findings provide insights into the potential active ingredients of these TCMs.

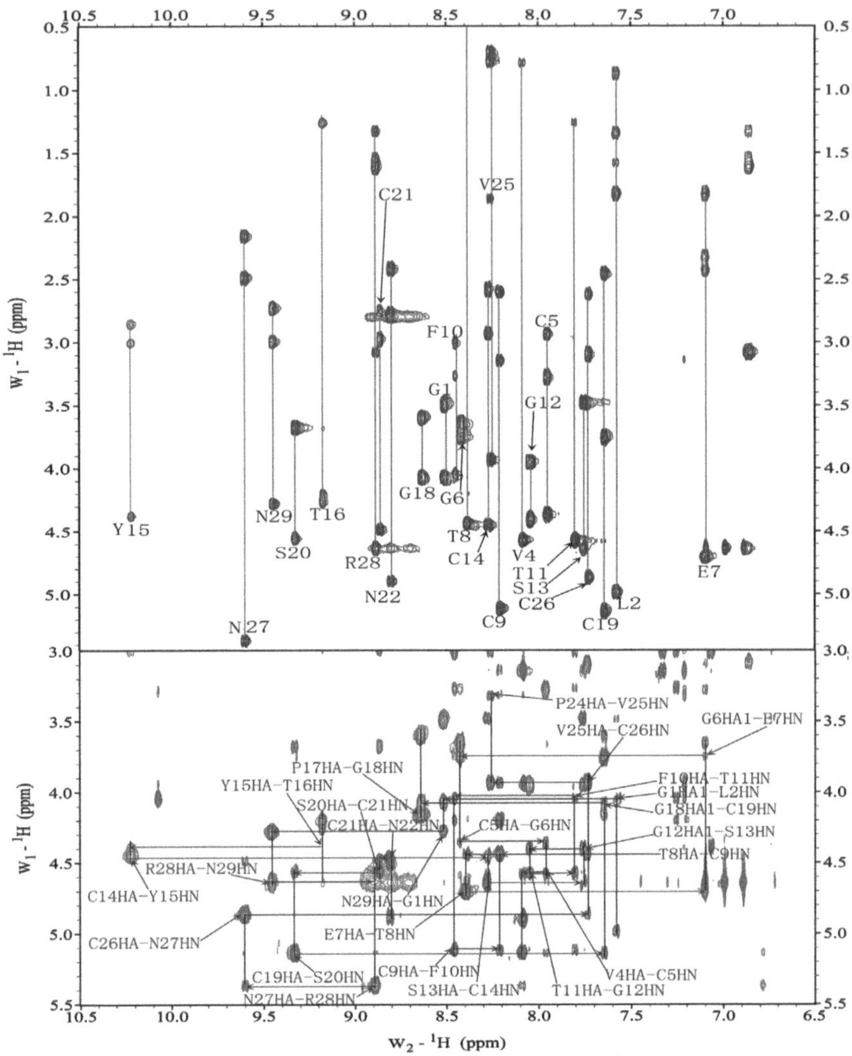

Figure 5 The TOCSY amide region and spin systems of Vitri D (above). The NOESY fingerprint region and sequential connectivity via H_α-H_N of Vitri D (below). *Reproduced with permission from Tang et al. (2010b).*

ACKNOWLEDGMENTS

This work was supported by the National Natural Science Foundation of China (3147028, 21202174).

REFERENCES

Cemazar, M., Daly, N. L., Haggblad, S., Lo, K. P., Yulyaningsih, E., & Craik, D. J. (2006). Knots in rings: the circular knotted protein *Momordica cochinchinensis* trypsin inhibitor-II folds via a stable two-disulfide intermediate. *Journal of Biological Chemistry, 281*, 8224–8232.

Chan, L. Y., He, W. J., Tan, N. H., Zeng, G. Z., Craik, D. J., & Daly, N. L. (2013). A new family of cystine knot peptides from the seeds of *Momordica cochinchinensis*. *Peptides, 39*, 29–35.

Chen, B., Colgrave, M. L., Daly, N. L., Rosengren, J. K., Gustafson, K. R., & Craik, D. J. (2005). Isolation and characterization of novel cyclotides from *Viola hederaceae* – solution structure and anti-HIV activity of vhl-1, a leaf-specific expressed cyclotide. *Journal of Biological Chemistry, 280*, 22395–22405.

Chen, B., Colurave, M. L., Wang, C., & Craik, D. J. (2006). Cycloviolacin H4, a hydrophobic cyclotide from *Viola hederaceae*. *Journal of Natural Products, 69*, 23–28.

Colgrave, M. L., Poth, A. G., Kaas, Q., & Craik, D. J. (2010). A new "era" for cyclotide sequencing. *Biopolymers Peptide Science, 94*, 592–601.

Craik, D. J. (2013). Joseph Rudinger memorial lecture: discovery and applications of cyclotides. *Journal of Peptide Science, 19*, 393–407.

Craik, D. J., Daly, N. L., Bond, T., & Waine, C. (1999). Plant cyclotides: a unique family of cyclic and knotted proteins that defines the cyclic cystine knot structural motif. *Journal of Molecular Biology, 294*, 1327–1336.

Editorial Committee of Chinese Flora. (1978). *Flora of China*. Beijing: Science Press.

Göransson, U., Luijendijk, T., Johansson, S., Bohlin, L., & Claeson, P. (1999). Seven novel macrocyclic polypeptides from *Viola arvensis*. *Journal of Natural Products, 62*, 283–286.

Gran, L. (1973). Oxytocic principles of *Oldenlandia affinis*. *Lloydia, 36*, 174–178.

Gran, L., Sandberg, F., & Sletten, K. (2000). *Oldenlandia affinis* (R&S) DC: a plant containing uteroactive peptides used in African traditional medicine. *Journal of Ethnopharmacology, 70*, 197–203.

He, W. J., Chan, L. Y., Zeng, G. Z., Daly, N. L., Craik, D. J., & Tan, N. H. (2011). Isolation and characterization of cytotoxic cyclotides from *Viola philippica*. *Peptides, 32*, 1719–1723.

Hernandez, J. F., Gagnon, J., Chiche, L., Nguyen, T. M., Andrieu, J. P., Heitz, A., et al. (2000). Squash trypsin inhibitors from *Momordica cochinchinensis* exhibit an atypical macrocyclic structure. *Biochemistry, 39*, 5722–5730.

Herrmann, A., Burman, R., Mylne, J. S., Karlsson, G., Gullbo, J., Craik, D. J., et al. (2008). The alpine violet, *Viola biflora*, is a rich source of cyclotides with potent cytotoxicity. *Phytochemistry, 69*, 939–952.

Ireland, D. C., Colgrave, M. L., & Craik, D. J. (2006). A novel suite of cyclotides from *Viola odorata*: sequence variation and the implications for structure, function and stability. *Biochemical Journal, 400*, 1–12.

Kaas, Q., & Craik, D. J. (2010). Analysis and classification of circular proteins in CyBase. *Biopolymers Peptide Science, 94*, 584–591.

Liu, M. Z., Yang, Y., Zhang, S. X., Tang, L., Wang, H. M., Chen, C. J., et al. (2014). A cyclotide against influenza A H1N1 virus from *Viola yedoensis*. *Acta Pharmaceutica Sinica, 49*, 905–912.

Mahatmanto, T., Mylne, J. S., Poth, A. G., Swedberg, J. E., Kaas, Q., Schaefer, H., et al. (2015). The evolution of *Momordica* cyclic peptides. *Molecular Biology and Evolution, 32*, 392–405.

Mulvenna, J. R., Sando, L., & Craik, D. J. (2005). Processing of a 22 kDa precursor protein to produce the circular protein tricyclon A. *Structure, 13*, 691–701.

Mylne, J. S., Chan, L. Y., Chanson, A. H., Daly, N. L., Schaefer, H., Bailey, T. L., et al. (2012). Cyclic peptides arising by evolutionary parallelism via asparaginyl-endopeptidase–mediated biosynthesis. *The Plant Cell, 24*, 2765–2778.

Nguyen, G. K. T., Lian, Y., Pang, E. W. H., Nguyen, P. Q. T., Tran, T. D., & Tam, J. P. (2013). Discovery of linear cyclotides in monocot plant *Panicum laxum* of Poaceae family provides new insights into evolution and distribution of cyclotides in plants. *Journal of Biological Chemistry, 288*, 3370–3380.

Nguyen, G. K. T., Zhang, S., Nguyen, N. T. K., Nguyen, P. Q. T., Chiu, M. S., Hardjojo, A., et al. (2011). Discovery and characterization of novel cyclotides originated from chimeric precursors consisting of albumin-1 chain a and cyclotide domains in the Fabaceae family. *Journal of Biological Chemistry, 286*, 24275–24287.

Poth, A. G., Colgrave, M. L., Lyons, R. E., Daly, N. L., & Craik, D. J. (2011). Discovery of an unusual biosynthetic origin for circular proteins in legumes. *Proceedings of the National Academy of Sciences of the United States of America, 108*, 10127–10132.

Poth, A. G., Colgrave, M. L., Philip, R., Kerenga, B., Daly, N. L., Anderson, M. A., et al. (2010). Discovery of cyclotides in the Fabaceae plant family provides new insights into the cyclization, evolution, and distribution of circular proteins. *ACS Chemical Biology, 6*, 345–355.

Poth, A. G., Mylne, J. S., Grassl, J., Lyons, R. E., Millar, A. H., Colgrave, M. L., et al. (2012). Cyclotides associate with leaf vasculature and are the products of a novel precursor in *Petunia* (Solanaceae). *Journal of Biological Chemistry, 287*, 27033–27046.

Saether, O., Craik, D. J., Campbell, I. D., Sletten, K., Juul, J., & Norman, D. G. (1995). Elucidation of the primary and three-dimensional structure of the uterotonic polypeptide kalata B1. *Biochemistry, 34*, 4147–4158.

Schöepke, T., Hasan Agha, M. I., Kraft, R., & Otto, A. (1993). Compounds with hemolytic activity from *Viola tricolor* L. and *Viola arvensis* Murray. *Scientia Pharmaceutica, 61*, 145–153.

Simonsen, S. M., Sando, L., Ireland, D. C., Colgrave, M. L., Bharathi, R., Göransson, U., et al. (2005). A continent of plant defense peptide diversity: cyclotides in Australian *Hybanthus* (Violaceae). *The Plant Cell, 17*, 3176–3189.

Sletten, K., & Gran, L. (1973). Some molecular properties of kalatapeptide B-1. A uterotonic polypeptide isolated from *Oldenlandia affinis* DC. *Meddelelser fra Norsk Farmaceutisk Selskap, 7–8*, 69–82.

Svangard, E., Göransson, U., Hocaoglu, Z., Gullbo, J., Larsson, R., Claeson, P., et al. (2004). Cytotoxic cyclotides from *Viola tricolor*. *Journal of Natural Products, 67*, 144–147.

Svangard, E., Goransson, U., Smith, D., Verma, C., Backlund, A., Bohlin, L., et al. (2003). Primary and 3-D modelled structures of two cyclotides from *Viola odorata*. *Phytochemistry, 64*, 135–142.

Tang, J., Wang, C. K., Pan, X. L., Yan, H., Zeng, G. Z., Xu, W. Y., et al. (2010a). Isolation and characterization of bioactive cyclotides from *Viola labradorica*. *Helvetica Chimica Acta, 93*, 2287–2295.

Tang, J., Wang, C. K., Pan, X. L., Yan, H., Zeng, G. Z., Xu, W. Y., et al. (2010b). Isolation and characterization of cytotoxic cyclotides from *Viola tricolor*. *Peptides, 31*, 1434–1440.

Tan, N. H., & Zhou, J. (2006). Plant cyclopeptides. *Chemical Reviews, 106*, 840–895.

Wang, C. K., Colgrave, M. L., Gustafson, K. R., Ireland, D. C., Göransson, U., & Craik, D. J. (2007). Anti-HIV cyclotides from the Chinese medicinal herb *Viola yedoensis*. *Journal of Natural Products, 71*, 47–52.

Xiang, B., Du, G. H., Wang, X. C., Zhang, S. X., Qin, X. Y., Kong, J. Q., et al. (2010). Elucidating the structure of two cyclotides of *Viola tianschanica* Maxim by MALDI TOF/TOF MS analysis. *Acta Pharmaceutica Sinica, 45*, 1402–1409.

Xu, W. Y., Tang, J., Ji, C. J., He, W. J., & Tan, N. H. (2008). Application of a TLC chemical method to detection of cyclotides in plants. *Chinese Science Bulletin, 53*, 1671–1674.

Zarrabi, M., Dalirfardouei, R., Sepehrizade, Z., & Kermanshahi, R. K. (2013). Comparison of the antimicrobial effects of semipurified cyclotides from Iranian *Viola odorata* against some of plant and human pathogenic bacteria. *Journal of Applied Microbiology, 115*, 367–375.

Zhang, J., Liao, B., Craik, D. J., Li, J. T., Hu, M., & Shu, W. S. (2009). Identification of two suites of cyclotide precursor genes from metallophyte *Viola baoshanensis*: cDNA sequence variation, alternative RNA splicing and potential cyclotide diversity. *Gene, 431*, 23–32.

Zhou, J., & Tan, N. H. (2000). Application of a new TLC chemical method for detection of cyclopeptides in plants. *Chinese Science Bulletin, 45*, 1825–1831.

CHAPTER FIVE

Primary Structural Analysis of Cyclotides

Michelle L. Colgrave
CSIRO Agriculture Flagship, Brisbane, QLD, Australia
E-mail: michelle.colgrave@csiro.au

Contents

Advances in Botanical Research, Volume 76
ISSN 0065-2296
http://dx.doi.org/10.1016/bs.abr.2015.08.004

Abstract

Cyclotides are topologically complex molecules owing to the unique combination of a head-to-tail cyclized backbone and a knotted arrangement of disulfide bonds. The so-called cyclic cystine knot imparts exceptional stability to this family of peptides but brings with it a great many analytical challenges. Standard peptide sequencing approaches such as Edman degradation are not useful for studying the native peptides owing to the blocked (cyclic) termini. Over the past 20 years the number of cyclotides appearing in the literature has expanded exponentially (4 in 1994; ~50 in 2004; >280 in 2014) from reported discoveries in the Violaceae and Rubiaceae to reports in the Cucurbitaceae, Fabaceae, Solanaceae, and Apocynaceae. The cystine knot and the site of cyclization are conserved, but significant sequence variation occurs in the loop regions, the backbone segments between the disulfide bonds. This variation may be as subtle as a single amino acid to variation in both loop size and composition. The gene architecture varies between plant families and the majority of cyclotide-containing plant species identified to date have not been the target of genome sequencing. For these reasons, standard proteomic approaches to peptide/protein discovery are not well suited to cyclotide sequencing. The most fruitful approach to cyclotide primary structure analysis has been the application of tandem mass spectrometry. This chapter will examine the first attempts to sequence cyclotides, review the tools and techniques that have been employed to characterize cyclotides, present the challenges and complexities with their analysis, and discuss the future prospects for cyclotide discovery.

1. CYCLOTIDE DISCOVERY—UNTANGLING THE CYSTINE KNOT

A polypeptide of molecular weight of about 4000 Da was first described in 1973 by the Norwegian physician Lorents Gran as an active agent in the plant "kalata-kalata," *Oldenlandia affinis* (Gran, 1973b). Extracts of this plant, prepared as a tea, were used by African natives to induce contractions and childbirth (Gran, 1970). The mode of use provided anecdotal evidence of a thermally stable, orally bioavailable compound. Phytochemical investigations on the plant revealed two uterotonic compounds, serotonin and an unknown peptide (Gran, 1973a) that was named kalata B1. The peptide was isolated, but its complete sequence and structural characterization was not undertaken for some 20 years (Saether et al., 1995). The molecular weight determination was refined to ~3000 Da by ultracentrifugation and the amino acid composition after acid hydrolysis was elucidated (Sletten & Gran, 1973). The high Cys content suggested a compact structure bound by disulfide bonds. The N-terminus was determined to be blocked as the peptide was found to be resistant to standard degradation procedures. After reduction and aminoethylation, trypsin cleavage yielded six peptide fragments that were analyzed by Edman degradation allowing partial elucidation of the sequence (27 of 29 amino acids) (Sletten & Gran, 1973).

The early 1990s saw the simultaneous discovery of cyclic plant peptides by a diverse number of groups guided by plant screening investigations for different biological activities: violapeptide I which was hemolytic (Schöpke, Hasan Agha, Kraft, Otto, & Hiller, 1993); cyclopsychotride A which inhibited neurotensin binding (Witherup et al., 1994); and the circulins A and B which showed anti-HIV activity (Gustafson et al., 1994). These early characterizations of a disulfide-bridged cyclic peptide structure were remarkable and required multiple, orthogonal approaches to untangle the knot and reveal the sequence. In the case of cyclopsychotride A (Witherup et al., 1994), acid hydrolysis revealed the amino acid composition (excluding Cys). The failure of Edman degradation suggested a blocked N-terminus. The native peptide was found to be impervious to enzymatic digestion. Partial acid hydrolysis yielded peptide fragments amenable to Edman degradation which showed overlapping primary sequence providing the first concrete evidence of the cyclic nature of the peptide backbone. The peptide was subsequently treated with iodoacetamide and the carbamidomethylated peptide was cleaved using the endoproteinase Glu-C to yield a linear peptide which was subsequently sequenced confirming a 31-amino acid cyclic peptide. The analysis of the circulins A and B from *Chassalia parvifolia* proved to be equally challenging (Gustafson et al., 1994). The active components were determined to be small peptides as they showed staining by Coomassie brilliant blue with a single band detected at ~3 kDa, the mass confirmed by fast atom bombardment (FAB) mass spectrometry. Attempts to sequence the peptides by Edman degradation failed. Linearization of the peptides by enzymatic digest after reduction enabled the sequences to be determined and was complemented by amino acid analysis (AAA).

In 1995, the prototypic cyclotide kalata B1 was investigated revealing both the primary and three-dimensional structure (Saether et al., 1995). Purified by HPLC and capillary electrophoresis, the amino acid composition (including Cys after performic acid oxidation) was elucidated and its mass determined by ion spray MS. The 2892 Da peptide was cleaved chemically (BNPS-skatole) and enzymatically (endo Glu-C, trypsin) to yield overlapping peptide fragments that were sequenced by Edman degradation confirming the earlier investigations of the amino acid composition and partial sequence characterization (Sletten & Gran, 1973). Furthermore, the 3-D structure elucidated by NMR spectroscopy revealed for the first time a knotted topology in which the disulfide bonds occupy the core of the molecule leading to surface-exposed hydrophobic patches (Saether et al., 1995).

Cyclotide was the name given in 1999 (Craik, Daly, Bond, & Waine, 1999) to describe the unique peptide family characterized by the combination of a cyclic peptide backbone and the knotted disulfide core. It is this combination called the cyclic cystine knot (CCK) that imparts their exceptional thermal, chemical and enzymatic stability (Colgrave & Craik, 2004). The cystine knot was demonstrated to be more important than the cyclic backbone to the peptides stability through the use of both acyclic and disulfide-deficient mutants. As a result, the disulfide bonds must be removed prior to enzyme digestion for peptide sequencing. This structure also results in the exclusion of hydrophobic residues, typically forming solvent-exposed patches which have been shown to be important determinants in a number of prescribed biological activities (reviewed in Craik, 2010; Daly, Rosengren, & Craik, 2009). Despite this hydrophobic patch, cyclotides are water soluble; for example, they are extracted in water during the preparation of medicinal tea (Gran, 1970) and in water and/or methanol in lab-based protocols (Craik, Henriques, Mylne, & Wang, 2012). The hydrophobic nature of cyclotides leads to late elution of typical cyclotides by reversed phase (RP) chromatography, a feature that has been exploited for their isolation and purification.

2. TOOLS FOR PEPTIDE SEQUENCE ASSIGNMENT

2.1 Gene Sequencing

Unlike many smaller cyclic peptides, cyclotides are ribosomally synthesized. Cyclotide genes encode linear precursor proteins that contain one or more putative cyclotide domains and these domains are excised from the precursor proteins to produce the mature cyclic peptides (Jennings, West, Waine, Craik, & Anderson, 2001). The gene architecture within the Rubiaceae family is quite variable (Figure 1) with the *oak1* gene that encodes kalata B1 containing a single copy of the mature peptide, whereas the *oak4* gene that encodes kalata B2 contains three copies. The *oak2* gene contains two peptide domains encoding both kalata B3 and kalata B6. The precursor

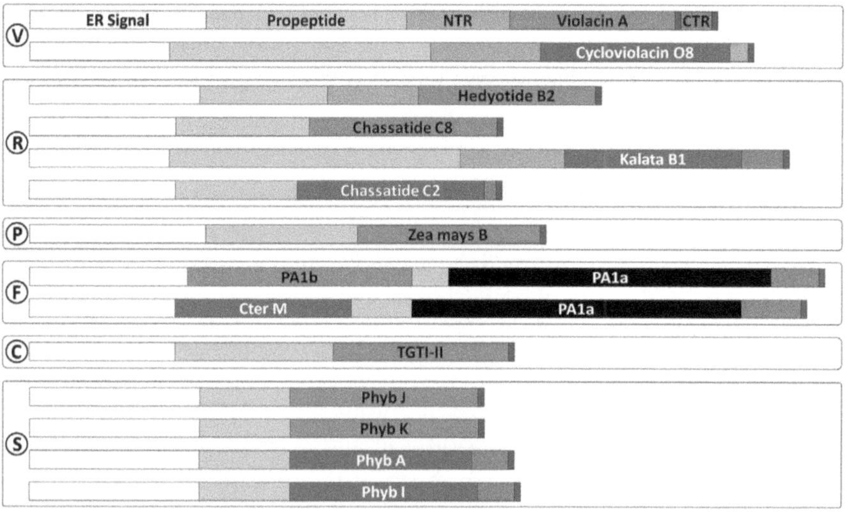

Figure 1 ***Variation in cyclotide and related knottin gene architecture among angiosperms.*** Signal sequence is shown in white boxes. Knottin and acyclotide domains are shown in light blue boxes. Cyclotide domains are shown in orange boxes. N-terminal prodomains are shown in green boxes. C-terminal repeats are shown in mauve boxes. *V*, Violaceae; *R*, Rubiaceae; *P*, Poaceae; *F*, Fabaceae; *C*, Cucurbitaceae; *S*, Solanaceae. *This research was originally published in Journal of Biological Chemistry. Poth et al. (2012), © the American Society for Biochemistry and Molecular Biology.* (See color plate)

protein in *O. affinis* consists of an endoplasmic reticulum (ER) signal sequence, a variable proregion, an N-terminal repeat region followed by the mature peptide domain (which may be repeated) and a short C-terminal tail. Four cyclotide genes from *Viola odorata*, a member of the Violaceae family were isolated and their overall gene structure was found to be conserved with those isolated from *O. affinis*, that is, conservation was observed in two plant families from different lineages of the phylogenetic tree (Dutton et al., 2004). It was noted that the mature cyclotide domains showed high homology, typically >80% within cyclotide subfamilies (Möbius and bracelet), but the flanking regions showed greater sequence variation.

The legume butterfly pea, *Clitoria ternatea*, from the Fabaceae family has a distinct gene architecture to the cyclotide genes initially described (Figure 1). In this instance, the mature cyclotide domain exists within an albumin-like precursor (Nguyen, Zhang, Nguyen, et al., 2011; Poth, Colgrave, Lyons, Daly, & Craik, 2011). The mature cyclotide domain immediately follows the ER signal sequence, presumably excised at the N-terminus by the action of a signal peptidase. A small linker region separates the cyclotide domain from the albumin a-chain. Comparison with PA1b reveals that the cyclotide domain has replaced the albumin b-chain that is present in typical albumin-1 genes. Despite, the different gene architecture and sequence variation in the composition of the encoded cyclotides, these peptides adopt a similar structure and possess activities reminiscent of their counterparts in the Violaceae and Rubiaceae families.

In petunia, a member of the Solanaceae family which contains such food crops as tomato and potato, both cyclotides and acyclotides (linear derivatives) were shown to exist. Interestingly, these were the products of another novel precursor structure (Poth et al., 2012). The genes in petunia were dedicated, only yielding a single mature cyclotide domain, but were more compact when compared to those from Violaceae and Rubiaceae. The precursors were noted to possess an ER target signal sequence, a mature cyclotide domain and, in most cases, a short (6-amino acid) C-terminal tail. Several of the expressed sequence tags (ESTs) discovered by homology searching (tBLASTn) of the NCBI database were punctuated by stop codons immediately following the cyclotide domain, suggestive that these would yield acyclic versions. Two of the three peptides sequenced by LC-MS/MS in this study were in fact acyclic and one of these contained an N-terminal Glu which was observed as the pyroglutaminated form, rendering it somewhat resistant to exoprotease activity and enhancing its stability in the absence of the cyclic backbone.

Despite the variation in gene architecture, all cyclotide precursors have typical ER signal sequences. The precursors enter the secretory pathway where folding and disulfide bond formation occurs before the cleavage and cyclization events that release the mature cyclic peptides. There are two key enzymes that appear to be involved in cyclotide biosynthesis. An asparaginyl endoproteinase (AEP) (Saska et al., 2007) has been implicated in the cyclization process, and a protein disulfide isomerase (Gruber et al., 2007) is purported to assist in the rapid folding of cyclotides. Further evidence for the role of AEP in the ligation reaction to produce cyclic products was provided by a study of the cyclotide-producing plant, *O. affinis*, when compared to transgenic *Arabidopsis thaliana* and *Nicotiana tabacum* (Gillon et al., 2008). Cyclic peptides were only produced in the transgenic plants when the conserved Asn was present at the C-terminal processing site and in the presence of a short C-terminal tail.

As exemplified in the studies undertaken in the Rubiaceae, Violaceae, Fabaceae, and Solanaceae families described herein, the amino acid sequence of a protein can be determined indirectly from the mRNA or the DNA that codes for the protein. One method of cyclotide sequencing involves the isolation of RNA from plant material and subsequent preparation of cDNA. The advantage of this approach is that it can be applied to relatively small amounts of plant tissue and information is gained not only on the predicted mature peptide sequence, but additionally of the precursor protein giving insight into the route of biosynthesis. RNA isolation may be accomplished using a range of commercial kits or using relatively simple protocols such as phenol—chloroform extraction followed by selective precipitation of RNA with lithium chloride as has recently been described (Craik et al., 2012). The RNA is then reverse transcribed to cDNA which is then amplified by polymerase chain reaction (PCR). The amplified cDNA copies are then sequenced.

2.2 Genome Mining

Gene or genome mining is the process of exploiting genomic information, such as DNA sequences of one genotype to isolate useful genes from related genotypes. Genome mining offers a useful alternative complementing peptide discovery by peptidomics. The number of available genomes has steadily increased over recent years as whole genome sequencing has become affordable, but there are many challenges that remain, for example, with the assembly of the sequenced genomes. Genome mining approaches applied to the discovery of ribosomally synthesized natural products including cyclotides has been recently reviewed (Velasquez & van der Donk, 2011). Cyclotide precursors are considered to be only moderately conserved with sequence conservation of ~60%. This is challenging when attempting to develop a single PCR strategy for the detection of novel peptides.

In the first applications of cDNA screening for cyclotide discovery, partial cDNA clones were isolated by RT-PCR using forward primers based on known sequences (Dutton et al., 2004; Jennings et al., 2001; Mulvenna, Sando, & Craik, 2005). These clones were either sequenced or used to design gene-specific primers for full-length clone isolation. In 2005, Simonsen et al. designed a new strategy making use of a forward primer that targets a conserved element upstream of the mature cyclotide domain. This element, AAFALPA, found in the ER region of genes from *O. affinis*, *V. odorata*, and *V. tricolor*, was then be used to rapidly reveal the precursor region encoding the mature cyclotide domain. Three of the resulting precursor proteins were confirmed at the peptide level by traditional LC-MS/MS sequencing.

In many cyclotide discovery programs the application of complementary techniques, such as parallel analysis of peptides by LC-MS/MS and of cDNA by RT-PCR, has been met with success. A peptide-based study of *V. odorata* revealed the first example of an acyclotide, named violacin A (Ireland, Colgrave, Nguyencong, Daly, & Craik, 2006). The unusual nature of this peptide prompted the sequencing of the cDNA clone in order to explain the acyclic nature of this peptide and revealed a stop codon that eliminated the conserved Asn required for production of the cyclic peptide. This is quite likely the result of a single nucleotide polymorphism in which the codon for a Lys (present in closely related peptides) is AAA or AAG, while a stop codon results from GAA or GAG.

In parallel with traditional peptide discovery approaches employing LC-MS/MS, a cDNA screening approach was utilized to uncover cyclotides in *Viola biflora* (Herrmann et al., 2008). In this approach mRNA was extracted and a primer was designed that targeted the conserved region in the ER signal (AAFALPA) as described previously (Simonsen et al., 2005). From 41 sequences, 6 unique clones were sequenced. Two of these encoded known cyclotide sequences and four were novel cyclotides, named vibi E and vibi I-K. Of the six precursors, peptide evidence was only found for the known peptide cycloviolacin O9 and the novel peptide vibi E and these were noted to be relatively minor components by LC-MS. Comparing the two approaches, 11 of the 15 cyclotide sequences were identified by peptide sequencing and 6 as expressed mRNA. The choice of primer is thus critical to cyclotide discovery as the diversity of cyclotide precursors in existence may result in lower affinity for the primer or in fact not contain codons that are targeted by the selected primer.

Discrepancies between mRNA and protein expression may exist for a number of reasons including the timing of plant material collection and extraction. Many cyclotides show tissue-specific expression, seasonal and/or geographical differences in peptide expression and the peptides may persist in plant material beyond the translation of the mRNA transcripts. In a similar manner, eight Melicytus species were screened at the peptide and mRNA level revealing 5 novel cyclotides, mra 1–5, and 17 nonredundant cDNA clones encoding cyclotide precursor proteins (Trabi, Mylne, Sando, & Craik, 2009). The degenerate AAFALPA primer was used to clone the full-length cyclotide precursors by 3′ RACE, but was used in combination with specific primers designed to the newly discovered cyclotide sequences. Of the 17 unique cDNA clones, 6 were found to encode single cyclotide domain genes, 7 represented 2 cyclotide domain genes and the remaining 4 encoded multidomain (>2) cyclotide genes. As was seen with the previous study, only two peptides of the five isolated were also detected as cDNA sequences. The three peptides that lacked cDNAs were again among the most abundant of the peptides isolated.

The vast majority of cDNA screening approaches applied to cyclotide discovery have examined plants from the Violaceae family. Some of these discoveries have been coincidental. For example, in an experiment aiming to uncover genes and/or proteins involved in defense mechanisms after cadmium exposure, a suite of five cyclotides were sequenced from cDNA in *Viola baoshanensis* (Zhang, Hu, et al., 2009).

Twelve novel cyclotides were isolated and sequenced from two Panamanian species (*Gleospermum blakeanum* or *G. pauciflorum*) using a combination of both protein isolation and cDNA screening (Burman et al., 2010). Protein isolation and analysis was carried out using the standard approach or reduction, alkylation and digestion with multiple complementary enzymes, and confirmatory AAA. In parallel, RNA extraction and cDNA sequencing making using of the AAFALPA protein sequence from the ER region resulted in 10 unique clones. Of the 10 precursor sequences, all but one contained a C-terminal Asn with Gpc1 containing a C-terminal Asp. Likewise the peptides sequenced by LC-MS/MS all contained Asn with the exception of Glopa D corresponding to the Gpc1 clone.

As described previously for the discovery of cyclotide-encoding ESTs from petunia (Poth et al., 2012), a similar bioinformatic approach was utilized to look for the presence of cyclotides across the plant kingdom (Mulvenna, Mylne, et al., 2006). Related sequences were uncovered in the Poaceae or grass family, specifically in rice, maize, and wheat, which are economically important food crops. Of the 22 precursors identified using regular expression searching, only one contained an Asn after the final conserved Cys and this precursor was noted to be in a member of the Rubiaceae family, *Hedyotis centranthoides*. The remaining sequences showed remarkable homology with mature cyclotide sequences and were postulated to be ancestral proteins related to cyclotides, but not likely to produce cyclic peptide products.

Recently, a suite of nine novel linear cyclotides, named panitides L1–9, from *Panicum laxum* of the Poaceae family were characterized using a combination of peptide sequencing and gene cloning (Nguyen et al., 2013). This study provided the first evidence of linear cyclotides at the protein level in a monocot plant. The cystine knot arrangement was determined to be similar to cyclotides by disulfide mapping of panitide L3 and the acyclic peptides were shown to exert high stability against heat and proteolytic degradation.

The confirmation of the presence of linear cyclotides (acyclotides) in both dicots and monocots supports their proposed ancient origin and their existence before the divergence of these two groups of flowering plants.

2.3 Next-Generation Sequencing

Relatively few plant species have been sequenced, especially when compared with the hundreds of thousands that exist globally. The model plant, *A. thaliana* (thale cress), was among the first plant genome to be sequenced (Arabidopsis-Genome-Initiative, 2000). Since then a number of commercially important plant species have been sequenced, including *Oryza sativa* (rice) (International-Rice-Genome-Sequencing-Project, 2005), *Carica papaya* (papaya) (Ming et al., 2008), *Zea mays* (maize) (Schnable et al., 2009), *Fragaria vesca* (strawberry) (Shulaev et al., 2011), *Solanum lycopersicum* (tomato) (The-Tomato-Genome-Consortium, 2012), and *Cajanus cajan* (pigeonpea) (Varshney et al., 2012).

Despite the advances in technology allowing high-throughput sequencing, the assembly and annotation of genomes remains the bottleneck. This is more challenging in plant genomes owing to the size and complexity of plant genomes (Schatz, Witkowski, & McCombie, 2012). Plant genomes can be 100 times larger than mammalian genomes (Pellicer, Fay, & Leitch, 2010). Nearly all mammals are diploid, that is, they contain two copies of each chromosome. However, polyploidy is far more common in plants with estimates of polyploidy occurrence as high as 80% of plant species (Meyers & Levin, 2006). Plants have higher rates of heterozygosity and repeat regions. Furthermore, genome duplication in plants leads to increased complexity with large gene families common.

To date, none of the cyclotide-containing plants have been subjected to genome sequencing. However, a study of the first cyclotide-containing plant, *O. affinis*, was undertaken in which the size of the genome was estimated to be 4.2 Gbp and over 1000 ESTs were identified (Qin et al., 2010). Of these, 31 were found to encode cyclotide precursors and interestingly, ESTs encoding the cyclotide precursors for kalata B1 and B2 were among the 20 most abundant ESTs. Another interesting finding of this study was the identification of an asparaginase (from 11 ESTs) postulated to be involved in excision of the mature cyclotide and ligation of the termini. Furthermore, cDNAs encoding cyclotides with C-terminal Asn, but not Asp were identified suggesting that the identification of Asp versions in *O. affinis* was the result of posttranslational modification (PTM) rather than existing as Asp *in planta*.

Although a number of cyclotides have been sequenced by cDNA screening (Burman et al., 2010; Dutton et al., 2004; Gillon et al., 2008; Gruber et al., 2008; Herrmann et al., 2008; Ireland, Colgrave, Nguyencong, et al., 2006; Simonsen et al., 2005; Trabi et al., 2009; Zhang, Liao, et al., 2009), the majority have been identified using mass spectrometry (MS).

2.4 Peptide Extraction and Purification

The vast array of extraction protocols that have been applied to the extraction of cyclotides has been reviewed (Goransson, Svangard, Claeson, & Bohlin, 2004). These techniques range from boiling plant material in water, as was used for the traditional preparation of the tea used in Africa, through to harsh extraction conditions employing salts that form complexes with amines (Reinecke's salt, sodium tetraphenylborate) and solvent–solvent partitioning. A fractionation protocol developed for extracting polypeptides from plant biomass (Claeson, Goransson, Johansson, Luijendijk, & Bohlin, 1998) provided the foundation for a range of extraction protocols employed for cyclotides. In fact, application of this protocol involving liquid/liquid extraction, filtration, size-exclusion, and solid phase (C18) chromatography yielded the identification of varv peptide A, a novel cyclotide from *Viola arvensis* (Claeson et al., 1998). The widely accepted extraction protocol (Craik et al., 2012) involves harvesting plant material, typically aerial parts, although cyclotides have also been found in the roots. The plant material is ground to a fine powder in a mortar and pestle under liquid nitrogen and then extracted using methanol (or alternatively ethanol) in a 1:1 dichloromethane:methanol extraction protocol. After filtration, the methanol layer is removed and concentrated by rotary evaporation. The extract is partially fractionated on a C18 column using stepwise elution, typically using 25%, 50%, 75%, and 100% acetonitrile (acidified with 0.05% trifluoroacetic acid, TFA). These crude fractions are then subjected to multiple rounds of RP chromatography employing a C18 column and a solvent gradient (typically acetonitrile based). The benefit of using dichloromethane is that potentially interfering compounds such as chlorophyll, polyphenols and low-molecular weight analytes may be separated from the cyclotide fraction. Salts and other reagents are removed in the <25% acetonitrile fraction eluted from the C18 column.

The high degree of homology between cyclotide sequences often results in co-eluting peptides, for example peptides that differ by a single conservative amino acid substitution such as Ser/Thr or Ile/Leu. Additionally, cyclotides with highly variable sequence may also co-elute, in part because of the generally hydrophobic nature of the cyclotides arising from the exclusion of the hydrophobic residues from the core of the molecule by the CCK resulting in the vast majority of cyclotides eluting in a small retention time window. These co-eluting peptides may be separated by using a shallower RP gradient, by using isocratic elution (Goransson et al., 2004), or by the addition of modifiers, such as tetraethylenepentamine (TEPA) (Ireland, Colgrave, & Craik, 2006).

A range of other separation and purification strategies have been employed. Adsorption chromatography using Sephadex gels is one such approach that has been shown to be complementary to RP HPLC separation. Sephadex LH-20 was used in the isolation of the kalata peptides from *O. affinis* by Gran (1973a). It was also utilized in the separation of circulins A and B from *C. parvifolia* (Gustafson et al., 1994) and the cycloviolins A-D from *Leonia cymosa* (Hallock et al., 2000).

Cyclotides are not suitable for protein-based separation strategies such as the use of one- or two-dimensional gels. In gel electrophoresis proteins are separated by their molecular weight and these techniques are efficient for the separation of proteins >10 kDa in size. The cyclotides typically have masses close to 3 kDa, but their compact structure resulting from the combination of the cyclic backbone and the knotted structure, means that they actually run smaller on a gel. Their size is also problematic in antibody assays, such as blotting applications involving membranes. The cyclotides pass through nitrocellulose membranes that are employed for the retention of proteins.

2.5 Acid Hydrolysis and Amino Acid Analysis

The amino acid composition of a protein can be determined by AAA. This is often undertaken prior to sequencing as it can inform which enzymes to use based on the amino acid frequency or is useful post-sequencing to detect errors in the sequencing process or to distinguish between ambiguous results. AAA involves hydrolyzing a known quantity of protein into its constituent amino acids and then separating the amino acids and in some cases quantifying them. Hydrolysis is typically achieved by heating the protein in 6 M hydrochloric acid to 100–110 °C for 24 h. However, under these conditions some amino acids (serine, threonine, tyrosine, tryptophan, glutamine, and cysteine) may be degraded. Cysteine and tryptophan are particularly susceptible and are usually analyzed separately after derivatization. The amino acids can be separated by ion-exchange chromatography or hydrophobic interaction chromatography. In ion-exchange chromatography, the amino acids are separated on the basis of their isoelectric points. RP chromatography is an example of hydrophobic interaction chromatography that is commonly used, wherein C8 and C18 silica columns are commonly used for the separation of amino acids in solution. The quantities of each amino acid can be determined by adding a reagent that will form a colored derivative. For example, ninhydrin can be used (gives a yellow color when reacted with proline and purple when reacted with other amino acids) to determine the concentration of each amino acid by measuring the absorbance of the resulting solution.

2.6 Edman Degradation

Edman degradation is a method of sequencing amino acids in a peptide (Edman, 1950) in which the amino-terminal (otherwise known as N-terminal) residue is labeled and cleaved from the peptide without hydrolyzing the rest of the peptide. Under mildly alkaline conditions, phenylisothiocyanate is reacted with the peptide to form a cyclical phenylthiocarbamoyl derivative. Subsequently, under acidic conditions, this derivative of the terminal amino acid is cleaved as a thiazolinone derivative which is then selectively extracted into an organic solvent and treated with acid to form the more stable phenylthiohydantoin (PTH)-amino acid derivative that can be identified by using chromatography or electrophoresis. These steps are repeated to identify each subsequent amino acid. Automated Edman sequencing is possible allowing peptides up to ~50 amino acids to be sequenced. Automated Edman sequencing was utilized in many of the early sequencing studies of cyclotides including the identification of cyclopsychotride A (Witherup et al., 1994), circulins A-F (Gustafson et al., 1994, 2000), kalata B1-B7 (Craik et al., 1999; Jennings et al., 2001; Saether et al., 1995), varv A-H (Claeson et al., 1998; Goransson, Luijendijk, Johansson, Bohlin, & Claeson, 1999), cycloviolacins O1-O12 (Craik et al., 1999), cycloviolins A-D (Hallock et al., 2000), vodo M-N (Svangard et al., 2003), and hypa A (Broussalis et al., 2001), however, in the latter case nanospray tandem mass spectrometry was also employed. Edman sequencing is not applicable to peptides with blocked N-termini, that is, peptides in which the N-terminus is modified such as by pyroglutamination or cyclic peptides that do not have N-termini. As such native cyclotides require chemical treatment to reduce (and alkylate) the Cys and cleavage either by acid hydrolysis or proteolytic digestion.

2.7 Chemical and Enzymatic Digestion

Specific cleavage can be achieved by chemical or enzymatic methods. For example, cyanogen bromide (CNBr) cleaves peptides only on the carboxyl side of methionine residues. A range of proteolytic enzymes are available that cleave peptide chains at the amide bond. Highly specific cleavage is also obtained with trypsin, a proteolytic enzyme isolated from the pancreas (of cows or pigs) that cleaves peptides on the carboxyl side of arginine and lysine residues. Endoproteinase Glu-C (also known as V8 protease) isolated from *Staphylococcus aureus* is particularly useful for sequencing cyclotides as it cleaves C-terminally to glutamic acid (and aspartic acid under some experimental conditions). Cyclotides contain a highly conserved Glu residue in loop 1 of the cyclotide structure which is typically the only Glu. This results in the cyclic peptide structure being opened at one site yielding a linear derivative suitable for sequence analysis. Chymotrypsin, another pancreatic enzyme, cleaves C-terminally to aromatic residues with high affinity for Trp, Phe, and Tyr and lower affinity for Leu and Met. The proteolytic enzymes typically used for cyclotide cleavage are given in Table 1 along with suggested conditions for digestion. Chymotrypsin is a useful enzyme in cyclotide-sequencing studies for delineating the isobaric amino acids isoleucine and leucine, which cannot be distinguished easily by MS/MS. The peptides obtained by specific chemical or enzymatic cleavage are typically separated by chromatography prior to sequence analysis by Edman degradation and/or mass spectrometry.

3. MASS SPECTROMETRY

Mass spectrometry (MS) is an analytical technique that measures the mass-to-charge ratio of charged compounds. It is primarily used for determining the mass of these compounds, but can also be used to determine the elemental composition of a compound or in the case of peptides, the amino acid composition. The general scheme applied to cyclotide sequencing is depicted in Figure 2.

3.1 Ionization

In order to detect and measure a peptide mass, the peptide must first be ionized to generate charged peptides. Ionization can be positive or negative, however, the vast majority of peptide research employs positive ionization making use of the presence of basic amino acids that accept and retain charge. Negative ionization techniques may be used for more specific peptide applications, such as the study of phosphopeptides which readily donate protons taking on a negative charge at the sites of phosphorylation. As there

Table 1 Cleavage Specificity of Proteolytic Enzymes and Commonly Employed Chemicals for Cyclotide Discovery

Enzyme	Cleavage	Except	Terminal	Optimum pH	Optimum Temp	Suitable Buffers
Trypsin	K, R	Before P	C-	7.8	37 °C	50–100 mM ammonium bicarbonate, 1 mM $CaCl_2$
Chymotrypsin	Y, F, W > L, M	Before P	C-	7.8–8.0	37 °C	50–100 mM ammonium bicarbonate or Tris HCl, 10 mM $CaCl_2$
Endoprotease Glu-C (protease V8)	E	Before P	C-	7.8	37 °C	50–100 mM ammonium bicarbonate or Tris HCl
Endoprotease Glu-C (protease V8)	E, D	Before P	C-	6.0–8.0	37 °C	100 mM phosphate buffer
Pepsin	F, L		C-	1.3	37 °C	10 mM HCl
Pepsin	F, L, W, Y, A		C-	2.2	37 °C	10 mM HCl
Thermolysin	A, F, I, L, M, V	After D, E	N-	8.0	65–85 °C	50 mM Tris HCl, 0.5 mM $CaCl_2$
CNBr	M		C-		25 °C	100 mM HCl or 70% aqueous formic acid

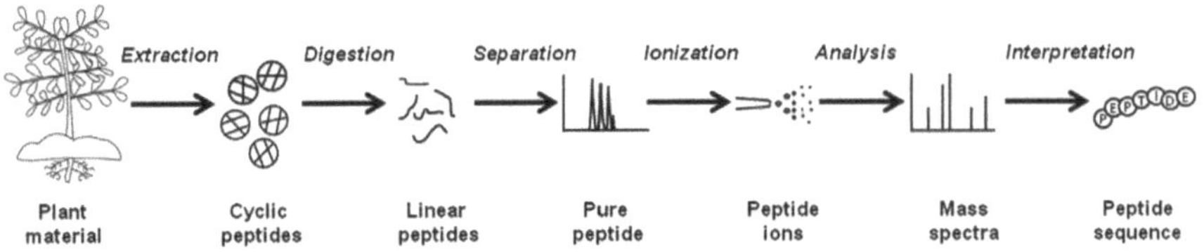

Figure 2 ***Experimental scheme for cyclotide sequencing.*** Plant material is extracted to yield a crude solution containing cyclotides and other peptides/analytes. Chemical treatment (including reduction and alkylation of Cys) and enzymatic digestion result in the linearization of the cyclic backbone. Peptide separation is performed to yield pure peptides (or simple mixtures) which are then introduced to the MS by ionization. MS analysis is undertaken by generating spectra for interpretation. This may be either automatic (database searching) or manual (de novo sequencing).

are no reports of phosphorylated cyclotides to date, the discussion will be limited to positive ionization.

There are a number of ionization techniques that have been developed over the years, including electron ionization, chemical ionization, and FAB, but the two techniques that have been widely applied to the study of peptides are electrospray ionization (ESI) and matrix-assisted laser desorption ionization (MALDI). These so-called soft ionization strategies are applicable to the study of biomolecules as very little fragmentation results from ionization.

3.1.1 Electrospray Ionization

The ESI technique was first reported by Masamichi Yamashita and John Fenn in 1984 (Yamashita & Fenn, 1984) and later resulted in Fenn receiving the Nobel Prize for this achievement. The ESI technique involves spraying of a solution of the sample through a highly charged needle (or capillary) which is at atmospheric pressure, transferring charge to the analytes. The resulting ions are transferred from the solution to gas phase in three steps: (1) dispersal of a fine spray of charge droplets, followed by (2) solvent evaporation, and (3) ion ejection from the highly charged droplets. A high voltage (2–5.5 kV) is applied to the capillary producing a mist (or aerosol) of highly charged droplets with the same polarity as the capillary voltage. The application of a nebulizing gas such as nitrogen assists in the streamlining of the sample flow. The charged droplets formed are solvated with solvent molecules. Increased temperature within the ionization source combined the application of a desolvation gas results in solvent evaporation. The desolvation process decreases the droplet size leading to an increase of surface charge density and a decrease of the droplet radius. This leads to Coulombic repulsion between the charges present in the droplet until it becomes unstable upon reaching its Rayleigh limit. At this point, the electrostatic repulsion of like charges, in a continually decreasing droplet size, becomes more powerful than the surface tension holding the droplet together. The droplet undergoes Coulomb fission, whereby the original droplet explodes creating many smaller, more stable droplets. The new droplets undergo further desolvation and subsequent Coulomb fissions. The process eventuates in the formation of individual gas phase analyte ions. These ions are then sampled by a skimmer cone and are then accelerated into the mass analyzer for subsequent analysis of molecular mass and measurement of ion intensity.

ESI can produce singly or multiply charged ions. The number of charges retained by a particular analyte depends on several factors such as the size, chemical composition, and higher-order structure of the analyte molecule, the solvent composition, the presence of cosolutes, and the instrument parameters. For small molecules (<2000 Da) ESI typically generates singly, doubly, or triply charged ions, while for large molecules (>2000 Da) ESI can produce a series of multiply charged ions. In this way ESI can be used to study a range of biomolecules from small peptides through to intact proteins.

3.1.2 Nanoelectrospray Ionization

Nanospray (or nanoelectrospray ionization) (Wilm & Mann, 1996) is a low flow rate version of ESI. In static nanospray, a small volume (1–2 μL) of the sample dissolved in a suitable volatile solvent is loaded into a capillary and a relatively low voltage (900–2000 V) is applied resulting in sample ionization which is directed into the MS. Static nanospray is particularly useful for pure compounds or simple mixtures and enables high quality data to be collected on the major components. The spectra generated are suitable for spectral interpretation by de novo sequencing (more later). Nanoflow LC-MS applications also exist in which the flow rate of solute and solvent using this procedure is very low (30–500 nL/min) consuming far less sample than standard ESI. A common application of this technique is tandem mass spectrometric (MS/MS) analysis of a protein digest. Workflows such as these are called independent data acquisition (IDA) in which the molecular masses of the components present are determined, generating a list of peptides to be analyzed "on the fly" by MS/MS.

3.1.3 Matrix-Assisted Laser Desorption/Ionization Mass Spectrometry

Matrix-assisted laser desorption/ionization mass spectrometry (MALDI-MS) was first introduced in 1988 by Hillenkamp and Karas (Karas & Hillenkamp,

1988) and later refined (Hillenkamp, Karas, Beavis, & Chait, 1991). MALDI is a technique suited to the study of peptides, proteins, and a range of other labile biomolecules. It can desorb intact analyte molecular ions with masses up to 300 kDa. In MALDI-MS analysis, the analyte is first cocrystallized with a larger excess of a matrix compound, usually a UV-absorbing weak organic acid such as α-cyano-4-hydroxycinnamic acid (CHCA), 2,5-dihydroxybenzoic acid (DBA), or sinapinic acid. Laser radiation of the dried droplets of the matrix-analyte preparation results in desorption of the matrix (and analyte) as a plume thus entering the gas phase. The matrix absorbs the laser energy indirectly resulting in vaporization of the analyte. The matrix acts to ionize the analyte by serving as a proton donor and acceptor, dependent on the type of ionization (positive, negative). MALDI typically produces ions with single charge.

3.2 Mass Analyzers

The main function of the mass analyzer is to separate, or resolve, the ions formed in the ionization source of the mass spectrometer according to their mass-to-charge (m/z) ratios. There are a number of mass analyzers which include quadrupoles, time-of-flight (TOF) analyzers, magnetic sectors, Fourier transform (FT), and quadrupole ion traps. Additionally, multiple mass analyzers have been combined to generate hybrid instruments with advantages in speed, sensitivity, and resolution.

These mass analyzers have different features, including the m/z range that can be covered, the mass accuracy, and the achievable resolution. The compatibility of different analyzers with different ionization methods varies. For example, all of the analyzers listed above can be used in conjunction with ESI, whereas MALDI is not usually coupled to a quadrupole analyzer, where TOF instruments are more commonly employed. The majority of cyclotide research has been accomplished on either ESI coupled to quadrupole, TOF or hybrid (Q-TOF) instruments or MALDI-TOF instruments and as such the discussion will focus on these types of mass analyzers.

3.2.1 Quadrupole Mass Analyzers

As the name suggests, quadrupoles consist of four cylindrical rods, set parallel to each other. Ions are separated in a quadrupole based on the stability of their trajectories in the oscillating electric fields that are applied to the rods. Each opposing rod pair is connected together electrically, and a radio frequency (RF) voltage is applied between one pair of rods and the other. A direct current voltage is then superimposed on the RF voltage. Ions travel down the quadrupole between the rods. Only ions of a certain mass-to-charge (*m*/*z*) ratio will reach the detector for a given ratio of voltages. The other ions will have unstable trajectories and will collide with the rods. This permits selection of an ion with a particular *m*/*z* or analysis of a range of *m*/*z* values by continuously varying the applied voltage.

Multiple quadrupoles can be combined to enable tandem mass spectrometry. For example, the arrangement of three quadrupoles in a linear fashion is known as the triple quadrupole (QQQ) mass spectrometer. The first (Q1) and third (Q3) quadrupoles act as mass filters, and the middle (Q2) quadrupole is employed as a collision cell. This collision cell is filled with an inert gas, such as Ar, He, or N_2 gas, enabling fragmentation of the analyte in a process known as collision-induced dissociation (CID). The fragment ions produced in Q2 are transmitted to Q3 for mass analysis.

This process allows for the study of fragments that are useful in the structural elucidation of a peptide by tandem mass spectrometry (MS/MS). For example, the first mass analyzer may be set to select a peptide of known mass, which is fragmented in the collision cell. The third quadrupole can then be set to scan the entire *m*/*z* range, giving information on the mass and intensities of the fragments, allowing interpretation of the peptide structure.

Quadrupole mass analyzers may also be combined with TOF mass analyzers in hybrid instruments, such as the quadrupole-time-of-flight (Q-TOF) MS. Quadrupole mass analyzers have relatively low resolution, but fast scanning speed. In contrast, TOF analyzers have high resolution, but are limited in their scanning speed. The combination provides benefits by utilizing the attributes of each mass analyzer.

3.2.2 TOF Mass Analyzers

There are two types of TOF mass analyzers: the linear TOF mass analyzer is the simplest. TOF analysis is based on accelerating the ions produced in the source to a detector. The ions are accelerated from the source with the same kinetic energy, but owing to their difference in mass, the ions reach the detector at different times. The so-called TOF is measured and converted to the *m*/*z*. The smaller ions reach the detector first because of their greater velocity while the larger ions take longer owing to their larger mass. The TOF reflectron combines TOF technology with an electrostatic analyzer, the reflectron. The reflectron is composed of a constant electrostatic field which acts to reflect the ion beam toward the detector. More energetic ions penetrate deeper into the reflectron and as such take a slightly longer path to the detector. Less energetic ions of the same *m/z* penetrate a shorter distance into the reflectron and, correspondingly, take a shorter path to the detector. In this way, variation in the kinetic energy distribution in the direction of ion flight can be corrected by using a reflectron. Moreover, resolution, defined as the mass of a peak divided by the width of a peak, is improved by increasing the TOF. This increased resolution, however, often comes at the expense of sensitivity and a relatively low mass range, typically <10,000 on reflectron TOF systems.

3.3 Peptide Sequencing

De novo is the latin term for "from the beginning" or "beginning again." Peptide de novo sequencing is the analytical process that derives a peptides amino acid sequence from its tandem mass spectrum (MS/MS) without the assistance of a sequence database. It stands in contrast to automated peptide identification known as database searching. In the latter approach, the MS/MS spectrum, or spectra, is searched against a given database in an attempt to identify the peptide sequence, but is only successful if the precise sequence of the peptide is present within the selected database. A clear advantage of de novo sequencing is that it works for both known and novel peptides.

The most widely used fragmentation methods today are CID and electron-transfer dissociation (ETD) (Syka, Coon, Schroeder, Shabanowitz, & Hunt, 2004). In CID, activation of the peptide ions proceeds via multiple collisions with the inert background gas, such as helium, argon, or nitrogen. When the internal energy reaches the fragmentation threshold, the most labile bonds are cleaved preferentially, irrespective of whether they are present in the peptide backbone or in the side chains, such as PTMs. CID typically generates b- and y-fragment ions via fragmentation of the amide bond (Steen & Mann, 2004), but this does not occur equally along the backbone. For instance, cleavage preceding Pro is preferred, followed by cleavage C-terminally to Glu/Asp and further abundant internal fragmentation and loss of PTMs are common. In contrast, ETD produces mostly c- and z-ions. In some comparisons with CID, ETD was demonstrated to generate a more complete series of ions and thus more extensive sequence information, with the additional advantage of leaving labile PTMs intact (Good, Wirtala, McAlister, & Coon, 2007). ETD is particularly useful for larger peptides (and even proteins) or for phosphorylated and glycosylated peptides. However, it is optimally applied to the analysis of ions with higher charge state (>2).

The fragmentation of peptides has been well documented (Biemann, 1990; Johnson & Biemann, 1989; Roepstorff & Fohlman, 1984). The protonated peptides fragment along the peptide backbone at either the NH—CH, CH—CO, and/or CO—NH bonds. In the case of singly protonated peptides, each bond breakage gives rise to two species, one neutral and the other one charged, where only the charged species is monitored by the mass spectrometer. For multiple protonated peptides, charge may be retained on both fragments, dependent on the chemistry and relative proton affinity of the two species. The six possible fragment ions for each amino acid residue are depicted in Figure 3. When the charge is retained on the N-terminus, the ions are termed a-, b-, and c-ions, whereas when the charge is retained on the C-terminal fragment, the ions are termed x-, y-, and z-ions. As described above, the most common cleavage sites in CID are at the CO—NH bonds which give rise to the b- and y-ions. The mass difference between two adjacent b-ions, or y-ions, is indicative of a particular amino acid residue. Additional to these sequence ions, immonium ions may also appear in the low *m/z* range of the MS/MS spectrum. Each amino

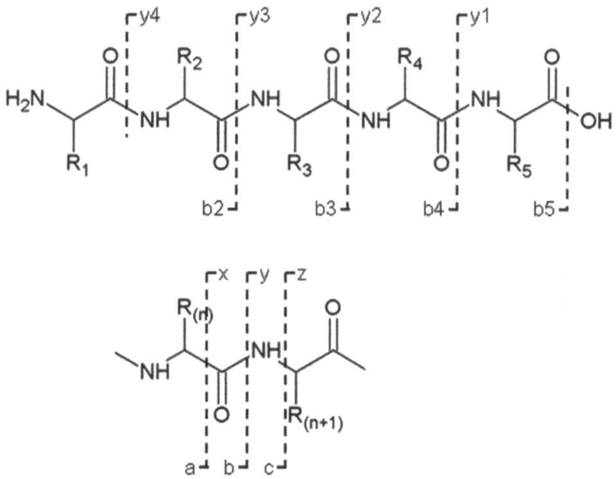

Figure 3 ***Peptide product (or fragment) ions produced in tandem mass spectrometry.*** Peptide fragmentation arising from collisions with an inert gas (as occurs in CID) results in breakage of the peptide backbone. The most labile bond is the amide bond which yields b- and y-ions as shown in the upper schematic. When the charge is retained on the amino-terminal fragment, b-ions are produced, whereas when the charge is retained on the carboxy-terminal fragment y-ions are formed. The possible backbone cleavages are a-, b-, and c-ions (charge retained on N-terminus); or x-, y-, and z-ions (charge retained on C-terminus) as depicted in the lower schematic. In ETD, the c- and z-ions dominate the MS/MS spectra.

acid residue leads to a diagnostic immonium ion, with the exception of the two isobaric pairs: leucine (L) and isoleucine (I); and lysine (K) and glutamine (Q), which produce immonium ions with the same *m/z* (86 for I/L; 101 for K/Q). The immonium ions are useful for detecting and confirming the amino acid composition of a peptide, although the absence of a particular immonium ion is not indicative that the amino acid is absent as some immonium ions are detected preferentially.

4. APPLICATION OF MASS SPECTROMETRY TO CYCLOTIDE SEQUENCING

4.1 MS Analysis of Cyclotides

MS is an ideal analytical method for the identification of peptides. Detection of peptides by MS relies on ionization by protonation of the peptide (facilitated by acidification of the peptide solution) which preferentially occurs at basic sites (Lys, Arg, and His) and at the N-terminus. The absence of termini and the relatively low number of basic residues in cyclotides is a challenge to their analysis by MS. Despite this, cyclotides do ionize, but take on lower charge states than typical linear peptides of the same size (2800–3500 Da) (Figure 4).

Mass spectrometry was initially applied to the confirmation of peptide mass only (Goransson et al., 1999; Gustafson et al., 1994, 2000; Hallock et al., 2000; Svangard et al., 2003). In the earliest studies, FAB MS was employed (Gustafson et al., 1994, 2000; Hallock et al., 2000). MALDI-TOF MS has been used to directly analyze cyclotides based on molecular weight (Claeson et al., 1998; Conlan et al., 2012; Goransson et al., 1999; Hallock et al., 2000), but has also been applied in a semi-quantitative manner (Colgrave, Jones, & Craik, 2005; Saska, Colgrave, Jones, Anderson, & Craik, 2008; Saska et al., 2007).

4.2 Tandem Mass Spectrometric (MS/MS) Analysis of Cyclotides

MS/MS sequencing is enabled by fragmentation of a peptide along the backbone primarily at the amide bonds in a process termed CID. A typical peptide will produce a suite of sequence ions, termed b- and y-ions, allowing the amino acid sequence to be elucidated from a tandem mass spectrum. Owing to their cyclic nature and the presence of disulfide bridges, fragmentation of native cyclotides is severely hindered (Figure 5(A)). Reduction of the disulfide bonds and alkylation of the resulting free sulfhydryl groups on

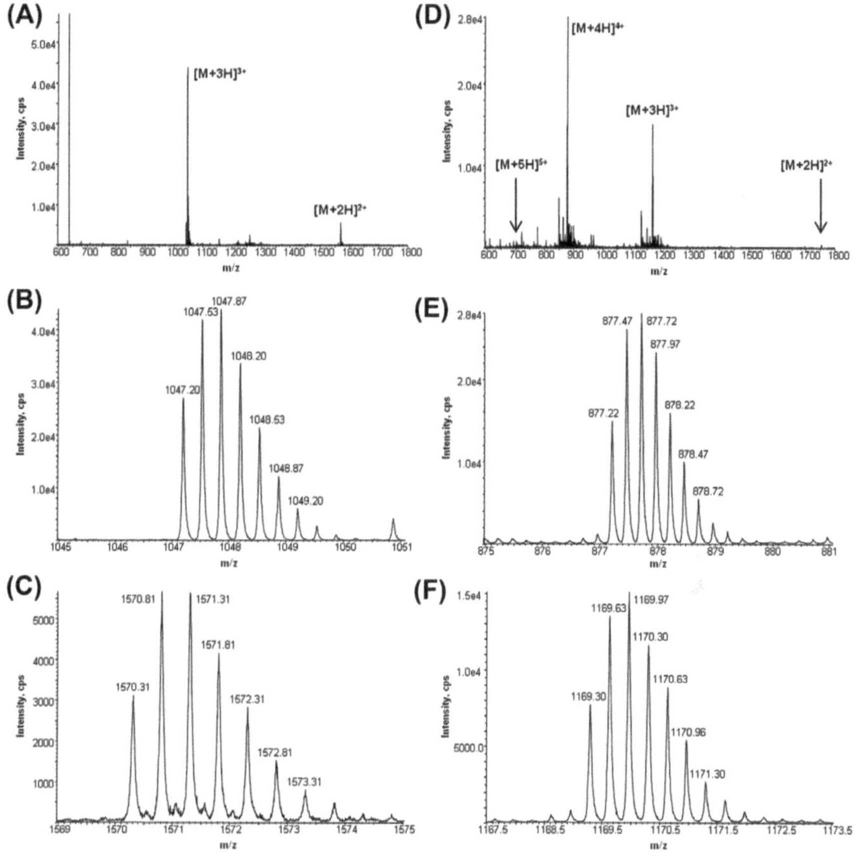

Figure 4 ***Native cyclotides adopt lower charge states than linear peptides in MS analysis.*** The shift in charge state from predominantly doubly and triply charged ions to triply and quadruply charged ions is exemplified by cycloviolacin O2. The native peptide (A) shows two peaks at *m/z* 1047.2^{3+} (B) and 1570.3^{2+} (C) despite containing three basic amino acids (2 Lys, 1 Arg). Upon reduction, alkylation and digestion (D), the linear derivative showed an increase in the charge state of the ion series with peaks observed at *m/z* 701.9^{5+}, 877.2^{4+} (E), 1169.2^{3+} (F), and 1753.3^{2+}.

Cys effectively removes the cystine knot, however the cyclic nature of the peptide backbone precludes efficient fragmentation by CID (Figure 5(B)). In some cases fragmentation of reduced and alkylated peptides does occur at labile bonds such as X-Pro, however, when multiple labile sites are present, it is possible to generate a mixed spectrum, that is, two series of overlapping fragment ions which render the spectrum difficult to interpret. Proteolytic digestion to cleave the peptides can be employed to produce linear derivatives amenable to MS/MS (Figure 5(C)).

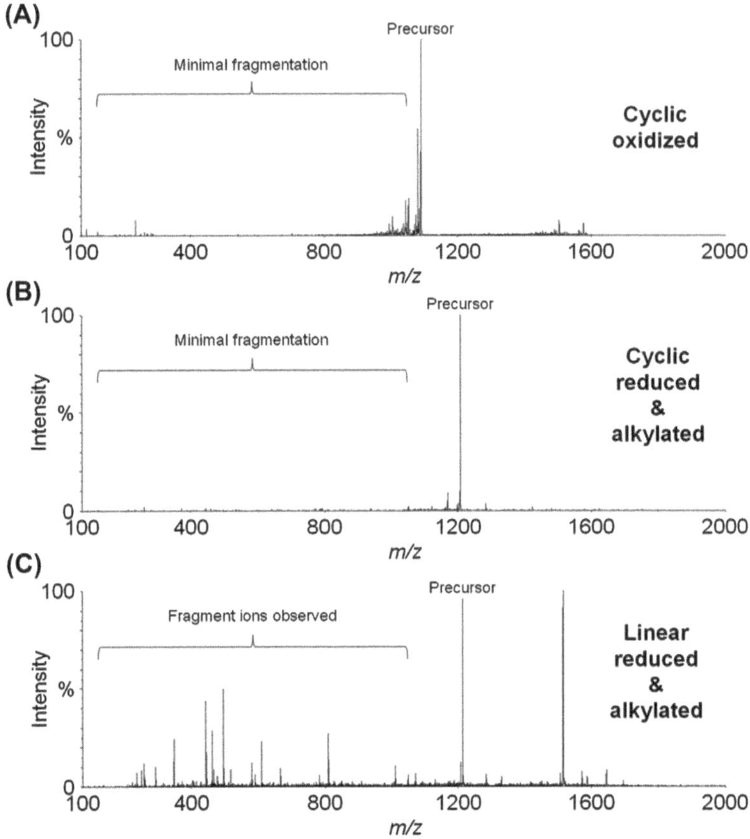

Figure 5 ***Native cyclotides do not readily fragment owing to presence of cyclic cystine knot.*** Nanospray tandem MS fragmentation patterns for "native" versus chemically modified Cter A at a collision energy setting of 50 V. (A) "Native" (cyclic oxidized) cyclotide precursor at *m/z* 1090.1^{3+}. (B) Cyclic reduced and alkylated precursor at *m/z* 1206.1^{3+}. (C) Digested/linearized reduced and alkylated precursor at *m/z* 1212.1^{3+}. These triply charged fragment ions correspond to species of molecular masses 3267.3, 3615.3, and 3633.3 Da, respectively. *Reprinted with permission from Poth et al. (2011). Copyright 2011, American Chemical Society.*

The determination of the primary structure of the vast majority of cyclotides discovered to date has been carried out by tandem mass spectrometry sequencing. Traditional peptide sequence analysis by mass spectrometry employs bottom-up peptide sequencing, in which enzymes are employed to cut proteins or peptides into smaller fragments amenable to MS/MS sequencing. In this approach cyclotides, isolated from crude extracts, are first reduced (in some cases alkylated) and then linearized by enzymatic digestion prior to MS/MS analysis being undertaken. One of the first examples of the use of MS/MS sequencing of cyclotides was the characterization of Hypa A from *Hybanthus parviflorus* (Broussalis et al., 2001). Since this time, MS/MS has been applied with increased frequency and great success. A great number of cyclotide discovery studies have employed bottom-up peptide sequencing (Chan et al., 2013; Chen et al., 2005; Chen, Colurave, Wang, & Craik, 2006; Gerlach, Burman, Bohlin, Mondal, & Goransson, 2010; Goransson & Craik, 2003; Gruber et al., 2008; Hashempour, Koehbach, Daly, Ghassempour, & Gruber, 2013; He et al., 2011; Ireland, Colgrave, & Craik, 2006; Ireland, Colgrave, Nguyencong, et al., 2006; Mulvenna et al., 2005; Nguyen, Zhang, Nguyen, et al., 2011; Pinto et al., 2012; Plan et al., 2007; Poth, Colgrave, Lyons, et al., 2011; Poth, Colgrave, Philip, et al., 2011; Poth et al., 2012; Simonsen et al., 2005; Tang et al., 2010; Xiang et al., 2010; Yeshak, Burman, Asres, & Goransson, 2011).

When applying MS/MS sequencing, the high sequence homology between cyclotides, the unusual gene structure, and the lack of genomic data for many of the cyclotide-containing plant species provide additional challenges in the discovery of cyclotides. A range of strategies have been used over the years to elucidate the sequences of cyclotides, including the loop sequencing method (Goransson, Broussalis, & Claeson, 2003). This protocol (Figure 6) involved modification of the six conserved cysteines by aminoethylation to produce pseudolysines that are sites for digestion by trypsin. In this way, each intercysteine loop can be sequenced by MS/MS. In order to maintain the structure of each loop, any lysines present in the sequence must be first modified, for example by acetylation, to prevent cleavage at these sites. To avoid cleavage at Arg, an alternate enzyme such as endoproteinase Lys-C may be employed. This method requires that the plant extract is subjected to extensive purification to single cyclotide components prior to analysis in order to be able to reconstruct the native sequences and as such has seen limited application (Chen et al., 2006; Gerlach et al., 2010; Goransson et al., 2003; Plan et al., 2007).

Another approach involving partial acid hydrolysis has been applied to cyclotide sequencing (Nair et al., 2006; Sze et al., 2009). In the first study (Nair et al., 2006), kalata B2 was initially sequenced by a standard protocol involving reduction, carbamidomethylation (of Cys) and digestion using endoproteinase Glu-C to yield a single linear derivative that was sequenced by high-resolution MS employing an FT-ICR. The same peptide was also subjected to microwave-assisted partial acid hydrolysis (using 6 M HCl) yielding peptide fragments with mass increases of either 18 or 36 Da

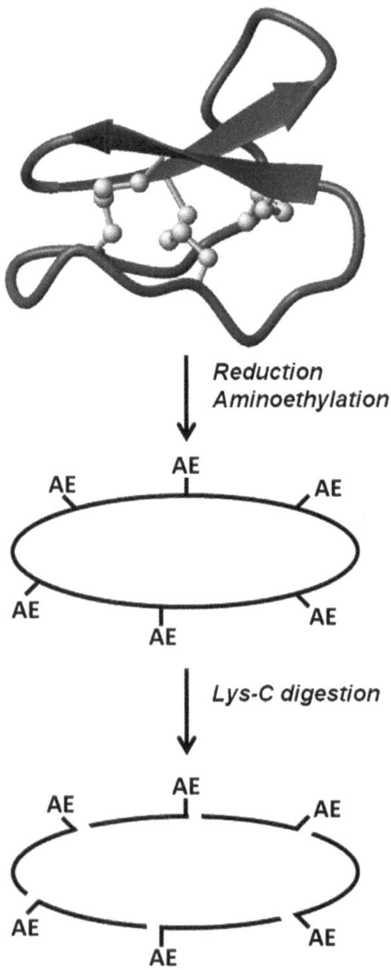

Figure 6 ***Intercysteine loop sequencing.*** The prototypic cyclotide kalata B1 shows the disulfide bonds in yellow (light gray in print versions) ball and stick arrangement and the β-sheets in purple (dark gray in print versions). The disulfide bonds are reduced before alkylation with bromoethylamine to produce the S-(β-aminoethyl) cysteine derivatives. The use of endoproteinase Lys-C (or trypsin) results in cleavage of the peptide backbone after each lysine and aminoethylated cysteine (and additionally arginine when trypsin is used). An additional experimental step can be included to avoid cleavage of lysines by first performing acetylation of the peptide with acetic anhydride. The resulting six loop-specific fragments may then be analyzed by MS/MS to reveal the peptide sequence. *This figure was adapted from Goransson et al. (2003).*

indicating acid hydrolysis at one or two backbone sites. HPLC analysis of the partially reduced (and NEM-alkylated) fragments revealed that the peptide was hydrolyzed at many sites and a number of the different fragments contained one to two intact disulfide bonds. Subsequent MS/MS sequencing facilitated the assignment of N- and C-terminal sequence tags (short stretches of sequence) preceding and/or succeeding the disulfide-bonded regions. Through interrogation of all peptide fragments resulting from partial acid hydrolysis combined with partial reduction and alkylation, the disulfide bond connectivity of kalata B2 was elucidated confirming connectivity proposed by NMR. It should be noted that these approaches also require at least partial peptide purification prior to analysis. In a subsequent study, partial acid hydrolysis (using 2 M HCl) and the use of sophisticated software tools have been implemented for the sequence analysis of a mixture of kalata B1 and varv A purified from *Viola yedoensis* (Sze et al., 2009). The two peptides differ by only a single conservative amino acid substitution (Ser—Thr) resulting in co-elution under standard HPLC conditions. After partial acid hydrolysis, the peptide fragments were analyzed by LC-MS/MS. The use of a computer program designed to interpret the MS/MS spectra of single disulfide-linked fragments facilitated the characterization of the disulfide bond connectivity of the two peptides. Acid hydrolysis combined with downstream LC-MS/MS analysis is a useful tool for disulfide connectivity, but in its current form is only applicable to single cyclotides or simple mixtures wherein the primary structure (sequence) is already known.

The combination of nano-LC with FT MS (Hashempour, Ghassempour, Daly, Spengler, & Rompp, 2011) offers significant advantages in terms of chromatographic resolution of peptides and high mass accuracy (with FT-MS capable of providing monoisotopic masses to five decimal places). In contrast to bottom-up proteomics described above, top-down sequencing (Kelleher, 2004) does not require the use of enzymes and as such the top-down strategy is not as time-consuming from a sample-processing perspective. A top-down peptide sequencing strategy was utilized for the identification of cyclotides from *Viola ignobilis* from a crude extract (Hashempour et al., 2011). Two peptides were identified on the basis of mass alone (cycloviolacin B2 and cycloviolacin O8), but a third peptide of mass 2876.1 Da corresponded to three known cyclotides in the CyBase database (Mulvenna, Wang, & Craik, 2006; Wang, Kaas, Chiche, & Craik, 2008): Varv peptides A, C, and D. In order to identify the correct peptide, reduction and alkylation followed by MS/MS sequencing was undertaken generating seven peptide fragments confirming the peptide identity as varv A. The success of this approach was based on matching experimentally determined masses with the theoretical masses derived from a cyclotide database. While this strategy was successful in identifying three known cyclotides, it cannot be readily applied to the discovery of novel cyclotides in the absence of an incomplete database.

Top-down sequencing was also employed in a study of hedyotide B2 from *Hedyotis biflora* (Nguyen, Zhang, Wang, et al., 2011). This study revealed the presence of four peptides, three of which were determined to be acyclic permutants resulting from the presence of a premature stop codon in the precursor protein sequence. Hedyotides 3 and 4 were identical in sequence to hedyotide 2, but N-terminally truncated forms of a potentially less stable linear peptide product. Interestingly, once the N-terminal glycine (G) and isoleucine (I) residues were removed, exposing an N-terminal glutamine (Q), pyroglutamination was observed. This type of modification imparts additional stability to linear peptides by precluding N-terminal exoprotease activity. The disulfide connectivity of the linear peptide hedyotide B2 was determined by top-down sequencing after partial reduction and alkylation employing two different alkylation reagents.

In an approach termed sequence fragment assembly (Hashempour et al., 2013), a suite of 13 cyclotides were recently characterized from *V. ignobilis* without the need for purification to individual components. Crude extracts were subjected to solid phase extraction, two rounds of HPLC to yield fractions containing simple mixtures of cyclotides. Each fraction was subjected to enzymatic digestion using single enzymes or combinations of trypsin, chymotrypsin, and endoproteinase Glu-C. The resulting peptide fragments were analyzed by MALDI-TOF-TOF and spectra were manually interpreted. Alignment and sequence fragment assembly was undertaken to yield the full-length sequences of 13 cyclotides, including 10 novel sequences.

4.3 Application of "Omics" Technologies to Cyclotide Sequencing

As discussed above, the successful application of MS/MS sequencing approaches has been limited to the analysis of purified peptides or simple mixtures resulting from partial purification strategies. This remains a bottleneck in the discovery of cyclotide sequences. The ability to sequence complex mixtures and/or crude extracts is highly desirable.

A proteomics approach in which four semipurified fractions from *V. odorata*, each of which contained tens of cyclotide masses was developed to accelerate the discovery of cyclotides (Colgrave, Poth, Kaas, & Craik,

2010). This approach was termed "ERA" referring to the bioinformatic strategy that Excised mature cyclotide domains from precursor proteins; Replicated these sequences; and finally Appended these sequences together in a database. This facilitated the database to be searched in an automatic search strategy commonly employed for true proteomic studies using any enzyme or combination of enzymes. While this approach will only yield results for peptides that exist in the database, using the Paragon algorithm (Shilov et al., 2007) within the ProteinPilot software tool enabled cyclotides with single amino acid substitutions to be identified fast tracking the discovery of highly homologous cyclotide sequences. Using this approach, 11 novel cyclotide sequences, among a plethora of known and modified sequences, were uncovered in what was thought to be a well-characterized cyclotide-containing plant species.

5. CHALLENGES AND COMPLEXITIES

5.1 Co-eluting, Isobaric Peptides

As previously noted, a huge abundance of natural diversity exists among cyclotide sequences, but furthermore, many of the peptide sequences are highly homologous. In an examination of plant material from a Viola species multiple co-eluting isobaric peptides were detected. These peptides were not able to be chromatographically resolved by RP HPLC, but by MS/MS analysis which was apparent that multiple isobaric peptides existed in what were initially thought to be pure peptide fractions (Figure 7). MS analysis revealed at least three distinct peptides in a single HPLC fraction with mass 2876, 2890 and 2904 Da. The difference in mass (14 Da) corresponds to a single amino acid substitution, for example a Ser—Thr or Val—Ile/Leu. The precursor at m/z 2877.1 corresponds to the mass of kalata S (Uniprot: P58458) and subsequent MS/MS analysis confirmed the lowest mass peptide to be kalata S (Figure 7(A)). It was postulated based on mass and similar properties, that the peptide with *m/z* 2891.1 would be kalata B1 (Ser—Thr substitution in loop 4). However, MS/MS analysis revealed that the dominant peptide was in fact cycloviolacin O12 (Uniprot: P83836) which differed from kalata S by a Val—Ile substitution in loop 6 (Figure 7(B)). Furthermore, at least one additional peptide shared the same mass, hydrophobicity (as judged by HPLC co-elution) and overlapping y-ion series in the MS/MS spectrum (Figure 7(C)). It is possible that these isobaric cyclotides may be separated using modified HPLC methodologies

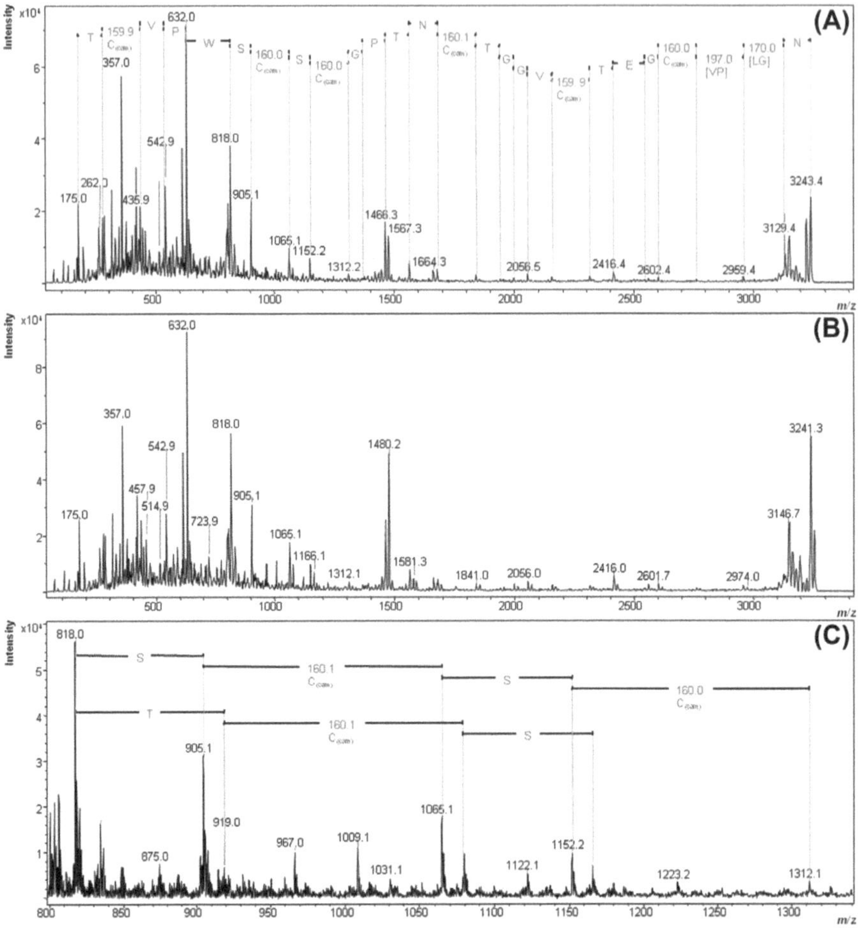

Figure 7 ***MS/MS analysis of cyclotides from* Viola *species shows the detection of isobaric, coeluting peptides.*** MALDI-TOF-MS spectra for a simple spectrum representing kalata S, precursor *m/z* 3243 (A); a mixed MS/MS spectrum containing multiple isobaric cyclotides, precursor *m/z* 3257 (B); and an expanded range of panel B showing selected y-ions through loops 4–5 of cycloviolacin O12, sequenced from the dominant peptide y-ion series (C).

(employing slower gradients or incorporating buffer additives), but it is clear that care must be taken when interpreting MS/MS spectra.

5.2 N/D Isoforms

A number of sequencing studies have identified pairs of coeluting cyclotides that differ in their sequence by a single amino acid at the site of cyclization, wherein the typical asparagine may either be an Asn (N) or Asp (D) immediately preceding the glycine. Examples of cyclotides are known to exist as both the N and D forms from gene sequencing studies (Jennings et al., 2001). For example, *oak1* encodes kalata B1 containing an Asn in loop 6, whereas *oak4* encodes three copies of kalata B2 which contains an Asp in loop 6. However, detection of these as peptides may arise from either deamidation of Asn to Asp as a result of sample handling and/or processing or alternatively from their existence *in planta*. The study of the legume *C. ternatea* revealed five pairs of N/D isoforms (Poth, Colgrave, Philip, et al., 2011). Whether these peptides arose from deamidation was investigated by extracting fresh plant material and immediately analyzing by MALDI-TOF MS (Figure 8). In this study, it was confirmed that both N- and D-isoforms existed *in planta* as evidenced by the skewed ^{13}C isotope ratio in the TOF-MS analysis of the freshly extracted plant material (Figure 8(A)). Further MS analysis revealed the cyclization site as the location of the amino acid substitution.

In contrast, in recent work in our laboratory examining a Viola species containing kalata S and cycloviolacin O12 among other cyclotides, masses corresponding only to the Asn isoform were detected in the freshly extracted plant material (Figure 9(A)). After chemical and enzymatic treatment, the masses of several cyclotides shifted by either 349 Da (reduced and alkylated peptides, Figure 9(B)) or 367 Da (after trypsin digestion, Figure 9(C)) rather than the expected molecular weight increases of 348 and 366 Da, respectively, indicating that complete deamidation of the loop 6 Asn had occurred. A similar phenomenon was observed in extracts of the Ethiopian plant species *Viola abyssinica* (Yeshak et al., 2011). While determining the sequences of the novel cyclotides, it was observed that the mass of Vaby C increased by an additional 2 Da after reduction and alkylation. MS/MS analyses of the enzymatic digests of Vaby C revealed that the site of modification was the Asn residues in loops 3 and 6. Non-enzymatic deamidation of Asn to Asp in peptides often occurs and is promoted by alkaline conditions, particularly when an Asn is followed by either a Gly or, to a lesser extent, Thr (Bischoff & Kolbe, 1994) as was the case for both Asn in Vaby C.

5.3 Unknown Genomes/Incomplete DBs

Automated database searching requires that the database to be interrogated includes the sequence of the peptides present in the sample. In the case of the vast majority of cyclotide-containing plants, the genomes remain to be sequenced or are, at best, only partially sequenced. For this reason,

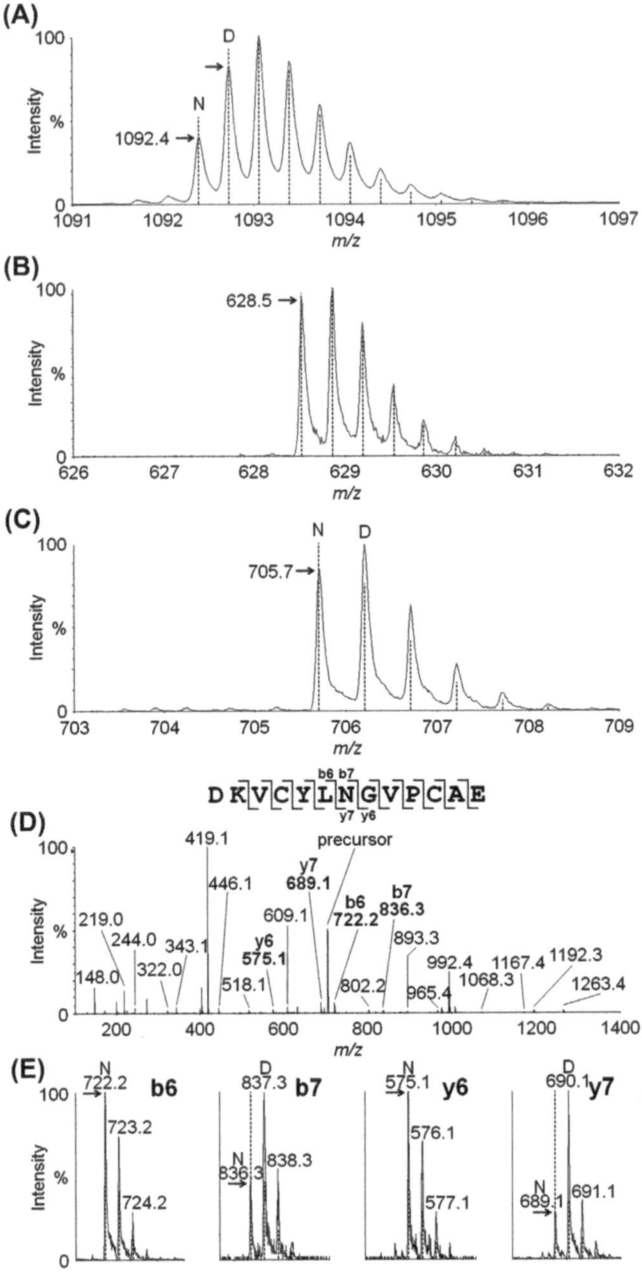

Figure 8 ***MS analysis of Cter B from*** **Clitoria ternatea** ***reveals the presence of both N- and D-isoforms.*** Nanospray spectra for reduced and digested (trypsin and endoproteinase Glu-C) Cter B. (A) TOF-MS spectrum of full-length linearized Cter B precursor 3274.2 Da. (B) TOF-MS spectrum of Cter B digest product with precursor 628.5^{3+} (1882.6 Da). (C) TOF-MS spectrum of Cter B digest product with precursor 705.7^{2+}

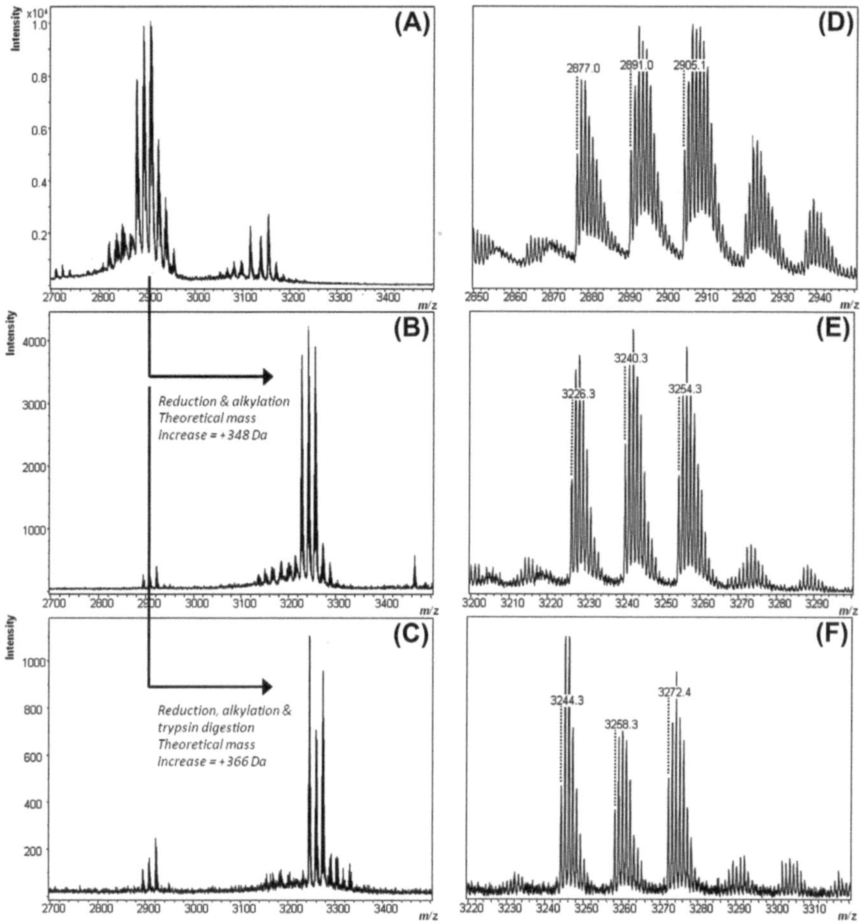

Figure 9 ***MS analysis of cyclotides from*** **Viola** ***species shows the conversion of Asn to Asp.*** MALDI-TOF-MS spectra for native (A); reduced and alkylated (B); and trypsin-digested (C) peptides. The expanded TOF-MS spectra are shown in panels (D–F), respectively, and show the increased mass indicating non-enzymatic deamidation of the Asn in loop 6.

(1409.4 Da). (D) Full product ion spectrum of precursor *m/z* 705.7^{2+} (1409.4 Da). Sequence ions shown in bold represent cleavage of the amide bonds either side of the amino acid at position 7. (E) Isotopic distributions of diagnostic fragment ions b6, b7, y6, and y7 indicate the presence of both Asn and Asp at position 7 and thus the heterogeneous nature of the selected precursor ion within the transmission window. Dotted lines illustrate the theoretical isotopic distributions for precursor and fragment ions, assuming that the residue at position 7 is an asparagine. Arrows (E) indicate the observed intensities of labeled monoisotopic peaks. *Reprinted with permission from Poth et al. (2011). Copyright 2011, American Chemical Society.*

automated database searching is only able to efficiently identify and characterize peptides that have been previously characterized (generated primarily by de novo sequencing efforts).

6. FUTURE PROSPECTS

6.1 Transcriptomics and Bioinformatics

As described above, the application of NGS has great potential in fast tracking the discovery of cyclotides by providing the foundation for automated peptide sequencing. However, the analysis of publicly available transcriptome data sets may act as intermediate in the absence of complete genomes. A recent study by Koehbach et al. (2013) utilized the transcriptome data sets provided by the 1 KP consortium that includes data from over 1000 plant species. Homology searching using a tBLASTn (nucleotide level) algorithm revealed nine precursor proteins from *Carapichea ipecacuanha* encoding five novel putative cyclotide sequences, named caripe 2–6. This approach was supported by traditional de novo sequencing which yielded nine cyclotide-containing plant species from a screen of 296 different plant species, all nine of which belonged to the Rubiaceae plant family. Interestingly, the two approaches employed failed to identify the same peptide sequence(s), a phenomenon often observed when comparing the results of RNA sequencing and peptide sequencing approaches. It is clear that NGS, transcriptomics and genome mining approaches offer great potential in cyclotide discovery, but traditional peptide sequencing is a complementary approach. Clear advantages of applying bioinformatic approaches include the ability to unambiguously identify isobaric amino acids (Ile/Leu, Gln/Lys), to distinguish similar amino acids (Asn/Asp) and to quickly identify candidate plant species for further analysis. Analytical procedures to detect, identify and characterize the peptide sequence provide answers to questions that cannot be gleaned from nucleotide level studies: Is the cyclotide expressed? Is the peptide processed to yield the cyclic product? Is the peptide modified? How much of each peptide is present?

Bioinformatic tools will undoubtedly assist in the characterization of cyclic peptides. CyPred, a recently developed bioinformatic tool (Kedarisetti, Mizianty, Kaas, Craik, & Kurgan, 2014), is able to predict whether a given peptide chain is cyclic. The model was designed based on the four well-populated cyclic peptide families: cyclotides, cyclic defensins, circular bacteriocins and trypsin inhibitors. Analysis with CyPred has indicated that over 3500 cyclic peptides may exist in the 642 proteomes examined, which is significantly more than the 600+ cyclic peptides listed in CyBase.

A recent study (Zhang et al., 2015) described the development of a novel time-efficient Perl program, CyPerl for searching cyclotides from predicted open reading frames (ORFs) of the 34 available plant genomes and existing plant protein sequences from GenBank databases. After sequences were isolated from the public databases, they were further analyzed by removing repeats, evaluating their cysteine-distributed regions, and comparing with CyBase-collected cyclotides. Genome-screening using CyPerl and CyExcel, a Microsoft Office-based comparison tool, yielded 186 ORFs containing 145 unique cyclotide analogues from 30 plant genomes tested from 10 plant families. The common bean, *Phaseolus vulgaris*, and corn, *Zea mays*, were the two richest species containing cyclotide analogues in the plants tested. Furthermore, screening of plant protein databases yielded 266 unique cyclotides and analogues from seven plant families. Overall, when combined with the 288 unique CyBase-listed cyclotides, 510 unique cyclotides and analogues were obtained from 13 plant families. In total, seven novel plant families containing cyclotide analogues and 202 novel cyclotide analogues were identified in this study. The results of this study provide the impetus for future peptide screening specifically targeting the potential cyclotide-containing plant species identified using the CyPerl and CyExcel tools.

6.2 Automation and Robotics

One of the major bottlenecks to peptide discovery is sample processing, requiring either extensive purification and/or multiple sample processing steps prior to analysis. These can be minimized through the use of robotic platforms that enable sample handling to occur in an automated and reproducible fashion reducing the need for manual handling. Raw materials, chemicals, and enzymes can be delivered by such robotic platforms greatly increasing the throughput for peptide extraction, processing and/or purification. Peptides may additionally be digested online using immobilized enzyme columns.

6.3 Advances in MS Hardware

Advances in hardware have seen high-resolution mass spectrometers (where resolution is the ability to separate two narrow width mass spectral peaks) becoming more affordable and available. Combined with high mass accuracy (the ability to measure the true mass of the analyte) the state-of-the-art mass spectrometers allow delineation of peptides of similar mass. For example, cyclotides possessing a phenylalanine or oxidized methionine (both bearing a residue mass of 147 amu) can be distinguished as is the case with vhl-1 from *Viola hederaceae* (Chen et al., 2005). Likewise, peptides that differ by a single mass unit, as would be the case for Asn/Asp isoforms, can be distinguished.

6.4 Advances in MS Software

Significant advances in both the manner in which data are acquired and the way it is interpreted will pave the way for accelerated discovery of natural compounds. For example, SWATH (sequential window acquisition of all theoretical fragment-ion spectra) (Gillet et al., 2012) is a data acquisition strategy that is capable of comprehensive spectral library generation. SWATH-MS is a data-independent acquisition method which aims to complement traditional mass spectrometry-based proteomics techniques. In essence, it allows a complete and permanent recording of all fragment ions of the detectable peptide precursors present in a biological sample. It combines the advantages of shotgun proteomics (high throughput) with those of targeted proteomics (high reproducibility and consistency). In SWATH analysis, the mass spectrometer steps within 2–4 s cycle time through a set of precursor acquisition windows designed to cover the *m/z* range 400–1200 (the typical mass range covered by a quadrupole mass analyzer). During each cycle, the mass spectrometer fragments all precursors within the defined quadrupole isolation window. For example, if a 25 Da window is selected then all precursors of *m/z* 400–425 will be fragmented generating a high-accuracy fragment-ion spectrum of all precursors in that isolation window before moving on to the next 25 Da window. The MS cycles through the SWATH windows during the chromatographic separation, providing a time-resolved recording of the fragment ions of all the peptide precursors that elute on the chromatography. The SWATH-MS data consist of highly multiplexed fragment ion maps that can be mined retrospectively. Data acquisition strategies such as SWATH-MS have not yet been applied to cyclotide sequencing and/or quantification.

The development of an automated database search methodology for cyclopeptides shows potential for the discovery of novel cyclic peptides, despite not yet being applied to the analysis of cyclotides (those of mass 2800–3500 Da). The tool called Cycloquest (Mohimani et al., 2011) uses a four step process to convert MS/MS spectra into candidate peptide sequences: (1) filtering the database by mass (as done for conventional search algorithms); (2) constructing the theoretical spectra for candidate peptides; (3) scoring the theoretical spectra against experimental spectra; and (4) listing the top scoring peptide spectrum matches. Cycloquest was successfully applied to the analysis and identification of two cyclic peptides from sunflower (SFTI-1 and SFT-L1), a bacterial cyclic peptide (sporulation killing factor) and a mammalian θ-defensin.

7. CONCLUDING REMARKS

The number of sequenced cyclotides has expanded dramatically from 4 in 1994, ~50 in 2004, >280 in 2014. This has been driven by improved methodologies for extraction and purification, but primarily by advances in sequencing approaches. Mass spectrometry has played a key role and will continue to into the future, but will be complemented by the "omics" technologies and bioinformatic tools. While cyclotides present a number of unique challenges in their analysis, many of these have been overcome by simple, but elegant sample processing workflows and sophisticated data acquisition and interpretation strategies.

REFERENCES

Arabidopsis-Genome-Initiative. (2000). Analysis of the genome sequence of the flowering plant *Arabidopsis thaliana*. *Nature, 408*(6814), 796–815.

Biemann, K. (January 01, 1990). Appendix 5. Nomenclature for peptide fragment ions (positive ions). In J. A. McCloskey (Ed.), *Methods in enzymology* (Vol. 193, pp. 886–887). San Diego: Elsevier.

Bischoff, R., & Kolbe, H. V. (1994). Deamidation of asparagine and glutamine residues in proteins and peptides: structural determinants and analytical methodology. *Journal of Chromatography B: Biomedical Sciences and Application, 662*(2), 261–278.

Broussalis, A. M., Goransson, U., Coussio, J. D., Ferraro, G., Martino, V., & Claeson, P. (2001). First cyclotide from *Hybanthus* (Violaceae). *Phytochemistry, 58*(1), 47–51.

Burman, R., Gruber, C. W., Rizzardi, K., Herrmann, A., Craik, D. J., Gupta, M. P., et al. (2010). Cyclotide proteins and precursors from the genus *Gloeospermum*: filling a blank spot in the cyclotide map of Violaceae. *Phytochemistry, 71*(1), 13–20.

Chan, L. Y., He, W., Tan, N., Zeng, G., Craik, D. J., & Daly, N. L. (2013). A new family of cystine knot peptides from the seeds of *Momordica cochinchinensis*. *Peptides, 39*, 29–35.

Chen, B., Colgrave, M. L., Daly, N. L., Rosengren, K. J., Gustafson, K. R., & Craik, D. J. (2005). Isolation and characterization of novel cyclotides from *Viola hederacea* –solution structure and anti-HIV activity of vhl-1, a leaf-specific expressed cyclotide. *Journal of Biological Chemistry, 280*(23), 22395–22405.

Chen, B., Colurave, M. L., Wang, C. N., & Craik, D. J. (2006). Cycloviolacin H4, a hydrophobic cyclotide from *Viola hederaceae*. *Journal of Natural Products, 69*(1), 23–28.

Claeson, P., Goransson, U., Johansson, S., Luijendijk, T., & Bohlin, L. (1998). Fractionation protocol for the isolation of polypeptides from plant biomass. *Journal of Natural Products, 61*(1), 77–81.

Colgrave, M. L., & Craik, D. J. (2004). Thermal, chemical, and enzymatic stability of the cyclotide kalata B1: the importance of the cyclic cystine knot. *Biochemistry, 43*(20), 5965–5975.

Colgrave, M. L., Jones, A., & Craik, D. J. (2005). Peptide quantification by matrix-assisted laser desorption ionisation time-of-flight mass spectrometry: Investigations of the cyclotide kalata B1 in biological fluids. *Journal of Chromatography A, 1091*(1–2), 187–193.

Colgrave, M. L., Poth, A. G., Kaas, Q., & Craik, D. J. (2010). A new "era" for cyclotide sequencing. *Biopolymers, 94*(5), 592–601.

Conlan, B. F., Colgrave, M. L., Gillon, A. D., Guarino, R., Craik, D. J., & Anderson, M. A. (2012). Insights into processing and cyclization events associated with biosynthesis of the cyclic Peptide kalata B1. *Journal of Biological Chemistry, 287*(33), 28037–28046.

Craik, D. J. (2010). Discovery and applications of the plant cyclotides. *Toxicon, 56*(7), 1092–1102.

Craik, D. J., Daly, N. L., Bond, T., & Waine, C. (1999). Plant cyclotides: a unique family of cyclic and knotted proteins that defines the cyclic cystine knot structural motif. *Journal of Molecular Biology, 294*(5), 1327–1336.

Craik, D. J., Henriques, S. T., Mylne, J. S., & Wang, C. K. (2012). Cyclotide isolation and characterization. In J. N. Abelson, & M. I. Simon (Eds.), *Methods in enzymology* (Vol. 516, pp. 37–62). San Diego: Elsevier.

Daly, N. L., Rosengren, K. J., & Craik, D. J. (2009). Discovery, structure and biological activities of cyclotides. *Advanced Drug Delivery Reviews, 61*(11), 918–930.

Dutton, J. L., Renda, R. F., Waine, C., Clark, R. J., Daly, N. L., Jennings, C. V., et al. (2004). Conserved structural and sequence elements implicated in the processing of gene-encoded circular proteins. *Journal of Biological Chemistry, 279*(45), 46858–46867.

Edman, P. (1950). Method for determination of the amino acid sequence in peptides. *Acta Chemica Scandinavica, 4*, 283–293.

Gerlach, S. L., Burman, R., Bohlin, L., Mondal, D., & Goransson, U. (2010). Isolation, characterization, and bioactivity of cyclotides from the Micronesian plant *Psychotria leptothyrsa*. *Journal of Natural Products, 73*(7), 1207–1213.

Gillet, L. C., Navarro, P., Tate, S., Rost, H., Selevsek, N., Reiter, L., et al. (2012). Targeted data extraction of the MS/MS spectra generated by data-independent acquisition: a new concept for consistent and accurate proteome analysis. *Molecular & Cellular Proteomics, 11*(6). O111.016717.

Gillon, A. D., Saska, I., Jennings, C. V., Guarino, R. F., Craik, D. J., & Anderson, M. A. (2008). Biosynthesis of circular proteins in plants. *Plant Journal, 53*(3), 505–515.

Good, D. M., Wirtala, M., McAlister, G. C., & Coon, J. J. (2007). Performance characteristics of electron transfer dissociation mass spectrometry. *Molecular & Cellular Proteomics, 6*(11), 1942–1951.

Goransson, U., Broussalis, A. M., & Claeson, P. (2003). Expression of Viola cyclotides by liquid chromatography-mass spectrometry and tandem mass spectrometry sequencing of intercysteine loops after introduction of charges and cleavage sites by aminoethylation. *Analytical Biochemistry, 318*(1), 107–117.

Goransson, U., & Craik, D. J. (2003). Disulfide mapping of the cyclotide kalata B1-Chemical proof of the cyclic cystine knot motif. *Journal of Biological Chemistry, 278*(48), 48188–48196.

Goransson, U., Luijendijk, T., Johansson, S., Bohlin, L., & Claeson, P. (1999). Seven novel macrocyclic polypeptides from *Viola arvensis*. *Journal of Natural Products, 62*(2), 283–286.

Goransson, U., Svangard, E., Claeson, P., & Bohlin, L. (2004). Novel strategies for isolation and characterization of cyclotides: the discovery of bioactive macrocyclic plant polypeptides in the Violaceae. *Current Protein & Peptide Science, 5*(5), 317–329.

Gran, L. (1970). An oxytocic principle found in *Oldenlandia affinis* DC. An indigenous, Congolese drug "Kalata-Kalata" used to accelerate delivery. *Meddelelser fra Norsk Farmaceutisk Selskap, 12*, 173–180.

Gran, L. (1973a). Isolation of oxytoxic peptides from *Oldenlandia affinis* by solvent-extraction of tetraphenylborate complexes and chromatography on sephadex LH-20. *Lloydia-The Journal of Natural Products, 36*(2), 207–208.

Gran, L. (1973b). Oxytocic principles of *Oldenlandia affinis*. *Lloydia, 36*(2), 174–178.

Gruber, C. W., Cemazar, M., Clark, R. J., Horibe, T., Renda, R. F., Anderson, M. A., et al. (2007). A novel plant protein-disulfide isomerase involved in the oxidative folding of cystine knot defense proteins. *Journal of Biological Chemistry, 282*(28), 20435–20446.

Gruber, C. W., Elliott, A. G., Ireland, D. C., Delprete, P. G., Dessein, S., Goransson, U., et al. (2008). Distribution and evolution of circular miniproteins in flowering plants. *Plant Cell, 20*(9), 2471–2483.

Gustafson, K. R., Sowder, R. C., Henderson, L. E., Parsons, I. C., Kashman, Y., Cardellina, J. H., et al. (1994). Circulin-A and circulin-B—novel HIV-inhibitory macrocyclic peptides from the tropical tree *Chassalia parvifolia*. *Journal of the American Chemical Society, 116*(20), 9337–9338.

Gustafson, K. R., Walton, L. K., Sowder, R. C., Jr., Johnson, D. G., Pannell, L. K., Cardellina, J. H., Jr., et al. (2000). New circulin macrocyclic polypeptides from *Chassalia parvifolia*. *Journal of Natural Products, 63*(2), 176–178.

Hallock, Y. F., Sowder, R. C., 2nd, Pannell, L. K., Hughes, C. B., Johnson, D. G., Gulakowski, R., et al. (2000). Cycloviolins A-D, anti-HIV macrocyclic peptides from *Leonia cymosa*. *Journal of Organic Chemistry, 65*(1), 124–128.

Hashempour, H., Ghassempour, A., Daly, N. L., Spengler, B., & Rompp, A. (2011). Analysis of cyclotides in *Viola ignobilis* by Nano liquid chromatography fourier transform mass spectrometry. *Protein and Peptide Letters, 18*(7), 747–752.

Hashempour, H., Koehbach, J., Daly, N. L., Ghassempour, A., & Gruber, C. W. (2013). Characterizing circular peptides in mixtures: sequence fragment assembly of cyclotides from a violet plant by MALDI-TOF/TOF mass spectrometry. *Amino Acids, 44*(2), 581–595.

He, W., Chan, L. Y., Zeng, G., Daly, N. L., Craik, D. J., & Tan, N. (2011). Isolation and characterization of cytotoxic cyclotides from *Viola philippica*. *Peptides, 32*(8), 1719–1723.

Herrmann, A., Burman, R., Mylne, J. S., Karlsson, G., Gullbo, J., Craik, D. J., et al. (2008). The alpine violet, *Viola biflora*, is a rich source of cyclotides with potent cytotoxicity. *Phytochemistry, 69*(4), 939–952.

Hillenkamp, F., Karas, M., Beavis, R. C., & Chait, B. T. (1991). Matrix-assisted laser desorption/ionization mass spectrometry of biopolymers. *Analytical Chemistry, 63*(24), 1193A–1203A.

International-Rice-Genome-Sequencing-Project. (2005). The map-based sequence of the rice genome. *Nature, 436*(7052), 793–800.

Ireland, D. C., Colgrave, M. L., & Craik, D. J. (2006). A novel suite of cyclotides from *Viola odorata*: sequence variation and the implications for structure, function and stability. *Biochemical Journal, 400*, 1–12.

Ireland, D. C., Colgrave, M. L., Nguyencong, P., Daly, N. L., & Craik, D. J. (2006). Discovery and characterization of a linear cyclotide from *Viola odorata*: Implications for the processing of circular proteins. *Journal of Molecular Biology, 357*(5), 1522–1535.

Jennings, C., West, J., Waine, C., Craik, D., & Anderson, M. (2001). Biosynthesis and insecticidal properties of plant cyclotides: the cyclic knotted proteins from *Oldenlandia affinis*. *Proceedings of the National Academy of Sciences of the United States of America, 98*(19), 10614–10619.

Johnson, R. S., & Biemann, K. (1989). Computer program (SEQPEP) to aid in the interpretation of high-energy collision tandem mass spectra of peptides. *Biomedical & Environmental Mass Spectrometry, 18*(11), 945–957.

Karas, M., & Hillenkamp, F. (1988). Laser desorption ionization of proteins with molecular masses exceeding 10,000 daltons. *Analytical Chemistry, 60*(20), 2299–2301.

Kedarisetti, P., Mizianty, M. J., Kaas, Q., Craik, D. J., & Kurgan, L. (2014). Prediction and characterization of cyclic proteins from sequences in three domains of life. *Biochimica et Biophysica Acta, 1844*(1 Pt B), 181–190.

Kelleher, N. L. (2004). Top-down proteomics. *Analytical Chemistry, 76*(11), 197A–203A.

Koehbach, J., Attah, A. F., Berger, A., Hellinger, R., Kutchan, T. M., Carpenter, E. J., et al. (2013). Cyclotide discovery in Gentianales revisited-identification and characterization of cyclic cystine-knot peptides and their phylogenetic distribution in Rubiaceae plants. *Biopolymers, 100*(5), 438–452.

Meyers, L. A., & Levin, D. A. (2006). On the abundance of polyploids in flowering plants. *Evolution, 60*(6), 1198–1206.

Ming, R., Hou, S., Feng, Y., Yu, Q., Dionne-Laporte, A., Saw, J. H., et al. (2008). The draft genome of the transgenic tropical fruit tree papaya (*Carica papaya* Linnaeus). *Nature, 452*(7190), 991–996.

Mohimani, H., Liu, W. T., Mylne, J. S., Poth, A. G., Colgrave, M. L., Tran, D., et al. (2011). Cycloquest: identification of cyclopeptides via database search of their mass spectra against genome databases. *Journal of Proteome Research, 10*(10), 4505–4512.

Mulvenna, J. P., Mylne, J. S., Bharathi, R., Burton, R. A., Shirley, N. J., Fincher, G. B., et al. (2006). Discovery of cyclotide-like protein sequences in graminaceous crop plants: ancestral precursors of circular proteins? *Plant Cell, 18*(9), 2134–2144.

Mulvenna, J. P., Wang, C., & Craik, D. J. (2006). CyBase: a database of cyclic protein sequence and structure. *Nucleic Acids Research, 34*, D192–D194.

Mulvenna, J. R., Sando, L., & Craik, D. J. (2005). Processing of a 22 kDa precursor protein to produce the circular protein tricyclon A. *Structure, 13*(5), 691–701.

Nair, S. S., Romanuka, J., Billeter, M., Skjeldal, L., Emmett, M. R., Nilsson, C. L., et al. (2006). Structural characterization of an unusually stable cyclic peptide, kalata B2 from *Oldenlandia affinis*. *Biochimica et Biophysica Acta-Proteins and Proteomics, 1764*(10), 1568–1576.

Nguyen, G. K., Lian, Y., Pang, E. W., Nguyen, P. Q., Tran, T. D., & Tam, J. P. (2013). Discovery of linear cyclotides in monocot plant *Panicum laxum* of Poaceae family provides new insights into evolution and distribution of cyclotides in plants. *Journal of Biological Chemistry, 288*(5), 3370–3380.

Nguyen, G. K., Zhang, S., Nguyen, N. T., Nguyen, P. Q., Chiu, M. S., Hardjojo, A., et al. (2011). Discovery and characterization of novel cyclotides originated from chimeric precursors consisting of albumin-1 chain a and cyclotide domains in the Fabaceae family. *Journal of Biological Chemistry, 286*(27), 24275–24287.

Nguyen, G. K., Zhang, S., Wang, W., Wong, C. T., Nguyen, N. T., & Tam, J. P. (2011). Discovery of a linear cyclotide from the bracelet subfamily and its disulfide mapping by top-down mass spectrometry. *Journal of Biological Chemistry, 286*(52), 44833–44844.

Pellicer, J., Fay, M. F., & Leitch, I. J. (2010). The largest eukaryotic genome of them all? *Botanical Journal of the Linnean Society, 164*(1), 10–15.

Pinto, M. F., Fensterseifer, I. C., Migliolo, L., Sousa, D. A., de Capdville, G., Arboleda-Valencia, J. W., et al. (2012). Identification and structural characterization of novel cyclotide with activity against an insect pest of sugar cane. *Journal of Biological Chemistry, 287*(1), 134–147.

Plan, M. R. R., Goransson, U., Clark, R. J., Daly, N. L., Colgrave, M. L., & Craik, D. J. (2007). The cyclotide fingerprint in *Oldenlandia affinis*: elucidation of chemically modified, linear and novel macrocyclic peptides. *Chembiochem, 8*(9), 1001–1011.

Poth, A. G., Colgrave, M. L., Lyons, R. E., Daly, N. L., & Craik, D. J. (2011). Discovery of an unusual biosynthetic origin for circular proteins in legumes. *Proceedings of the National Academy of Sciences of the United States of America, 108*(25), 10127–10132.

Poth, A. G., Colgrave, M. L., Philip, R., Kerenga, B., Daly, N. L., Anderson, M. A., et al. (2011). Discovery of cyclotides in the fabaceae plant family provides new insights into the cyclization, evolution, and distribution of circular proteins. *ACS Chemical Biology, 6*(4), 345–355.

Poth, A. G., Mylne, J. S., Grassl, J., Lyons, R. E., Millar, A. H., Colgrave, M. L., et al. (2012). Cyclotides associate with leaf vasculature and are the products of a novel precursor in petunia (Solanaceae). *Journal of Biological Chemistry, 287*(32), 27033–27046.

Qin, Q., McCallum, E. J., Kaas, Q., Suda, J., Saska, I., Craik, D. J., et al. (2010). Identification of candidates for cyclotide biosynthesis and cyclisation by expressed sequence tag analysis of *Oldenlandia affinis*. *BMC Genomics, 11*, 111.

Roepstorff, P., & Fohlman, J. (1984). Proposal for a common nomenclature for sequence ions in mass spectra of peptides. *Biomedical Mass Spectrometry, 11*(11), 601.

Saether, O., Craik, D. J., Campbell, I. D., Sletten, K., Juul, J., & Norman, D. G. (1995). Elucidation of the primary and 3-dimensional structure of the uterotonic polypeptide kalata B1. *Biochemistry, 34*(13), 4147–4158.

Saska, I., Colgrave, M. L., Jones, A., Anderson, M. A., & Craik, D. J. (2008). Quantitative analysis of backbone-cyclised peptides in plants. *Journal of Chromatography B Analytical Technologies in the Biomedical and Life Sciences, 872*(1–2), 107–114.

Saska, I., Gillon, A. D., Hatsugai, N., Dietzgen, R. G., Hara-Nishimura, I., Anderson, M. A., et al. (2007). An asparaginyl endopeptidase mediates in vivo protein backbone cyclization. *Journal of Biological Chemistry, 282*(40), 29721–29728.

Schatz, M. C., Witkowski, J., & McCombie, W. R. (2012). Current challenges in de novo plant genome sequencing and assembly. *Genome Biology, 13*(4), 243.

Schnable, P. S., Ware, D., Fulton, R. S., Stein, J. C., Wei, F., Pasternak, S., et al. (2009). The B73 maize genome: complexity, diversity, and dynamics. *Science, 326*(5956), 1112–1115.

Schöpke, T., Hasan Agha, M. I., Kraft, R., Otto, A., & Hiller, K. (1993). Hämolytisch aktive komponenten aus *Viola tricolor L.* und *Viola arvensis Murray*. *Scientia Pharmaceutica, 61*, 145–153.

Shilov, I. V., Seymour, S. L., Patel, A. A., Loboda, A., Tang, W. H., Keating, S. P., et al. (2007). The Paragon Algorithm, a next generation search engine that uses sequence temperature values and feature probabilities to identify peptides from tandem mass spectra. *Molecular & Cellular Proteomics, 6*(9), 1638–1655.

Shulaev, V., Sargent, D. J., Crowhurst, R. N., Mockler, T. C., Folkerts, O., Delcher, A. L., et al. (2011). The genome of woodland strawberry (*Fragaria vesca*). *Nature Genetics, 43*(2), 109–116.

Simonsen, S. M., Sando, L., Ireland, D. C., Colgrave, M. L., Bharathi, R., Goransson, U., et al. (2005). A continent of plant defense peptide diversity: cyclotides in Australian Hybanthus (Violaceae). *Plant Cell, 17*(11), 3176–3189.

Sletten, K., & Gran, L. (1973). Some molecular properties of kalatapeptide B-1. *Meddelelser fra Norsk Farmaceutisk Selskap, 35*, 69–82.

Steen, H., & Mann, M. (2004). The ABC's (and XYZ's) of peptide sequencing. *Nature Reviews Molecular Cell Biology, 5*(9), 699–711.

Svangard, E., Goransson, U., Smith, D., Verma, C., Backlund, A., Bohlin, L., et al. (2003). Primary and 3-D modelled structures of two cyclotides from *Viola odorata*. *Phytochemistry, 64*(1), 135–142.

Syka, J. E., Coon, J. J., Schroeder, M. J., Shabanowitz, J., & Hunt, D. F. (2004). Peptide and protein sequence analysis by electron transfer dissociation mass spectrometry. *Proceedings of the National Academy of Sciences of the United States of America, 101*(26), 9528–9533.

Sze, S. K., Wang, W., Meng, W., Yuan, R. D., Guo, T. N., Zhu, Y., et al. (2009). Elucidating the structure of cyclotides by partial acid hydrolysis and LC-MS/MS analysis. *Analytical Chemistry, 81*(3), 1079–1088.

Tang, J., Wang, C. K., Pan, X., Yan, H., Zeng, G., Xu, W., et al. (2010). Isolation and characterization of cytotoxic cyclotides from Viola tricolor. *Peptides, 31*(8), 1434–1440.

The-Tomato-Genome-Consortium. (2012). The tomato genome sequence provides insights into fleshy fruit evolution. *Nature, 485*(7400), 635–641.

Trabi, M., Mylne, J. S., Sando, L., & Craik, D. J. (2009). Circular proteins from *Melicytus* (Violaceae) refine the conserved protein and gene architecture of cyclotides. *Organic & Biomolecular Chemistry, 7*(11), 2378–2388.

Varshney, R. K., Chen, W., Li, Y., Bharti, A. K., Saxena, R. K., Schlueter, J. A., et al. (2012). Draft genome sequence of pigeonpea (*Cajanus cajan*), an orphan legume crop of resource-poor farmers. *Nature Biotechnology, 30*(1), 83–89.

Velasquez, J. E., & van der Donk, W. A. (2011). Genome mining for ribosomally synthesized natural products. *Current Opinion in Chemical Biology, 15*(1), 11–21.

Wang, C. K., Kaas, Q., Chiche, L., & Craik, D. J. (2008). CyBase: a database of cyclic protein sequences and structures, with applications in protein discovery and engineering. *Nucleic Acids Research, 36*(Database issue), D206–D210.

Wilm, M., & Mann, M. (1996). Analytical properties of the nanoelectrospray ion source. *Analytical Chemistry, 68*(1), 1–8.

Witherup, K. M., Bogusky, M. J., Anderson, P. S., Ramjit, H., Ransom, R. W., Wood, T., et al. (1994). Cyclopsychotride-A, a biologically-active, 31-residue cyclic peptide isolated from *Psychotria longipes*. *Journal of Natural Products-Lloydia, 57*(12), 1619–1625.

Xiang, B., Du, G. H., Wang, X. C., Zhang, S. X., Qin, X. Y., Kong, J. Q., et al. (2010). Elucidating the structure of two cyclotides of *Viola tianschanica* maxim by MALDI TOF/TOF MS analysis. *Yao Xue Xue Bao, 45*(11), 1402–1409.

Yamashita, M., & Fenn, J. B. (1984). Electrospray ion source. Another variation on the free-jet theme. *The Journal of Physical Chemistry, 88*(20), 4451–4459.

Yeshak, M. Y., Burman, R., Asres, K., & Goransson, U. (2011). Cyclotides from an extreme habitat: characterization of cyclic peptides from *Viola abyssinica* of the Ethiopian highlands. *Journal of Natural Products, 74*(4), 727–731.

Zhang, J., Hu, M., Li, J. T., Guan, J. P., Yang, B., Shu, W. S., et al. (2009). A transcriptional profile of metallophyte *Viola baoshanensis* involved in general and species-specific cadmium-defense mechanisms. *Journal of Plant Physiology, 166*(8), 862–870.

Zhang, J., Hua, Z., Huang, Z., Chen, Q., Long, Q., Craik, D. J., et al. (2015). Two Blast-independent tools, CyPerl and CyExcel, for harvesting hundreds of novel cyclotides and analogues from plant genomes and protein databases. *Planta, 4*(241), 929–940.

Zhang, J., Liao, B., Craik, D. J., Li, J. T., Hu, M., & Shu, W. S. (2009). Identification of two suites of cyclotide precursor genes from metallophyte *Viola baoshanensis*: cDNA sequence variation, alternative RNA splicing and potential cyclotide diversity. *Gene, 431*(1–2), 23–32.

CHAPTER SIX

Structural Studies of Cyclotides

Norelle L. Daly*,[1], K. Johan Rosengren§,[1]
*Centre for Biodiscovery and Molecular Development of Therapeutics, AITHM, James Cook University, Cairns, QLD, Australia
§The University of Queensland, School of Biomedical Sciences, Brisbane, QLD, Australia
[1]Corresponding authors: E-mail: norelle.daly@jcu.edu.au; j.rosengren@uq.edu.au

Contents

Abstract

Over the last two decades, the cyclotides have generated widespread interest both in the plant biology and peptide communities for a number of reasons, including their physiological roles in plants, evolution, and wide range of biological activities with potential therapeutic applications. But perhaps most of all it is their unique structure with a cyclic cystine knot motif that provide remarkable features such as resistance to chemical and biological degradation, which have caught researchers' attention. In this chapter, we summarize the key studies that have contributed to our understanding of the cyclotide fold, the structural features of these peptides, and the role and utilization of these features for native and engineered biological functions.

Advances in Botanical Research, Volume 76
ISSN 0065-2296
http://dx.doi.org/10.1016/bs.abr.2015.09.003

1. THE CYCLIC CYSTINE KNOT

The cyclic cystine knot (CCK) is a unique structural motif found only in the plant cyclotides (Craik, Daly, Bond, & Waine, 1999). It comprises an inhibitor cystine knot (ICK) motif (Pallaghy, Nielsen, Craik, & Norton, 1994), combined with a cyclic peptide backbone, as highlighted in Figure 1(A). In contrast to the CCK motif, which is only found in cyclotides, the ICK motif is present in a range of species, including venomous creatures, fungi, and plants (Craik, Daly, & Waine, 2001). Both the CCK and ICK motifs comprise three disulfide bonds, two of which (CysI–CysIV and CysII–CysV) form a ring through which the third bond (CysIII–CysVI) threads. The knotted arrangement of the disulfide bonds coupled with the cyclic backbone generally results in CCK peptides having well-defined structures (Rosengren, Daly, Plan, Waine, & Craik, 2003). This structural motif appears to have significant advantages in terms of stability and function, but has led to challenges in structure elucidation, as outlined in this section.

The original cyclotide discovered, kalata B1, was isolated in the 1970s (Gran, 1973) but it was not until 1995 that the full sequence and structure were published (Saether et al., 1995). The structure was determined using NMR spectroscopy; a technique that is highly effective for the structural analysis of disulfide-rich peptides such as the cyclotides, which are often highly soluble and with well-defined structures that result in high-quality

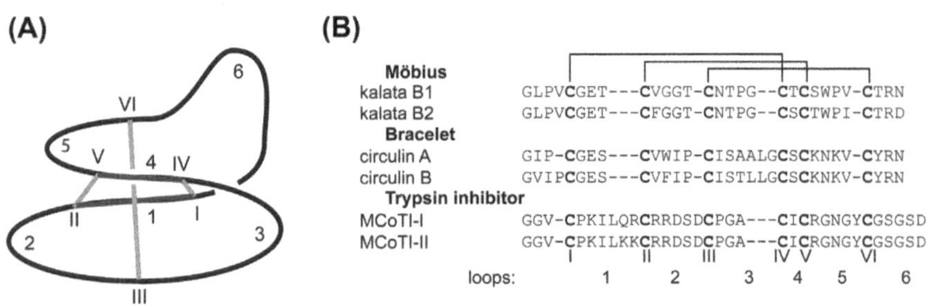

Figure 1 ***Schematic representation of the (A) Cyclic Cystine Knot (CCK) motif and (B) representative sequences of cyclotides from the different subclasses.*** The CysI–CysIV and CysII–CysV disulfide bonds create, together with their interconnecting peptide backbone, a ring that is penetrated by the CysIII–CysVI disulfide bond. Sequences are shown for typical Möbius, bracelet, and trypsin inhibitor cyclotides, illustrating the differences in size and character of the various "loops." Cysteine residues are labeled with Roman numbers and the intercystine segments referred to as "loops" are labeled with Arabic numbers.

NMR data (Marx, Daly, & Craik, 2006). However, when the structure of kalata B1 was first published the disulfide connectivity could not be conclusively determined (Saether et al., 1995). Two approaches were used to analyze the disulfide connectivity in this initial study. The first approach involved analysis of a family of structures that were calculated with Nuclear Overhauser Effect (NOE)-derived distance restraints but no disulfide bond restraints. The second approach involved calculating families of structures with the covalent geometry of each of the 15 theoretically possible disulfide connectivities enforced. Analysis of the distances between sulfur atoms, and how well the structures satisfy the experimental restraints suggested that the most likely connectivity was CysI–CysIV, CysII–CysV, and CysIII–CysVI. This connectivity was consistent with the ICK motif present in peptides such as the squash trypsin inhibitors (Nielsen, Alewood, Andrews, Kent, & Craik, 1994). However, three other connectivities were also considered to be possible, due to the clustering of the disulfide bonds at the core of the molecule making it difficult to distinguish between the connectivities.

Later an independent study on the structure of kalata B1 suggested the alternative disulfide connectivity CysI–CysVI, CysII–CysV, and CysIII–CysIV (Skjeldal, Gran, Sletten, & Volkman, 2002). The CysII–CysV bond is consistent with the Saether et al. (1995) study, but the other two bonds combine to form a topologically distinct "ladder" arrangement of the disulfide bonds. The study by Skjeldal (Skjeldal et al., 2002) also involved the use of NMR spectroscopy, albeit at a higher field than the Saether et al. (1995) study and involved a similar approach for analyzing the disulfide bonds. Structures were calculated either without disulfide bond restraints, or with the 15 possible disulfide connectivities. The target function used to analyze how well the structures satisfied the distance restraints was significantly lower for the CysI–CysVI, CysII–CysV, and CysIII–CysIV connectivity compared to the other 14 possibilities. Furthermore, analysis of distances between sulfur atoms in the structures calculated without disulfide bond restraints was also consistent with this alternative connectivity. These results highlighted the challenges involved in assigning disulfide connectivities in cyclotides using indirect methods such as solution NMR. Given that two different connectivities had been proposed for kalata B1, additional studies were carried out in an attempt to resolve this discrepancy.

Mapping disulfide bonds in tightly folded, disulfide-rich peptides can be difficult as these peptides are generally resistant to enzymatic cleavage and therefore small fragments required for analysis cannot be easily obtained. To circumvent this problem, chemical and structural studies have been used in the analysis of the disulfide bonds of kalata B1 and related cyclotides. A mass spectrometry study on circulins A and B (Derua, Gustafson, & Pannell, 1996), which involved partial acid hydrolysis, suggested that the CCK motif was indeed present in these peptides rather than the alternative connectivity suggested by Skjeldal et al. (2002) for kalata B1. A synthetic study on circulin A and cyclopsychotride A involved the explicit formation of a single disulfide bond (Tam, Lu, Yang, & Chiu, 1999). However, this bond was between CysII–CysIV, which is present in both suggested connectivities, and therefore did not provide support for either connectivity. Direct chemical evidence of the CCK motif in kalata B1 was found when Göransson and Craik (2003) used selective reduction and stepwise alkylation of the cysteine residues. The reduction was done at low pH to prevent disulfide bond shuffling. Peptides that contained either a single disulfide bond or two disulfide bonds were present. The selectively reduced species were alkylated, cleaved, and then analyzed by mass spectrometry. This analysis unequivocally showed that the CCK motif was present in kalata B1.

The CCK motif was also confirmed by subsequent structural studies using both NMR spectroscopy and X-ray crystallography. The use of side chain dihedral angle restraints allowed the determination of a high-resolution structure of kalata B1 using NMR spectroscopy (Rosengren et al., 2003). Structures were calculated using both the CCK and laddered arrangement of the disulfide bonds. Several of the experimentally derived angle restraints were not consistent with the laddered arrangement, in contrast to the CCK arrangement. The final piece of evidence for the CCK fold came from the determination of the crystal structure of varv F, a cyclotide isolated from *Viola arvensis*, which confirmed the cystine knot motif was present (Wang, Hu, et al., 2009). Although it has been suggested that the disulfide connectivity might vary between different cyclotides (Skjeldal et al., 2002), the studies on kalata B1 and related cyclotides indicate that the overall fold and the disulfide connectivity is conserved in the plant cyclotides.

2. STRUCTURAL FEATURES OF CYCLOTIDES

2.1 Structural Studies

The unique features of cyclotides have generated significant interest in their structures, and since the initial work by Saether et al. considerable amounts of structural data have been generated for a wide range of naturally occurring cyclotides, and designed analogues. As of August 2015, more than 40 structures of native and engineered cyclotides have been deposited to the protein data bank and these are summarized in Table 1. From here it is clear that by far the most commonly used technique for studying cyclotides is solution NMR spectroscopy. Their highly ordered structure generally results in both sharp lines and excellent signal dispersion, allowing complete resonance assignments based solely on two-dimensional proton NMR data (Daly, Rosengren, Henriques, & Craik, 2011). Consequently, most studies have been performed directly on material isolated from plants or from synthetic sources, without the need for isotopic labeling. Not requiring isotopic labeling is a distinct advantage, as recombinant production of cyclotides and isotopic labeling in *Escherichia coli* are not straightforward because of the cyclic backbone. It has been shown that it is possible to grow cyclotide-producing plants on agar supplemented with ^{15}N-enriched salts to generate peptides for NMR studies (Mylne & Craik, 2008), but this is a relatively expensive approach that is not generally applicable to low-abundance cyclotides. NMR spectroscopy has the advantage that different solvent systems, including ligands, membrane-mimicking micelle systems, and a wide range of pH and temperatures, can be utilized. It also provides details about flexibility and disorder in structures, and in the sections below we will provide examples of where these different approaches have been utilized.

More recently X-ray crystallography has also started to contribute to our understanding of these molecules. As noted above, the first crystallographic study of a cyclotide was of varv F (Wang, Hu, et al., 2009). Applying both crystallography and solution NMR to the same peptide allowed an assessment of the validity of both techniques in terms of how well they represent the true bioactive structure of cyclotides. The solution and crystal structures are remarkably similar, with minimal differences observed. This highlights the rigidity of the fold, which due to its conformational restriction appears to be unaffected by crystal packing forces (Wang, Hu, et al., 2009). Small disulfide-rich proteins are generally difficult to crystallize, and significant efforts into crystallizing various cyclotides were required before the final success with varv F. However, recent advances utilizing racemic and quasi-racemic strategies promises to significantly improve this (Wang, King, Northfield, Ojeda, & Craik, 2014). Racemic crystallization is based on the idea that a racemic mixture of synthetic mirror-image proteins made from either all L or all D amino acids are more likely to crystallize as they have access to a larger range of crystal space groups. Using this approach

Table 1 Cyclotide and Kalata Related Structures in the Protein Data Bank (August 2015)

Cyclotide	Family	Data	PDB	Comment	References
Kalata B1	Möbius	NMR	1KAL	First cyclotide structure and proof of the cyclic backbone of kalata B1.	Saether et al. (1995)
Kalata B1	Möbius	NMR	1JJZ	Structure highlighting ambiguities and questioning CCK disulfide connectivities.	Skjeldal et al. (2002)
Kalata B1	Möbius	NMR	1NB1	High-resolution structure of kalata B1 and NMR evidence of the CCK motif.	Rosengren et al. (2003)
Kalata B1	Möbius	NMR	1ZNU	First structure in membrane-like environment—DPC micelles.	Shenkarev et al. (2006)
Kalata B1 (all-D)	Möbius	NMR	2JUE	Synthetic all-D amino acid mirror image variant of kalata B1.	Sando et al. (2011)
Kalata B1 (des 24–28)	Möbius	NMR	1ORX	Synthetic linear variant of kalata B1 with loop 6 removed.	Barry, Daly, Clark, Sando, and Craik (2003)
Kalata B (linear)	Möbius	NMR	2KHB	Synthetic variant of kalata B1 with loop 6 broken used for quartz crystal microbalance and neutron scattering.	Wang et al. (2012)
Kalata B1 (C1,15A)	Möbius	NMR	1N1U	Synthetic 2-disulfide variant of kalata B1 with the Cys1–Cys15 bond removed.	Daly, Clark, and Craik (2003)
Kalata B1 (P20D, V21K)	Möbius	NMR	2F2I	Engineered variant of kalata B1 with the Möbius defining *cis*-Pro removed.	Clark et al. (2006)
Kalata B1 (W19K, P20N, V21K)	Möbius	NMR	2F2J	Engineered variant with a bracelet loop 5 inserted into kalata B1.	Clark et al. (2006)
Kalata B1 (23–28; GHRW)	Möbius	NMR	2LUR	Engineered cyclotide with melanocortin agonist activity.	Eliasen et al. (2012)
Kalata B1 (W23WW)	Möbius	NMR	2MN1	Engineered cyclotide for studying membrane targeting in anticancer applications.	Henriques, Huang, Chaousis, Wang, and Craik (2014)

Kalata B1 (30–33; TGGG)	Möbius	NMR	2MH1	Structure of a variant of kalata B1 produced through sortase-mediated cyclization.	Jia et al. (2014)
Kalata B1	Möbius	X-ray	4TTM	X-ray structure of kalata B1 generated from racemic crystallization.	Wang, King, et al. (2014)
Kalata B1 (G6A)	Möbius	X-ray	4TTN	X-ray structure of variant of kalata B1 from quasi-racemic crystallization.	Wang, King, et al. (2014)
Kalata B1 (V25A)	Möbius	X-ray	4TTO	X-ray structure of variant of kalata B1 from quasi-racemic crystallization.	Wang, King, et al. (2014)
Kalata B2	Möbius	NMR	1PT4	Second structure of Möbius cyclotide confirming the *cis*-Pro bond.	Jennings et al. (2005)
Kalata B2	Möbius	NMR	2KCH	Kalata B2 in DPC micelles confirming differences in membrane binding mode.	Wang, Colgrave, et al. (2009)
Kalata B7	Möbius	NMR	2JWM	Kalata B7 in a complex with DPC micelles and Mn^{2+}.	Shenkarev et al. (2008)
Kalata B7	Möbius	NMR	2M9O	Kalata B7 in solution and usage as template for GPCR ligand design.	Koehbach et al. (2013)
Kalata B8	Bracelet	NMR	2B38	Structure of a "hybrid" cyclotide containing both Möbius and bracelet-type loops.	Daly et al. (2006)
Kalata B12	Möbius	NMR	2KVX	Structure of a native cyclotide lacking the conserved Glu in loop 1.	Wang et al. (2011)
Varv F	Möbius	X-ray	3E4H	First crystal structure of a cyclotide, confirming the rigidity of the fold.	Wang, Hu, et al. (2009)
Varv F	Möbius	NMR	2K7G	Solution structure of varv F for direct comparison with the crystal structure.	Wang, Hu, et al. (2009)
Violacin A	Möbius	NMR	2FQA	Structure of a "linear-cyclotide" in which the Asn important for cyclization is absent.	Ireland, Colgrave, Nguyencong, Daly, et al. (2006)

(*Continued*)

Table 1 Cyclotide and Kalata Related Structures in the Protein Data Bank (August 2015)—cont'd

Cyclotide	Family	Data	PDB	Comment	References
Kalata B5	Bracelet	NMR	2KUX	Cyclotide structure confirming role of Glu but with weaker h-bond network interactions.	Plan et al. (2010)
vhl-1	Bracelet	NMR	1ZA8	Structure of a cyclotide with leaf-specific expression.	Chen et al. (2005)
vhl-2	Bracelet	NMR	2KUK	Structure of a cyclotide with leaf-specific expression.	Daly, Chen, Nguyencong, and Craik (2010)
vhr-1	Bracelet	NMR	1VB8	Structure of a cyclotide with root-specific expression.	Trabi and Craik (2004)
Cter-M	Möbius	NMR	2LAM	Structure of a cyclotide with an unusual biosynthetic origin from an albumin.	Poth, Colgrave, Lyons, Daly, and Craik (2011)
Cycloviolacin O1	Bracelet	NMR	1NBJ	High-resolution structure of prototypic bracelet cyclotide cycloviolacin O1.	Rosengren et al. (2003)
Cycloviolacin O2	Bracelet	NMR	2KNM	Structure of the cytotoxic cyclotide cycloviolacin O2.	Göransson et al. (2009)
Cycloviolacin O2 (Glu6-methyl)	Bracelet	NMR	2KNN	Structure of methylated cycloviolacin O2, confirming a structural role of Glu6.	Göransson et al. (2009)
Cycloviolacin O14	Bracelet	NMR	2GJ0	Structure of a cyclotide from the violet *Viola odorata*.	Ireland, Colgrave, and Craik (2006)
Palicourein	Bracelet	NMR	1R1F	Structure of the largest known cyclotide.	Barry et al. (2004)
Tricyclon A	Bracelet	NMR	1YP8	Structure of cyclotide with unusually long loop 6 and lack of hydrophobic patch.	Mulvenna et al. (2005)

MCoTI-II	TI	NMR	1IB9	Structure of CCK trypsin inhibitor related to linear squash trypsin inhibitors.	Felizmenio-Quimio et al. (2001)
MCoTI-II	TI	NMR	1HA9	Structure of CCK trypsin inhibitor related to linear squash trypsin inhibitors.	Heitz et al. (2001)
MCoTI-II-trypsin	TI	X-ray	4GUX	Crystal structure of a macrocyclic trypsin inhibitor in complex with trypsin.	Daly et al. (2013)
MCoTI-V	TI	NMR	2LJS	Structure of a linear trypsin inhibitor from *Momordica*.	Mylne et al. (2012)
MCoTi-II (C4,21A)	TI	NMR	2PO8	Structure of 2-disulfide folding intermediate of a trypsin inhibitor.	Cemazar, Joshi, Daly, Mark, and Craik (2008)
MCoTi-II (linear)	TI	NMR	2IT8	Structure of linearized trypsin inhibitor investigating role of circular backbone.	Heitz et al. (2008)
Mco-PMI	TI	X-ray	2M86	Cyclotide grafted with sequence mimicking p53 in complex with Hdm2.	Ji et al. (2013)

CCK, cyclic cystine knot.

the crystal structure of kalata B1 was resolved and confirmed the fold of the high-resolution NMR structure (Rosengren et al., 2003; Wang, King, et al., 2014). Analogues or mutants can also be crystallized in conjunction with a wild-type peptide mirror-image template in a quasi-racemic fashion, which was also highlighted by the pairing of two alanine-mutated variants of the L-form of kalata B1 with the D-form of the native peptide (Wang, King, et al., 2014).

2.2 Möbius, Bracelets, and Trypsin Inhibitors

The original study on kalata B1, although not unequivocally proven, hinted that the Trp19-Pro20 amide bond may be in a *cis* rather than *trans* conformation, which was confirmed by subsequent studies (Skjeldal et al., 2002). The peptide backbone, due to its planar amide bonds, can be considered a ribbonlike entity, which when cyclized head-to-tail can be imagined to be in the form of a bracelet. However a *cis* peptide bond, with an omega dihedral angle of $\sim 0°$ rather than $\sim 180°$ can be considered a twist in the ribbon creating a conceptual molecular Möbius strip (Craik et al., 1999). As more and more cyclotide sequences were determined, it became obvious that they could be classified in two different groups, where sequence homology in the various loops was notable. In the first group, which includes kalata B1, all peptides comprise a loop 5 with a Trp—Pro or homologous motif, and in all cases where structural studies have been undertaken a *cis* bond has been confirmed (Jennings et al., 2005). Consequently this family is referred to as the Möbius family, while the second family lacking this feature is referred to as the bracelet family. A third small subfamily of peptides also comprising a CCK motif is the trypsin inhibitory peptides. These cyclotides have quite different sequences to the other cyclotides, and share more homology with a group of linear squash trypsin inhibitors. Figure 1(B) presents a few examples of sequences from the different families and highlights the disulfide array and the various intercysteine segments referred to as loops. Screening ventures using extraction, separation, and mass spectrometry analyses, as well as nucleic acid based approaches, have provided novel sequences at an ever-increasing rate. The database CyBase comprises a collection of several hundred sequences; however, current estimates suggest that family has tens of thousands of members (Kaas & Craik, 2010). Figure 2 presents a diversity wheel highlighting aspects of the vast sequence diversity, as well as some more conserved features.

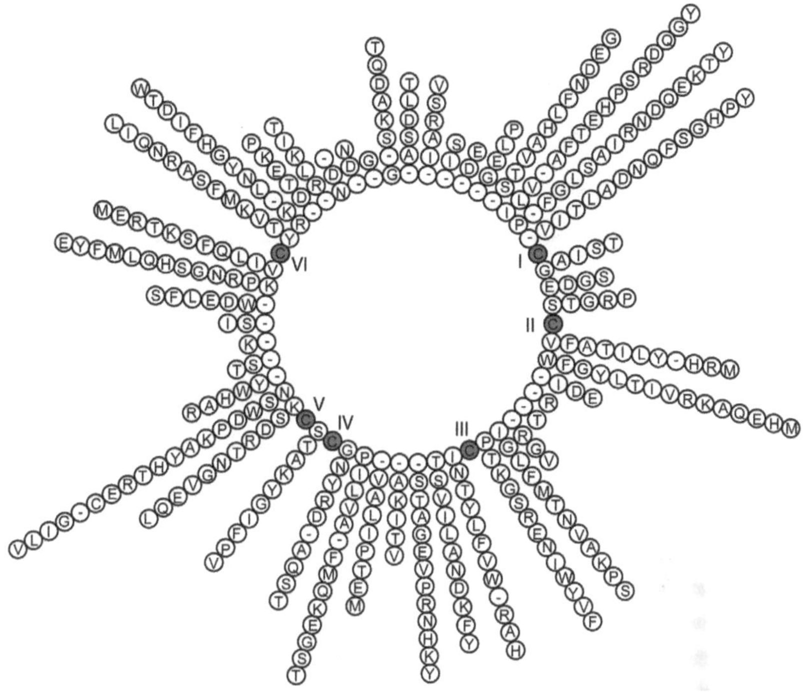

Figure 2 ***Representation of the sequence diversity observed in the Möbius and bracelet cyclotides.*** The consensus sequence is shown in the inner circle and the known sequence variations shown protruding out from the inner circle. *The data were obtained from Cybase (Kaas & Craik, 2010).*

2.3 Secondary Structure and Hydrogen Bonds

Cyclotides can be considered true mini-proteins in the sense that they are folded into functional shapes characterized by regular elements of secondary structure. The CCK motif is classically associated with a central antiparallel β-sheet, which is cross-braced by the disulfide bonds (Craik et al., 1999). Figure 3 shows a set of structures that represent the diversity of the cyclotide fold. The selected structures highlight the structural conservation with all sharing a hairpin feature comprising loops 4, 5, and part of loop 6. Associated with this is what can be considered a more disordered third strand comprising parts of loop 1 and loop 6 (Saether et al., 1995). The bracelet cyclotides comprise a longer loop 3, and this loop tends to adopt a 3_{10} helical conformation (Rosengren et al., 2003). The trypsin inhibitor family has a different shape, but still comprises a hairpin and a number of tight turns (Felizmenio-Quimio, Daly, & Craik, 2001). The ability of the cyclotide

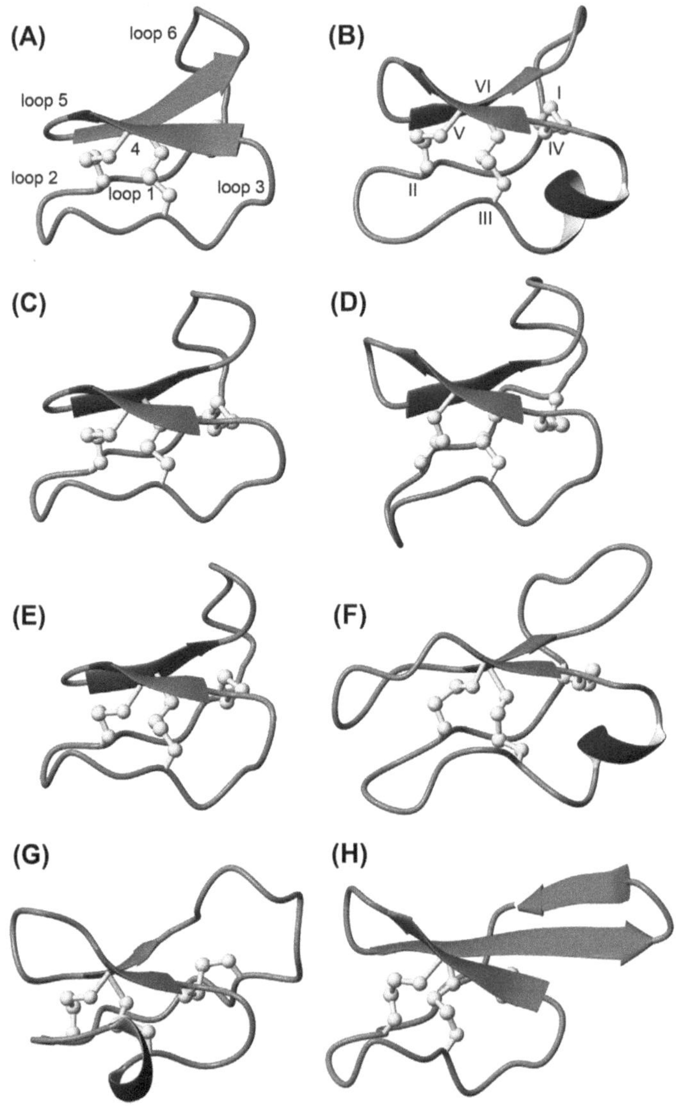

Figure 3 ***Representative structures highlighting the diversity of the cyclotide family.*** (A) Solution NMR structure of kalata B1 (PDB code 1NBI). (B) Solution NMR structure of cycloviolacin O1 (1NBJ). (C) Crystal structure of varv F (3E4H). (D) Solution NMR structure of kalata B8 (2B38). (E) Solution NMR structure of kalata B7 in DPC micelles and in complex with Mn^{2+} ions (2JWM). (F) Solution NMR structure of palicourein (1R1F). (G) Crystal structure of MCoTi-II in complex with trypsin (4GUX). (H) NMR solution structure of tricyclon A (1YP8). B-sheet secondary structure is shown in red and helical secondary structure in blue. Disulfides are shown in yellow in ball-and-stick representation. (See color plate)

fold to accommodate many different loop sizes and sequences is a key for their usage as natural templates for drug development (Daly, Rosengren, & Craik, 2009). Figure 3 also highlights some of the more unusual cyclotides with longer loops and structures. Palicourein, for example, contains 37 residues and is the largest cyclotide identified to date (Barry, Daly, Bokesch, Gustafson, & Craik, 2004), while tricyclon A has an unusually long loop 6 (Mulvenna, Sando, & Craik, 2005).

The structural motifs and overall fold of the cyclotides are stabilized by an extensive network of hydrogen bonds. Figure 4 illustrates the backbone hydrogen bond network for prototypical members of the different classes, and the majority of these hydrogen bonds are highly conserved, as can be

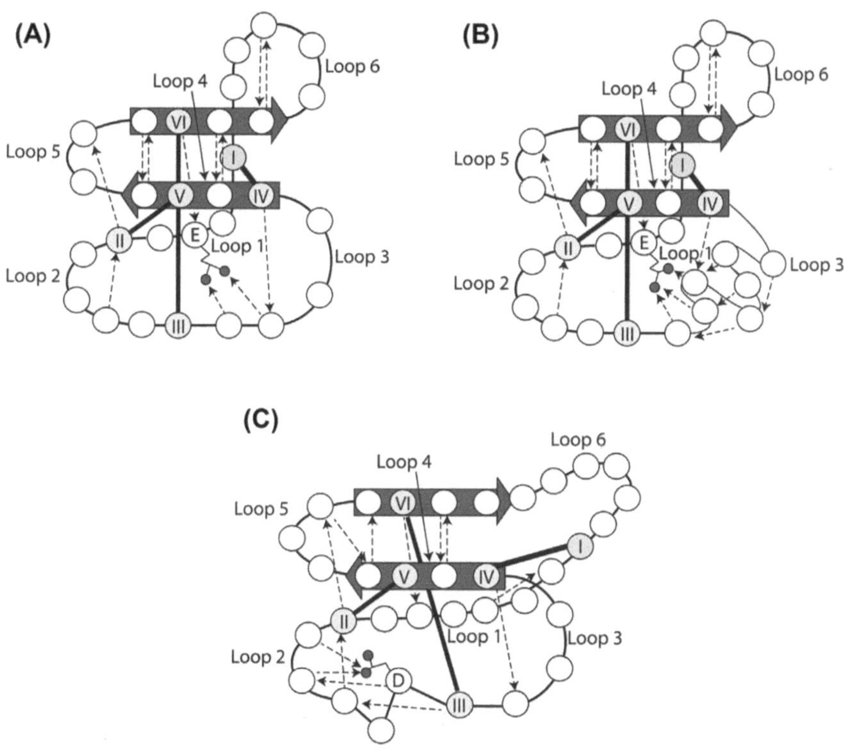

Figure 4 ***Conserved hydrogen bonding networks in cyclotides.*** Schematic illustration of the (A) Möbius, (B) bracelet, and (C) trypsin inhibitor folds, and key hydrogen bonds that stabilize the different folds. Backbone hydrogen bonds involving amide protons and carbonyls are shown as dashed arrows from the donator to the acceptor. In addition, backbone amide hydrogen bonds to key side chain carboxyl groups are highlighted. Hydrogen bonds were identified in crystal structures or by analysis of deuterium exchange and temperature coefficient data (Rosengren et al., 2003). (See color plate)

confirmed by NMR temperature titration experiments and deuterium exchange experiments that readily identify amide protons involved in hydrogen bonds as they are protected from the solvent (Göransson, Herrmann, Burman, Haugaard-Jönsson, & Rosengren, 2009; Plan, Rosengren, Sando, Daly, & Craik, 2010; Rosengren et al., 2003).

2.4 The Role of the Conserved Loop 1 Glutamic Acid

Apart from the disulfide array and the Asn/Asp and Gly residues involved in the cyclization of the backbone a single residue, Glu7 in kalata B1 is particularly highly conserved throughout both the Möbius and bracelet families. High conservation generally means an important role for structure and/or function, and this is indeed the case for this Glu. Although it was not recognized in the earlier structural studies, the high-resolution NMR structures of kalata B1 and cycloviolacin O1 highlighted a key structural role (Rosengren et al., 2003). An advantage of solution NMR is the ability to change solvent conditions, including pH and monitoring structural changes as a result of different conditions. During pH titrations of kalata B1 and cycloviolacin O1, it was established that the pKa of the glutamic acid was 3.0–3.5, which is 1.0–1.5 pH units less than the ~4.5 expected for this residue (Rosengren et al., 2003). This suggests that the negatively charged deprotonated form is stabilized by favorable interactions. Indeed the structures reveal the side chain of the Glu stretches across from loop 1 over to loop 3 placing the carboxyl group in a position to act as an acceptor for multiple hydrogen bonds, involving both backbone amides and side chain groups. In kalata B1 and typical members of the Möbius family, the Glu stabilizes a type II β-turn by hydrogen bonding the backbone amides of Asn15 and Thr16, as well as the hydroxyl proton of the side chain of Thr16. In cycloviolacin O1 the hydrogen bonds are accepted from the backbone amides of Thr15, Val16, and Thr17, as well as the Thr17 side chain hydroxyl proton (Rosengren et al., 2003). This interaction provides an N-terminal Cap for the short helical segment present in cycloviolacin O1 and other bracelet cyclotides with the larger loop 3. It would appear from sequence analysis that a hydroxyl bearing residue is always present at either position two or three of loop 3, thus allowing hydrogen bonds that can stabilize either a turn or helical conformation.

The importance of the Glu7 interactions has been studied by various modification strategies. Methylation of the Glu in cycloviolacin O2 resulted in a large drop in cytotoxic activity (Herrmann et al., 2006), and subsequent structural studies revealed the helical motif in loop 3 was completely disrupted without the Glu stabilizing "cap" (Göransson et al., 2009). Kalata B12 is an unusual natural variant comprising a Glu7 to Asp7 mutation. Although the change does not substantially affect the fold, the structure of kalata B12 appears more flexible, presumably because of the disruption to the hydrogen bonds (Wang et al., 2011).

2.5 The Hydrophobic Patch

One of the features of cyclotides is their generally rather hydrophobic surface, which is evident from the late elution on reversed-phase high-performance liquid chromatography. Generally proteins are stabilized by a hydrophobic core where the hydrophobic side chains are buried from the solvent. However, in the cyclotides the core is occupied by the cystine knot forcing hydrophobic groups to the surface (Craik et al., 1999). Hydrophobic patches are recognized as key elements for interacting with biological membranes and it has been suggested that potency in some cases are linked to degree of hydrophobicity (Wang et al., 2008). It is interesting to note that the position of the hydrophobic patch is distinctly different between different cyclotides, suggesting if it is important for activity the mode of action will differ between different cyclotides. The trypsin inhibitor cyclotides lack a hydrophobic patch and are highly polar, consistent with their target being an enzyme rather than a membrane (Heitz et al., 2001).

3. IMPORTANCE OF THE CYCLIC BACKBONE

Understanding the role of the cyclic backbone in cyclotides has been a question that has intrigued researchers since the initial characterization of this peptide class. The first study aimed at elucidating the role of the cyclic backbone involved the chemical synthesis of "acyclic permutants" of kalata B1 (Daly & Craik, 2000). A suite of six acyclic permutants were synthesized such that an N and C-termini were introduced into each of the six intercysteine loops. Analysis of the structures of these peptides using NMR spectroscopy revealed that the native fold was present in all of the permutants with the exception of the peptides opened in loops 1 and 4. It appears likely that these peptides did not fold into the native conformation because loops 1 and 4 are directly involved in the cystine knot motif. The peptides that had native folds were tested in a hemolytic activity and were shown to be devoid of activity, in contrast to the native peptide, which has mild activity. It was suggested that the lack of activity was related to the absence of the cyclic backbone but the cyclic backbone also appeared to have a role in the stability of the structure. An increase in the amide exchange rates was observed for the folded acyclic permutants suggesting that they are more flexible and less stable than the native peptide. Overall, this initial study suggested a role for the cyclic backbone in both activity and stability.

Studies on naturally occurring homologues of cyclotides that do not contain a cyclic backbone have also provided insight into the role of the cyclic backbone. Violacin A was the first linear cyclotide discovered (Ireland, Colgrave, Nguyencong, Daly, & Craik, 2006). The gene encoding violacin A has a stop codon that results in a C-terminal truncation, which prevents cyclization. This discovery added support to the importance of a conserved Asp/Asn residue in the cyclization process (Jennings, West, Waine, Craik, & Anderson, 2001). Similarly to the synthetic acyclic permutants of kalata B1, violacin A has a native cyclotide fold based on NMR spectroscopy data and lacks hemolytic activity. Furthermore, the structures were more disordered than that observed for cyclic counterparts suggesting that the cyclic backbone appeared to have a role in reducing flexibility. Despite the absence of a cyclic backbone, violacin A was resistant to cleavage by trypsin and thermolysin, consistent with the cystine knot being involved with enzymatic resistance to endoproteases rather than the cyclic backbone. As expected, the peptide was susceptible to cleavage by an exoprotease in contrast to cyclotides.

Following the discovery of violacin A, additional linear cyclotides were discovered including linear versions of kalata B9/B10 (Plan et al., 2007), psyle C (Gerlach, Burman, Bohlin, Mondal, & Göransson, 2010), and recently linear cyclotides from *Chassalia chartacea* (Rubiaceae family) (Nguyen, Lim, Nguyen, & Tam, 2012) and *Panicum laxum*, a monocot from the Poaceae family (Nguyen et al., 2013), have been discovered. Interestingly, the linear cyclotides such as those isolated from *C. chartacea* have been shown to have more potent antimicrobial, cytotoxic, and hemolytic activity than cyclic versions, indicating that the cyclic backbone is not involved in bioactivity for these peptides (Nguyen et al., 2012). The suggestion that the cyclic backbone is not involved in bioactivity is supported by an alanine-scan on kalata B1, which revealed that individual point mutations can dramatically decrease the hemolytic activity (Simonsen et al., 2008). Therefore, it appears likely that the lack of hemolytic activity in the acyclic permutants was not a direct consequence of opening the cyclic backbone but rather a result of more localized changes in sequence/structure.

To complicate matters it appears that the cyclic backbone might actually play a role in bioactivity for the trypsin inhibitor cyclotides. Linear homologues of the trypsin inhibitor cyclotides have less potent trypsin inhibitory activity than the cyclic versions (Avrutina et al., 2005). The crystal structure of MCoTI-II bound to trypsin provides an explanation for why the cyclic backbone might be involved in bioactivity (Daly et al., 2013). Residues present in the cyclization loop (loop 6) interact with trypsin and it appears likely that these interactions stabilize the peptide:protein complex and result in enhanced trypsin inhibitory activity. The importance of the cyclic backbone on protease inhibition has also been shown for engineered protease inhibitors. A hybrid of MCoTI-II and the trypsin inhibitor EETI was generated using recombinant expression and shown to have threefold higher potency against tryptase in the cyclic form compared to the linear, knotted form (Thongyoo, Bonomelli, Leatherbarrow, & Tate, 2009).

In addition to conflicting data on the role of the cyclic backbone in bioactivity, there is also conflicting reports on the role in stability. Although structure studies indicate that the cyclic backbone results in more well-defined structures, molecular dynamics of linear and cyclic trypsin inhibitors have shown that cyclization does not enhance stability or modify flexibility (Heitz et al., 2008). Experimental evidence in support of this suggestion comes from analysis of the stability of cyclotides and acyclic permutants to chaotropic agents, temperature, acid, and proteases (Colgrave & Craik, 2004). These studies indicated that the cystine knot appears to be more important for stability than the cyclic backbone.

Overall, the studies to date suggest that the answer to the question regarding the role of the cyclic backbone in cyclotides is not straightforward. The cyclic backbone is not required to maintain the cystine knot fold but it is likely that the role of the cyclic backbone can be related to bioactivity, stability, or both and that the extent to which it plays a role is sequence and target specific.

4. DYNAMICS OF CYCLOTIDE STRUCTURES

The structures of cyclotides such as kalata B1 and cycloviolacin O1 (Rosengren et al., 2003) highlight the well-defined nature of the CCK motif. The crystal structure of varv F also emphasizes the rigid nature of this structural scaffold (Wang, Hu, et al., 2009). However, despite the presence of a cyclic backbone and a cystine knot motif, it is possible for individual loops to be highly mobile. One of the first indications that the cyclotide family can have structural flexibility came from the determination of the structure of MCoTI-II. The cyclization loop (loop 6) is disordered in the structures (Felizmenio-Quimio et al., 2001), and the disorder appears to be related to isomerization involving an Asp/Gly sequence (Hernandez et al., 2000). This type of isomerization is prevalent in flexible regions of protein structures, consistent with the three-dimensional structure of MCoTI-II. The discovery of kalata B8 and its subsequent structure determination indicated that disorder is not restricted to the trypsin inhibitor cyclotides. Kalata B8, isolated from *Oldenlandia affinis*, is a hybrid between the Möbius and bracelet cyclotides and also has disorder in loop 6 (Daly, Clark, Plan, & Craik, 2006). Interestingly, Asp/Gly isomerization in loop 6 was also observed for kalata B8. The site for the Asp/Gly isomerization in MCoTI-II and kalata B8 corresponds to the in vivo ligation site to form the mature cyclic peptide but this site is not always associated with disorder in cyclotide structures.

The structural disorder observed in the MCoTI-II and kalata B8 structures, although likely to be a result of flexibility, is a direct consequence of limited experimental restraints to define the cyclization loop. Consequently, it was of interest to carry out more intensive studies to determine if the disorder observed was indeed a result of structural flexibility. Molecular motion over different timescales can be quantified using NMR relaxation experiments. Such experiments were used by Puttamadappa, Jagadish, Shekhtman, and Camarero (2010) to study the backbone dynamics of MCoTI-I, a closely related peptide to MCoTI-II. MCoTI-I and MCoTI-II differ by only two residues, one of which is an Arg to Lys substitution. The backbone dynamics of MCoTI-I in solution and when bound to trypsin were analyzed using T1, T2, and heteronuclear NOE experiments. Based on this study it was concluded that MCoTI-I has a well-defined structure in a free state in solution, but that there is increased flexibility in loop 6 and the active site loop compared to the rest of the molecule. They also reported the unexpected finding that the flexibility of MCoTI-I increases upon binding to trypsin.

We have examined the backbone dynamics of MCoTI-II in a free state (Daly et al., 2013) and found flexibility in loop 6 and the active site loop, consistent with the study on MCoTI-I (Puttamadappa et al., 2010). We also determined the structure of MCoTI-II bound to trypsin using X-ray crystallography. Analysis of the crystal structure indicated that even when bound to trypsin there was a degree of flexibility in loop 6 compared to the rest of the molecule. By contrast, loop 1, the active site loop had a single well-defined conformation as a result of interactions with trypsin. Indeed, the majority of the interactions between the peptide and trypsin are associated with loop 1 as shown in Figure 5. Given that MCoTI-I/II are highly potent trypsin inhibitors with subnanomolar Kis (Avrutina et al., 2005; Quimbar et al., 2013), it was not unexpected to find that such a well-defined complex was formed. However, the well-defined complex observed for MCoTI-II bound to trypsin appears to be in contradiction with the study on MCoTI-I, which showed an increase in flexibility in the active site loop upon binding to trypsin. A possible explanation for the discrepancies between the two studies might be related to the cleavage of MCoTI-II that is observed after extended incubation with trypsin, as a hydrolyzed form would be expected to have quite different dynamic properties compared to the cyclic form. Further studies are required to fully understand the differences between the studies.

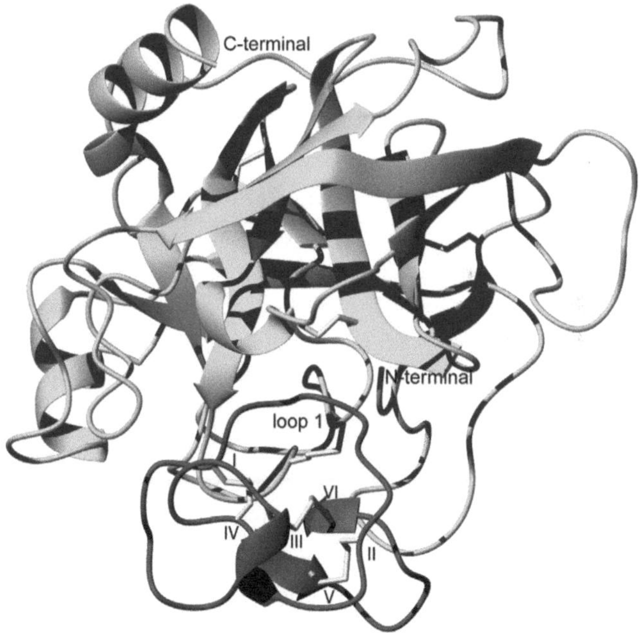

Figure 5 ***Three-dimensional structure of MCoTI-II bound to trypsin (PDB ID code 4GUX).*** Both proteins are show in ribbon style, with trypsin in green and MCoTI-II in red. Disulfide bonds are shown in yellow. The main interactions between the peptide and trypsin are associated with loop 1 of MCoTI-II. The diagram was made using Pymol (The PyMOL Molecular Graphics System, Schrödinger, LLC). (See color plate)

5. SELF-ASSOCIATION OF CYCLOTIDES

It was recognized in the early studies on cyclotides that the different physicochemical properties resulted in different solubilities and different behavior during NMR studies. For example, kalata B1 gave rise to excellent quality NMR data in aqueous solution (Saether et al., 1995), while the NMR studies of the closely related kalata B2 required the use of acetonitrile as a cosolvent to provide a monomeric form with sharp lines (Jennings et al., 2005). Nourse, Trabi, Daly, and Craik (2004) used analytical ultracentrifugation experiment to characterize these differences between kalata B1 and B2 in more detail. From the ultracentrifugation experiments it was clear that although kalata B1 exists primarily as a monomer at millimolar concentration in phosphate buffer, kalata B2 at a concentration of 1.6 mM exists in an equilibrium of monomers (~30%), tetramers (~42%), and octamers (~25%). The difference in behavior was postulated to be a result of both hydrophobic and electrostatic interactions, taking into account the nature of the five amino acid residue substitutions between the two peptides. The ability to form specific oligomeric states was particularly interesting from the perspective of biological function, and led to speculations that these oligomeric species could be further enhanced in environments where cyclotides exert their activity, such as membranes, and indeed represented the bioactive states (Nourse et al., 2004).

Mutational studies in the form of an alanine-scan of kalata B1 aimed at further describing the key features for bioactivity identified a patch of residues, centered around the conserved glutamic acid, in which mutations were found to be detrimental for biological function (Simonsen et al., 2008). Furthermore, a lysine-scan identified a second patch in which activity could be enhanced by introduction of positive charges (Huang, Colgrave, Clark, Kotze, & Craik, 2010). These regions, referred to as the "bioactive face" and the "amenable face" were suggested to at least in part represent interfaces between monomers during multimer formation and that the sensitivity of these regions to mutations were related to induced changes in propensity to form bioactive multimers (Simonsen et al., 2008).

More recently, the nature of the oligomeric forms of kalata B2 was revisited using solution NMR spectroscopy (Rosengren, Daly, Harvey, & Craik, 2013). We explored different peptide and solvent conditions and found that when using a similar buffer to the one reported for the ultracentrifugation studies, it was possible to manipulate the quality of the NMR data by altering the peptide concentration (Rosengren et al., 2013). At low concentrations (0.15 mM), very narrow and well-dispersed lines, consistent with a fully monomeric form, were observed. The data could easily be assigned and were found to closely match the previously reported shifts for kalata B2 (Jennings et al., 2005). However, increasing the peptide concentration to 0.7 mM had a detrimental effect on the quality of the NMR data with significant broadening of almost all resonances to beyond detection, suggesting chemical exchange between monomers and multimers on an intermediate NMR time scale. A further increase in concentration to 1.4 mM resulted in improvement in line shape due to the multimeric form becoming more dominant. The data recorded at the high concentration were well resolved and fully assignable, allowing for the first time a detailed comparison between the forms based on high-field NMR data (Rosengren et al., 2013).

Chemical shifts are sensitive to the environment surrounding the nuclei and comparing chemical shifts derived from spectra recorded with high or low concentrations of kalata B2 revealed that all resonances affected by the multimerization could be mapped exclusively to the hydrophobic face, not to the bioactive or amenable faces (Rosengren et al., 2013). At all concentrations, only a single set of resonances could be observed and the fact that all resonances from each individual monomer within the multimeric states perfectly overlap confirms a fully symmetric arrangement. A model in which residues within each monomer interacts with the corresponding residues of neighboring molecules, thus preventing identification of intermolecular NOEs due to overlap with intra molecular NOEs, could be derived, as shown in Figure 6(A). Two of these tetramers likely come together face to face to form the octameric form (Rosengren et al., 2013). Importantly these models are logical from the perspective of it being beneficial to bury hydrophobic residues when in aqueous solution, but suggest that the oligomers present under these conditions are not related to biological function. It would seem unlikely that the peptide would bury its hydrophobic residues when interacting with hydrophobic biological membranes (Rosengren et al., 2013). The role of well-defined oligomers in the biological function of cyclotides, if indeed any, remains to be clarified.

6. MEMBRANE INTERACTIONS OF CYCLOTIDES

The wide range of bioactivities observed for cyclotides suggested a common mechanism of action could be interaction with and disruption of

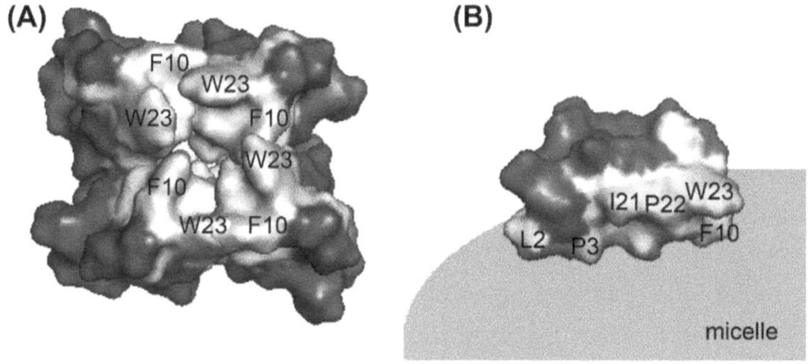

Figure 6 ***Aggregation and membrane interaction of kalata B2.*** (A) Proposed tetramer, which is a key oligomeric species of kalata B2 in aqueous buffer. The model is based on chemical shift changes occurring during aggregation, highlighted in pink, and the symmetry evident from the single set of resonances. An octamer is proposed to form by face-to-face association of two tetramers. Kalata B2 also binds and inserts into micelles according to the schematic view in (B), which is based on quenching of resonances using paramagnetic probes. Both interactions rely on the exposed hydrophobic patch. (See color plate)

biological membranes. In light of this it was of interest to study the structures of cyclotide not only in solution but also in membrane-mimicking environments. The first cyclotides to be characterized by NMR spectroscopy in micelles were kalata B1 (Shenkarev et al., 2006) and kalata B7 (Shenkarev et al., 2008). Shenkarov et al. utilized the versatility of NMR spectroscopy and added dodecylphosphocholine (DPC) micelles to the samples. The micelles were spiked with either 5- or 16-doxylstearate, which is a paramagnetic probe that quenches NMR signals from protons that are within proximity of the probe. By utilizing two different probes with the paramagnetic group positioned either at the surface or within the micelle monitoring the degree of quenching of different signals can provide information about the positioning of the peptide within the micelle. Both kalata B1 and B7 were found to be interacting with the micelles, and that they insert their hydrophic patch into the hydrophobic core of the micelles, as evident from the effect of the paramagnetic probe (Shenkarev et al., 2008). The structure of kalata B1 in micelles was found to be near identical to the solution structure from aqueous solution, again highlighting the rigidity of the fold and solution structures that are good representations of the active forms of the peptides. Interestingly, the orientations of kalata B1 and B7 in the micelle were found to be different, which can be attributed to the sequence differences, in particular an additional positive charge in loop 6 of kalata B7 (Shenkarev et al., 2008).

Kalata B1 and B7 are both Möbius cyclotides with similar positioning of the hydrophobic patch. However, the bracelet family has a distinctly different sequence, which is dominated by a highly hydrophobic longer loop 3, while loop 5 is highly charged. Consequently, Wang, Colgrave, Ireland, Kaas, and Craik (2009) used a similar approach to investigate how the bracelet cyclotide cycloviolacin O2 interacts with DPC micelles, and compared it to kalata B2. Again the data showed a clear interaction between the peptide and the micelles and the orientation of the peptide in the micelle was determined. Cycloviolacin O2 also utilizes its hydrophobic patch to drive the interaction, which intriguingly means it binds in a completely different orientation to kalata B2 and the other Möbius peptides (Wang, Colgrave, et al., 2009). It appears the conserved CCK motif does not determine the mode of interaction with membranes, but rather the interaction is guided by the surface features of the individual peptides. Figure 6(B) illustrates how kalata B2 interacts with membranes using its hydrophobic patch. Notably this is the same face that is responsible for self-association in solution.

Although the structural studies on cyclotides binding to micelles conclusively highlight the importance of the hallmark hydrophobic patch in directing the insertion into the membrane, that alone is not sufficient to explain all structure–activity relationship data available for cyclotides. For example, although a mirror image form of kalata B1, all-D kalata B1, retains significant activity, it does not show the full potency of the L-form, suggesting that rather than a fully nonspecific disruption of membranes, some component of the interaction is stereo-specific (Sando et al., 2011). Extensive studies exploring different membrane compositions revealed that kalata B1 has a distinct preference for membranes that comprise phosphatidylethanolamine (PE) lipids (Henriques et al., 2011). Again, solution NMR spectroscopy was utilized to provide further detail on this interaction. PE headgroups were added at increasing concentration to an NMR sample of kalata B1 and the chemical shift changes analyzed during the titration (Henriques et al., 2011). During this experiment the groups that did move were found to be located in the region described as the bioactive patch. Thus, the reason this region is highlighted as a key determinant of bioactivity by the alanine-scan is because of its role in specific interactions with the lipid headgroup, not because of mediating self-association. Further works by both Burman et al. and Henriques et al. suggest that the preference for PE lipids and the mechanism of membrane action is not unique to kalata B1. When the bracelet cyclotide cycloviolacin O2 was exposed to membranes, it appeared to be able to selectively extract PE lipids (Burman, Strömstedt, Malmsten, & Göransson, 2011). Furthermore, using surface plasmon resonance experiments on a range of cyclotides and different membrane compositions, it appears that all tested cyclotides have a preference for PE lipids (Henriques et al., 2012). Given the wide range of cyclotides that are often produced by a single plant it would perhaps not have been surprising to see different cyclotides having evolved a preferences for different lipids but from the current data available it appears not to be the case, thus the significance of PE lipids and any potential synergy from multiple peptides remain an enigma.

Cyclotide interactions with membranes have also been studied using different biophysical techniques including isothermal titration calorimetry and neutron reflectometry (Wang, Wacklin, & Craik, 2012). Wang et al. (2012) highlight using kalata B1 and B2, as well as analogues of B1, that the interaction is primarily entropy driven as a result of the hydrophobic interactions between peptide and lipid, but that the enthalpy changes in the free energy of binding resulting from the specific interaction with PE headgroups is also evident. The nature of the interaction between cyclotides and membranes is concentration dependant. At lower concentrations, kalata B1 associates with the membrane and initially inserts into the outer leaflet. As the concentration increases, the peptide starts to penetrate the membrane and induces the formation of pores (Wang et al., 2012).

7. GRAFTING OF NON-NATIVE SEQUENCES

The stability conferred by the well-defined structure of the cyclotides, combined with their natural sequence diversity, has led to the hypothesis that the cyclotides can be used as scaffolds for grafting nonnative sequences. Early studies showed that it was possible to synthesize kalata B1 and that with the right folding conditions, the native structure could be obtained in high yields (Daly, Love, Alewood, & Craik, 1999). Following on from these studies, mutants of kalata B1 were made where the charge distribution was altered (Clark, Daly, & Craik, 2006). The structures of the analogues maintained the CCK fold despite the mutations, providing support for the hypothesis that the CCK fold could accommodate nonnative sequences.

One of the first grafting studies on cyclotides involved grafting a poly arginine sequence, a peptide epitope involved in VEGF-A antagonism, into four of the intercysteine loops of kalata B1 (Gunasekera et al., 2008).

Loops 1 and 4 were not used for grafting based on the acyclic permutant study, which showed that these loops were critical for folding (Daly & Craik, 2000). The peptides were synthesized using solid phase methods, and the cyclization facilitated by a C-terminal thioester. The structures of the grafted analogues were analyzed using NMR spectroscopy. Two were found to have the native-like fold (i.e., the CCK fold), but two were misfolded. The peptides that were misfolded were redesigned to incorporate residues flanking the poly arginine sequence. The flanking residues were chosen based on the sequence of the wild-type peptide. The presence of these flanking residues was enough to rescue the folding and allow the CCK fold to form. The grafted peptides were analyzed in an in vitro VEGF-A antagonism assay, and one was found to be active at low micromolar concentrations. Interestingly, the poly arginine sequence in the active peptide was disordered in contrast to the other analogues. As a consequence of the disorder, the active peptide was more susceptible to proteolytic degradation than the other grafted analogues, but was more stable than the peptide epitope alone. The disorder/flexibility in the active peptide might be facilitating binding to the target whereas in the other analogues, the epitope might be too well constrained to allow binding. This study was a promising start to prove that it is possible to use the cyclotide scaffold to confer a nonnative bioactivity and improve the stability of a small bioactive sequence.

A range of grafting studies has now been carried out, with different cyclotides, targets, and potential applications. Several of the studies have been done on the trypsin inhibitor cyclotides; a subfamily of cyclotides with different cysteine spacings compared to the Möbius and bracelet subfamilies. Although loop 1 of the Möbius cyclotide kalata B1 appears to be important for folding into the native conformation and has not been extensively used for grafting, the equivalent loop in MCoTI-II has been successfully used in grafting studies. Thongyoo et al. (2009) have modified loop 1 of MCoTI-II to generate analogues that are potent inhibitors of β-tryptase and human leukocyte elastase; serine proteases with potential therapeutic applications for conditions such as asthma and rheumatoid arthritis. This study also provided insight into the structural plasticity of the MCoTI-II framework as one of the analogues that was highly potent against tryptase was truncated in loop 6 to just four residues but was still able to be cyclized and form the CCK fold based on NMR analysis.

The trypsin inhibitor cyclotides have also been used for the design of novel peptides with bioactivities not related to protease inhibitory activity. We have used the MCoTI-II scaffold to graft bioactive sequences from extracellular matrix proteins to generate novel angiogenic peptides (Chan et al., 2011). The development of potent angiogenic peptides has potential in wound healing or cardiovascular diseases. A chorioallantoic membrane assay was used to assess the activity and the most potent analogue had a region of osteopontin sequence grafted into loop 6. This study also explored the potential of another plant-derived cyclic peptide, SFTI-1. SFTI-1 is a backbone cyclic trypsin inhibitor isolated from the seeds of sunflowers but unlike MCoTI-II only has a single disulfide bond. Grafting the osteopontin epitope into the SFTI-1 scaffold resulted in a more potent analogue, but analysis of the binding to an integrin indicated that MCoTI had better binding properties and is potentially a better candidate for the design of angiogenic peptides. MCoTI-I has also been used in grafting studies such as the design of CXCR4 antagonists with anti-HIV activity (Aboye et al., 2012). One of the unexpected findings from this study was the improved stability of a grafted cyclotide compared to the parent peptide MCoTI-I as assessed by stability in human serum. Prior to this study the naturally occurring cyclotides were either equal or more stable than the engineered/grafted peptides.

The majority of the grafting studies have focused on extracellular targets, but the discovery that MCoTI-II can penetrate into cells opened up the possibility of pursuing intracellular targets. Ji et al. (2013) have shown that MCoTI-I can be used as a scaffold for targeting intracellular protein–protein interactions. An α-helical PMI peptide was grafted into loop 6 of MCoTI-I. The PMI peptide is able to bind to the p53 binding domains of Hdm2 and HdmX with low nanomolar affinity. This study suggests that grafting into the CCK scaffold could be used for a wide range of applications.

Several studies have shown that grafting into the CCK scaffold can increase the in vitro stability of the bioactive sequence, but one of the most pertinent questions in the field is whether enhanced in vivo stability can be achieved. The testing of grafted cyclotides in animal models is now emerging in the literature. Grafted cyclotides have been shown to be active in a mouse model of multiple sclerosis (Wang, Gruber, et al., 2014). Furthermore, Tam and colleagues have shown that grafting bradykinin receptor antagonist sequences into kalata B1 resulted in peptides that were orally active in a mouse model for inflammatory pain (Wong et al., 2012). In general, the bioavailability of cyclotides and grafted analogues has not been quantified and further analysis is required to determine if the cyclotide scaffold can be used in the development of orally active peptide drugs.

8. CONCLUSIONS

The structural studies on cyclotides have been critical in defining this family; the largest family of cyclic peptides to be characterized. These studies have revealed a highly conserved structural motif, which can be involved in both stability and bioactivity. Despite this structural conservation, subtle differences in the surface features of cyclotides can have a significant impact on membrane interaction and on the dynamics of the CCK fold. The role that membrane interactions and dynamics play in the bioactivity of cyclotides is only just beginning to be understood and further study will be required to understand how cyclotides interact with molecular targets. To date the only well-characterized protein target for a cyclotide is trypsin, but there is emerging evidence that other protein targets might exist such as the oxytocin and vasopressin V1a receptors (Koehbach et al., 2013). Analysis of the structures of cyclotide complexes with protein targets will provide significant advances in the field. Structural studies will also continue to play a substantial role in the engineering studies on cyclotides, as it is critical to be able to establish if the native, stable fold is maintained in the grafted analogues.

ACKNOWLEDGMENTS

NLD and KJR are Australian Research Council Future Fellows.

REFERENCES

Aboye, T. L., Ha, H., Majumder, S., Christ, F., Debyser, Z., Shekhtman, A., et al. (2012). Design of a novel cyclotide-based CXCR4 antagonist with anti-human immunodeficiency virus (HIV)-1 activity. *Journal of Medicinal Chemistry, 55*(23), 10729–10734.

Avrutina, O., Schmoldt, H. U., Gabrijelcic-Geiger, D., Le Nguyen, D., Sommerhoff, C. P., Diederichsen, U., et al. (2005). Trypsin inhibition by macrocyclic and open-chain variants of the squash inhibitor MCoTI-II. *Biological Chemistry, 386*(12), 1301–1306.

Barry, D. G., Daly, N. L., Bokesch, H. R., Gustafson, K. R., & Craik, D. J. (2004). Solution structure of the cyclotide palicourein: implications for the development of a pharmaceutical framework. *Structure, 12*(1), 85–94.

Barry, D. G., Daly, N. L., Clark, R. J., Sando, L., & Craik, D. J. (2003). Linearization of a naturally occurring circular protein maintains structure but eliminates hemolytic activity. *Biochemistry, 42*(22), 6688–6695.

Burman, R., Strömstedt, A. A., Malmsten, M., & Göransson, U. (2011). Cyclotide-membrane interactions: defining factors of membrane binding, depletion and disruption. *Biochimica et Biophysica Acta, 1808*(11), 2665–2673.

Cemazar, M., Joshi, A., Daly, N. L., Mark, A. E., & Craik, D. J. (2008). The structure of a two-disulfide intermediate assists in elucidating the oxidative folding pathway of a cyclic cystine knot protein. *Structure, 16*(6), 842–851.

Chan, L. Y., Gunasekera, S., Henriques, S. T., Worth, N. F., Le, S. J., Clark, R. J., et al. (2011). Engineering pro-angiogenic peptides using stable, disulfide-rich cyclic scaffolds. *Blood, 118*(25), 6709–6717.

Chen, B., Colgrave, M. L., Daly, N. L., Rosengren, K. J., Gustafson, K. R., & Craik, D. J. (2005). Isolation and characterization of novel cyclotides from *Viola hederaceae*: solution structure and anti-HIV activity of vhl-1, a leaf-specific expressed cyclotide. *Journal of Biological Chemistry, 280*(23), 22395–22405.

Clark, R. J., Daly, N. L., & Craik, D. J. (2006). Structural plasticity of the cyclic-cystine-knot framework: implications for biological activity and drug design. *Biochemical Journal, 394*(Pt 1), 85–93.

Colgrave, M. L., & Craik, D. J. (2004). Thermal, chemical, and enzymatic stability of the cyclotide kalata B1: the importance of the cyclic cystine knot. *Biochemistry, 43*(20), 5965–5975.

Craik, D. J., Daly, N. L., Bond, T., & Waine, C. (1999). Plant cyclotides: a unique family of cyclic and knotted proteins that defines the cyclic cystine knot structural motif. *Journal of Molecular Biology, 294*(5), 1327–1336.

Craik, D. J., Daly, N. L., & Waine, C. (2001). The cystine knot motif in toxins and implications for drug design. *Toxicon, 39*(1), 43–60.

Daly, N. L., Chen, B., Nguyencong, P., & Craik, D. J. (2010). Structure and activity of the leaf-specific cyclotide vhl-2. *Australian Journal of Chemistry, 63*, 771–778.

Daly, N. L., Clark, R. J., & Craik, D. J. (2003). Disulfide folding pathways of cystine knot proteins. Tying the knot within the circular backbone of the cyclotides. *Journal of Biological Chemistry, 278*(8), 6314–6322.

Daly, N. L., Clark, R. J., Plan, M. R., & Craik, D. J. (2006). Kalata B8, a novel antiviral circular protein, exhibits conformational flexibility in the cystine knot motif. *Biochemical Journal, 393*(Pt 3), 619–626.

Daly, N. L., & Craik, D. J. (2000). Acyclic permutants of naturally occurring cyclic proteins. Characterization of cystine knot and beta -sheet formation in the macrocyclic polypeptide kalata B1. *Journal of Biological Chemistry, 275*(25), 19068–19075.

Daly, N. L., Love, S., Alewood, P. F., & Craik, D. J. (1999). Chemical synthesis and folding pathways of large cyclic polypeptides: studies of the cystine knot polypeptide kalata B1. *Biochemistry, 38*(32), 10606–10614.

Daly, N. L., Rosengren, K. J., & Craik, D. J. (2009). Discovery, structure and biological activities of cyclotides. *Advanced Drug Delivery Reviews, 61*(11), 918–930.

Daly, N. L., Rosengren, K. J., Henriques, S. T., & Craik, D. J. (2011). NMR and protein structure in drug design: application to cyclotides and conotoxins. *European Biophysics Journal, 40*(4), 359–370.

Daly, N. L., Thorstholm, L., Greenwood, K. P., King, G. J., Rosengren, K. J., Heras, B., et al. (2013). Structural insights into the role of the cyclic backbone in a squash trypsin inhibitor. *Journal of Biological Chemistry, 288*(50), 36141–36148.

Derua, R., Gustafson, K. R., & Pannell, L. K. (1996). Analysis of the disulfide linkage pattern in circulin A and B, HIV-inhibitory macrocyclic peptides. *Biochemical and Biophysical Research Communications, 228*, 632–638.

Eliasen, R., Daly, N. L., Wulff, B. S., Andresen, T. L., Conde-Frieboes, K. W., & Craik, D. J. (2012). Design, synthesis, structural and functional characterization of novel melanocortin agonists based on the cyclotide kalata B1. *Journal of Biological Chemistry, 287*(48), 40493–40501.

Felizmenio-Quimio, M. E., Daly, N. L., & Craik, D. J. (2001). Circular proteins in plants: solution structure of a novel macrocyclic trypsin inhibitor from *Momordica cochinchinensis*. *Journal of Biological Chemistry, 276*(25), 22875–22882.

Gerlach, S. L., Burman, R., Bohlin, L., Mondal, D., & Göransson, U. (2010). Isolation, characterization, and bioactivity of cyclotides from the Micronesian plant *Psychotria leptothyrsa*. *Journal of Natural Products, 73*(7), 1207–1213.

Göransson, U., & Craik, D. J. (2003). Disulfide mapping of the cyclotide kalata B1. Chemical proof of the cyclic cystine knot motif. *Journal of Biological Chemistry, 278*(48), 48188–48196.

Göransson, U., Herrmann, A., Burman, R., Haugaard-Jönsson, L. M., & Rosengren, K. J. (2009). The conserved glu in the cyclotide cycloviolacin O_2 has a key structural role. *ChemBioChem, 10*(14), 2354–2360.

Gran, L. (1973). Oxytocic principles of *Oldenlandia affinis*. *Lloydia, 36*, 174–178.

Gunasekera, S., Foley, F. M., Clark, R. J., Sando, L., Fabri, L. J., Craik, D. J., et al. (2008). Engineering stabilized vascular endothelial growth factor-A antagonists: synthesis, structural characterization, and bioactivity of grafted analogues of cyclotides. *Journal of Medicinal Chemistry, 51*(24), 7697–7704.

Heitz, A., Avrutina, O., Le-Nguyen, D., Diederichsen, U., Hernandez, J. F., Gracy, J., et al. (2008). Knottin cyclization: impact on structure and dynamics. *BMC Structural Biology, 8*, 54.

Heitz, A., Hernandez, J. F., Gagnon, J., Hong, T. T., Pham, T. T., Nguyen, T. M., et al. (2001). Solution structure of the squash trypsin inhibitor MCoTI-II. A new family for cyclic knottins. *Biochemistry, 40*(27), 7973–7983.

Henriques, S. T., Huang, Y. H., Castanho, M. A., Bagatolli, L. A., Sonza, S., Tachedjian, G., et al. (2012). Phosphatidylethanolamine binding is a conserved feature of cyclotide-membrane interactions. *Journal of Biological Chemistry, 287*(40), 33629–33643.

Henriques, S. T., Huang, Y. H., Chaousis, S., Wang, C. K., & Craik, D. J. (2014). Anticancer and toxic properties of cyclotides are dependent on phosphatidylethanolamine phospholipid targeting. *ChemBioChem, 15*(13), 1956–1965.

Henriques, S. T., Huang, Y. H., Rosengren, K. J., Franquelim, H. G., Carvalho, F. A., Johnson, A., et al. (2011). Decoding the membrane activity of the cyclotide kalata B1: the importance of phosphatidylethanolamine phospholipids and lipid organization on hemolytic and anti-HIV activities. *Journal of Biological Chemistry, 286*(27), 24231–24241.

Hernandez, J. F., Gagnon, J., Chiche, L., Nguyen, T. M., Andrieu, J. P., Heitz, A., et al. (2000). Squash trypsin inhibitors from *Momordica cochinchinensis* exhibit an atypical macrocyclic structure. *Biochemistry, 39*(19), 5722–5730.

Herrmann, A., Svangård, E., Claeson, P., Gullbo, J., Bohlin, L., & Göransson, U. (2006). Key role of glutamic acid for the cytotoxic activity of the cyclotide cycloviolacin O_2. *Cellular and Molecular Life Sciences, 63*(2), 235–245.

Huang, Y. H., Colgrave, M. L., Clark, R. J., Kotze, A. C., & Craik, D. J. (2010). Lysine-scanning mutagenesis reveals an amendable face of the cyclotide kalata B1 for the optimization of nematocidal activity. *Journal of Biological Chemistry, 285*(14), 10797–10805.

Ireland, D. C., Colgrave, M. L., & Craik, D. J. (2006). A novel suite of cyclotides from *Viola odorata*: sequence variation and the implications for structure, function and stability. *Biochemical Journal, 400*(1), 1–12.

Ireland, D. C., Colgrave, M. L., Nguyencong, P., Daly, N. L., & Craik, D. J. (2006). Discovery and characterization of a linear cyclotide from *Viola odorata*: implications for the processing of circular proteins. *Journal of Molecular Biology, 357*(5), 1522–1535.

Jennings, C. V., Rosengren, K. J., Daly, N. L., Plan, M., Stevens, J., Scanlon, M. J., et al. (2005). Isolation, solution structure, and insecticidal activity of kalata B2, a circular protein with a twist: do Mobius strips exist in nature? *Biochemistry, 44*(3), 851–860.

Jennings, C., West, J., Waine, C., Craik, D., & Anderson, M. (2001). Biosynthesis and insecticidal properties of plant cyclotides: the cyclic knotted proteins from *Oldenlandia affinis*. *Proceedings of the National Academy of Sciences of the United States of America, 98*(19), 10614–10619.

Ji, Y., Majumder, S., Millard, M., Borra, R., Bi, T., Elnagar, A. Y., et al. (2013). In vivo activation of the p53 tumor suppressor pathway by an engineered cyclotide. *Journal of the American Chemical Society, 135*(31), 11623–11633.

Jia, X., Kwon, S., Wang, C. I., Huang, Y. H., Chan, L. Y., Tan, C. C., et al. (2014). Semi-enzymatic cyclization of disulfide-rich peptides using Sortase A. *Journal of Biological Chemistry, 289*(10), 6627–6638.

Kaas, Q., & Craik, D. J. (2010). Analysis and classification of circular proteins in CyBase. *Biopolymers, 94*(5), 584–591.

Koehbach, J., O'Brien, M., Muttenthaler, M., Miazzo, M., Akcan, M., Elliott, A. G., et al. (2013). Oxytocic plant cyclotides as templates for peptide G protein-coupled receptor ligand design. *Proceedings of the National Academy of Sciences of the United States of America, 110*(52), 21183–21188.

Marx, U. C., Daly, N. L., & Craik, D. J. (2006). NMR of conotoxins: structural features and an analysis of chemical shifts of post-translationally modified amino acids. *Magnetic Resonance in Chemistry, 44*. Spec No: S41–S50.

Mulvenna, J. P., Sando, L., & Craik, D. J. (2005). Processing of a 22 kDa precursor protein to produce the circular protein tricyclon A. *Structure, 13*(5), 691–701.

Mylne, J. S., Chan, L. Y., Chanson, A. H., Daly, N. L., Schaefer, H., Bailey, T. L., et al. (2012). Cyclic peptides arising by evolutionary parallelism via asparaginyl-endopeptidase-mediated biosynthesis. *Plant Cell, 24*(7), 2765–2778.

Mylne, J. S., & Craik, D. J. (2008). 15N cyclotides by whole plant labeling. *Biopolymers, 90*(4), 575–580.

Nguyen, G. K., Lian, Y., Pang, E. W., Nguyen, P. Q., Tran, T. D., & Tam, J. P. (2013). Discovery of linear cyclotides in monocot plant *Panicum laxum* of Poaceae family provides new insights into evolution and distribution of cyclotides in plants. *Journal of Biological Chemistry, 288*(5), 3370–3380.

Nguyen, G. K., Lim, W. H., Nguyen, P. Q., & Tam, J. P. (2012). Novel cyclotides and uncyclotides with highly shortened precursors from *Chassalia chartacea* and effects of methionine oxidation on bioactivities. *Journal of Biological Chemistry, 287*(21), 17598–17607.

Nielsen, K. J., Alewood, D., Andrews, J., Kent, S. B. H., & Craik, D. J. (1994). An ^{1}H NMR determination of the three-dimensional structures of mirror-image forms of a Leu-5 variant of the trypsin inhibitor from *Ecballium elaterium* (EETI-II). *Protein Science, 3*, 291–302.

Nourse, A., Trabi, M., Daly, N. L., & Craik, D. J. (2004). A comparison of the self-association behavior of the plant cyclotides kalata B1 and kalata B2 via analytical ultracentrifugation. *Journal of Biological Chemistry, 279*(1), 562–570.

Pallaghy, P. K., Nielsen, K. J., Craik, D. J., & Norton, R. S. (1994). A common structural motif incorporating a cystine knot and a triple-stranded β-sheet in toxic and inhibitory polypeptides. *Protein Science, 3*, 1833–1839.

Plan, M. R., Göransson, U., Clark, R. J., Daly, N. L., Colgrave, M. L., & Craik, D. J. (2007). The cyclotide fingerprint in *Oldenlandia affinis*: elucidation of chemically modified, linear and novel macrocyclic peptides. *ChemBioChem, 8*(9), 1001–1011.

Plan, M. R., Rosengren, K. J., Sando, L., Daly, N. L., & Craik, D. J. (2010). Structural and biochemical characteristics of the cyclotide kalata B5 from *Oldenlandia affinis*. *Biopolymers, 94*(5), 647–658.

Poth, A. G., Colgrave, M. L., Lyons, R. E., Daly, N. L., & Craik, D. J. (2011). Discovery of an unusual biosynthetic origin for circular proteins in legumes. *Proceedings of the National Academy of Sciences of the United States of America, 108*(25), 10127–10132.

Puttamadappa, S. S., Jagadish, K., Shekhtman, A., & Camarero, J. A. (2010). Backbone dynamics of cyclotide MCoTI-I free and complexed with trypsin. *Angewandte Chemie, International Edition in English, 49*(39), 7030–7034.

Quimbar, P., Malik, U., Sommerhoff, C. P., Kaas, Q., Chan, L. Y., Huang, Y. H., et al. (2013). High-affinity cyclic peptide matriptase inhibitors. *Journal of Biological Chemistry, 288*(19), 13885–13896.

Rosengren, K. J., Daly, N. L., Harvey, P. J., & Craik, D. J. (2013). The self-association of the cyclotide kalata B2 in solution is guided by hydrophobic interactions. *Biopolymers, 100*(5), 453–460.

Rosengren, K. J., Daly, N. L., Plan, M. R., Waine, C., & Craik, D. J. (2003). Twists, knots, and rings in proteins. Structural definition of the cyclotide framework. *Journal of Biological Chemistry, 278*(10), 8606–8616.

Saether, O., Craik, D. J., Campbell, I. D., Sletten, K., Juul, J., & Norman, D. G. (1995). Elucidation of the primary and three-dimensional structure of the uterotonic polypeptide kalata B1. *Biochemistry, 34*(13), 4147–4158.

Sando, L., Henriques, S. T., Foley, F., Simonsen, S. M., Daly, N. L., Hall, K. N., et al. (2011). A synthetic mirror image of kalata B1 reveals that cyclotide activity is independent of a protein receptor. *ChemBioChem, 12*(16), 2456–2462.

Shenkarev, Z. O., Nadezhdin, K. D., Lyukmanova, E. N., Sobol, V. A., Skjeldal, L., & Arseniev, A. S. (2008). Divalent cation coordination and mode of membrane interaction in cyclotides: NMR spatial structure of ternary complex kalata B7/Mn^{2+}/DPC micelle. *Journal of Inorganic Biochemistry, 102*(5–6), 1246–1256.

Shenkarev, Z. O., Nadezhdin, K. D., Sobol, V. A., Sobol, A. G., Skjeldal, L., & Arseniev, A. S. (2006). Conformation and mode of membrane interaction in cyclotides. Spatial structure of kalata B1 bound to a dodecylphosphocholine micelle. *FEBS Journal, 273*(12), 2658–2672.

Simonsen, S. M., Sando, L., Rosengren, K. J., Wang, C. K., Colgrave, M. L., Daly, N. L., et al. (2008). Alanine scanning mutagenesis of the prototypic cyclotide reveals a cluster of residues essential for bioactivity. *Journal of Biological Chemistry, 283*(15), 9805–9813.

Skjeldal, L., Gran, L., Sletten, K., & Volkman, B. F. (2002). Refined structure and metal binding site of the kalata B1 peptide. *Archives of Biochemistry and Biophysics, 399*(2), 142–148.

Tam, J. P., Lu, Y. A., Yang, J. L., & Chiu, K. W. (1999). An unusual structural motif of antimicrobial peptides containing end-to-end macrocycle and cystine-knot disulfides. *Proceedings of the National Academy of Sciences of the United States of America, 96*(16), 8913–8918.

Thongyoo, P., Bonomelli, C., Leatherbarrow, R. J., & Tate, E. W. (2009). Potent inhibitors of beta-tryptase and human leukocyte elastase based on the MCoTI-II scaffold. *Journal of Medicinal Chemistry, 52*(20), 6197–6200.

Trabi, M., & Craik, D. J. (2004). Tissue-specific expression of head-to-tail cyclized miniproteins in Violaceae and structure determination of the root cyclotide *Viola hederacea* root cyclotide1. *Plant Cell, 16*(8), 2204–2216.

Wang, C. K., Clark, R. J., Harvey, P. J., Rosengren, K. J., Cemazar, M., & Craik, D. J. (2011). The role of conserved Glu residue on cyclotide stability and activity: a structural and functional study of kalata B12, a naturally occurring Glu to Asp mutant. *Biochemistry, 50*(19), 4077–4086.

Wang, C. K., Colgrave, M. L., Gustafson, K. R., Ireland, D. C., Göransson, U., & Craik, D. J. (2008). Anti-HIV cyclotides from the Chinese medicinal herb *Viola yedoensis*. *Journal of Natural Products, 71*(1), 47–52.

Wang, C. K., Colgrave, M. L., Ireland, D. C., Kaas, Q., & Craik, D. J. (2009). Despite a conserved cystine knot motif, different cyclotides have different membrane binding modes. *Biophysical Journal, 97*(5), 1471–1481.

Wang, C. K., Gruber, C. W., Cemazar, M., Siatskas, C., Tagore, P., Payne, N., et al. (2014). Molecular grafting onto a stable framework yields novel cyclic peptides for the treatment of multiple sclerosis. *ACS Chemical Biology, 9*(1), 156–163.

Wang, C. K., Hu, S. H., Martin, J. L., Sjögren, T., Hajdu, J., Bohlin, L., et al. (2009). Combined X-ray and NMR analysis of the stability of the cyclotide cystine knot fold that underpins its insecticidal activity and potential use as a drug scaffold. *Journal of Biological Chemistry, 284*(16), 10672–10683.

Wang, C. K., King, G. J., Northfield, S. E., Ojeda, P. G., & Craik, D. J. (2014). Racemic and quasi-racemic X-ray structures of cyclic disulfide-rich peptide drug scaffolds. *Angewandte Chemie, International Edition in English, 53*(42), 11236–11241.

Wang, C. K., Wacklin, H. P., & Craik, D. J. (2012). Cyclotides insert into lipid bilayers to form membrane pores and destabilize the membrane through hydrophobic and phosphoethanolamine-specific interactions. *Journal of Biological Chemistry, 287*(52), 43884–43898.

Wong, C. T., Rowlands, D. K., Wong, C. H., Lo, T. W., Nguyen, G. K., Li, H. Y., et al. (2012). Orally active peptidic bradykinin B1 receptor antagonists engineered from a cyclotide scaffold for inflammatory pain treatment. *Angewandte Chemie, International Edition in English, 51*(23), 5620–5624.

CHAPTER SEVEN

Natural Functions and Structure–Activity Relationships of Cyclotides

Georgianna Kae Oguis, Meng-Wei Kan, David J. Craik[1]
Institute for Molecular Bioscience, The University of Queensland, Brisbane, QLD, Australia
[1]Corresponding author: E-mail: d.craik@imb.uq.edu.au

Contents

Abstract

Cyclotides have a diverse range of biological activities that have both pharmaceutical and agricultural applications. Here we examine the range of activities seen in natural cyclotides and describe the structure–activity relationship (SAR) studies that have helped to understand the mechanisms associated with these activities. These SAR

Advances in Botanical Research, Volume 76
ISSN 0065-2296
http://dx.doi.org/10.1016/bs.abr.2015.10.001

studies have also led to the development of "designer" cyclotides with an introduced range of pharmaceutical bioactivities. Overall, cyclotides continue to be of interest for their potential in drug design and crop protection.

1. INTRODUCTION

The first three chapters (Overview on the Discovery and Applications of Cyclotides, Cyclotides in the Violaceae, Cyclotides in the Rubiaceae) of this book described the history of the discovery of cyclotides and their distribution in some of the major families of angiosperms, including the Violaceae, Rubiaceae, Solanaceae, Fabaceae, and Cucurbitaceae. Chapter 4 Cyclotides from Chinese Plants provided a description of cyclotides from Chinese traditional medicines and outlined some of their medicinal uses. The background provided in these chapters was followed by a description of state-of-the-art methods for sequencing novel cyclotides and for analyzing their structures. In particular, Chapter 5 Primary Structural Analysis of Cyclotides described mass spectrometry sequencing methods and Chapter 6 Structural Studies of Cyclotides showed how NMR spectroscopy has been used to determine the three-dimensional structures of a wide range of cyclotides.

With this background on the discovery and structures of cyclotides complete, in this chapter we focus on the biological activities of cyclotides. We first examine their natural functions, which are primarily thought to be associated with host defense and pesticidal activities. In addition to having pesticidal functions, cyclotides have been screened for pharmaceutical activities; in particular, anti-HIV, anti-neurotensin, and antimicrobial activities were examined in early studies, following their first described bioactivity as uterotonic agents. In this chapter we provide a detailed survey of the known biological activities of naturally occurring cyclotides as well as some introduced activities in pharmaceutically modified cyclotides. This survey includes a detailed analysis of structure—activity relationships (SARs) of cyclotides that have been enabled by the development of chemical or biological approaches to make modified cyclotide derivatives (see Chapter 8 Biosynthesis of cyclotides for information on the biosynthesis of cyclotides and Chapter 9 Chemical and Biological Production of Cyclotides for more details on chemical and biological approaches to synthesizing cyclotides).

Table 1 shows the sequences of cyclotides for which extensive biological activity studies have been reported (Bokesch et al., 2001; Broussalis et al., 2001; Chen et al., 2005; Craik, Daly, Bond, & Waine, 1999; Daly, Clark,

Table 1 Sequences of Cyclotides with Reported Biological Activities*

Cyclotide	Sequence												Species of Origin	References
Loop		**1**		**2**		**3**		**4**		**5**		**6**		
Kalata B1	**C**	GET	**C**	VGGT	**C**	NTPG	**C**	T	**C**	SWPV	**C**	TRNGLPV	*Oldenlandia affinis*	Saether et al. (1995)
Kalata B2	**C**	GET	**C**	FGGT	**C**	NTPG	**C**	S	**C**	TWPI	**C**	TRDGLPV	*O. affinis*	Craik et al. (1999)
Kalata B6	**C**	GET	**C**	FGGT	**C**	NTPG	**C**	S	**C**	SSWPI	**C**	TRNGLPT	*O. affinis*	Craik et al. (1999)
Kalata B7	**C**	GET	**C**	TLGT	**C**	YTQG	**C**	T	**C**	SWPI	**C**	KRNGLPV	*O. affinis*	Craik et al. (1999)
Kalata B8	**C**	GET	**C**	LLGT	**C**	YTTG	**C**	T	**C**	NKYRV	**C**	TKDGSVLN	*O. affinis*	Daly et al. (2006)
Circulin A	**C**	GES	**C**	VWIP	**C**	ISAALG	**C**	S	**C**	KNKV	**C**	YRNGIP	*Chassalia parviflora*	Gustafson et al. (1994)
Circulin B	**C**	GES	**C**	VFIP	**C**	ISTLLG	**C**	S	**C**	KNKV	**C**	YRNGVIP	*C. parviflora*	Gustafson et al. (1994)
Circulin C	**C**	GES	**C**	VFIP	**C**	ITSVAG	**C**	S	**C**	KSKV	**C**	YRNGIP	*C. parviflora*	Gustafson et al. (2000)
Circulin D	**C**	GES	**C**	VWIP	**C**	VTSIFN	**C**	K	**C**	ENKV	**C**	YHDKIP	*C. parviflora*	Gustafson et al. (2000)
Circulin E	**C**	GES	**C**	VWIP	**C**	LTSVFN	**C**	K	**C**	ENKV	**C**	YHDKIP	*C. parviflora*	Gustafson et al. (2000)
Circulin F	**C**	GES	**C**	VWIP	**C**	ISAAIG	**C**	S	**C**	KNKV	**C**	YRAIP	*C. parviflora*	Gustafson et al. (2000)
Psyle A	**C**	GES	**C**	VFLG	**C**	FIPG	**C**	S	**C**	KSKV	**C**	YFNGIA	*Psychotria leptothyrsa*	Gerlach, Burman, et al. (2010)
Psyle C	**C**	GET	**C**	FKFK	**C**	YTPG	**C**	S	**C**	SYFP	**C**	KKL	*P. leptothyrsa*	Gerlach, Burman, et al. (2010)
Psyle E	**C**	GES	**C**	VFIP	**C**	ISSVLG	**C**	S	**C**	KNKV	**C**	YRDGVIP	*P. leptothyrsa*	Gerlach, Burman, et al. (2010)
Parigidin-br1	**C**	GES	**C**	VFIP	**C**	ITSLAG	**C**	S	**C**	KNKV	**C**	YYDGGSVP	*Palicourea rigida*	Pinto et al. (2012)
Palicourein	**C**	GET	**C**	RVIPV	**C**	TYSAALG	**C**	T	**C**	DDRSDGL	**C**	KRNGDPTF	*Palicourea condensata*	Bockesh et al. (2001)
Cyclopsychotride_A	**C**	GES	**C**	VFIP	**C**	TVTALLG	**C**	S	**C**	KSKV	**C**	YKNSIP	*Palicourea jungiana*	Witherup et al. (1994)
Hedyotide B1	**C**	GET	**C**	FVLP	**C**	WSAKFG	**C**	Y	**C**	QKGF	**C**	YRNGTR	*Hybanthus biflora*	Wong et al. (2011)
Hedyotide 5	**C**	GES	**C**	VYLP	**C**	LTTVVG	**C**	S	**C**	QNSV	**C**	YHNGAVP	*H. biflora*	Ding et al. (2014)
Hedyotide 6	**C**	GET	**C**	VYLP	**C**	ITAAIG	**C**	S	**C**	QNKV	**C**	YRDGAVP	*H. biflora*	Ding et al. (2014)

(*Continued*)

Table 1 Sequences of Cyclotides with Reported Biological Activities*—cont'd

Cyclotide	Sequence												Species of Origin	References
Loop		**1**		**2**		**3**		**4**		**5**		**6**		
Hedyotide 7	C	AET	C	VLLP	C	LSSVIG	C	S	C	TCQNKR	C	YKNGIS	*H. biflora*	Ding et al. (2014)
Hedyotide 8	C	GES	C	AFLP	C	LTSLLG	C	T	C	QNKV	C	YRDGIP	*H. biflora*	Ding et al. (2014)
Hedyotide 9	C	GES	C	VYLP	C	ITTVVG	C	S	C	QNSV	C	YHNGNP	*H. biflora*	Ding et al. (2014)
Vitri A	C	GES	C	VWIP	C	ITSAIG	C	S	C	KSKV	C	YRNGIP	*Viola tricolor*	Svangård et al. (2004)
Vitri F	C	GES	C	VWIP	C	ISSVVG	C	A	C	KSKV	C	YKDGLTP	*V. tricolor*	Tang et al. (2010a)
Cycloviolacin O1	C	AES	C	VYIP	C	TVTALLG	C	S	C	SNRV	C	YNGIP	*Viola odorata*	Rosengren et al. (2003)
Cycloviolacin O2	C	GES	C	VWIP	C	ISSAIG	C	S	C	KSKV	C	YRNGIPC	*V. odorata*	Wang et al. (2009)
Cycloviolacin O3	C	GES	C	VWIP	C	LTSAIG	C	S	C	KSKV	C	YRNGIP	*V. odorata*	Craik et al. (1999)
Cycloviolacin O8	C	GES	C	VWIP	C	ISSVVG	C	S	C	KSKV	C	YKNGTLP	*V. odorata, Viola adunca*	Dutton et al. (2004)
Cycloviolacin O13	C	GES	C	VWIP	C	ISAAIG	C	S	C	KSKV	C	YRNGIP	*V. odorata*	Dutton et al. (2004)
Cycloviolacin O14	C	GES	C	FKGK	C	YTPG	C	S	C	SKYPL	C	AKNGSIPA	*V. odorata*	Ireland et al. (2006)
Cycloviolacin O15	C	GET	C	FTGK	C	YTPG	C	S	C	SYPI	C	KKNGLVP	*V. odorata*	Ireland et al. (2006)
Cycloviolacin O16	C	GET	C	FTGK	C	YTPG	C	S	C	SYPI	C	KKINGLP	*V. odorata*	Ireland et al. (2006)
Cycloviolacin O24	C	GET	C	FGGT	C	NTPG	C	T	C	DPWPV	C	THNGLPT	*V. odorata*	Ireland et al. (2006)
Cycloviolacin H3	C	GET	C	FGGT	C	NTPG	C	I	C	DPWPV	C	CTRNGLPV	*Viola hederacea*	Chen et al. (2005)
Cycloviolacin Y1	C	GET	C	FLGT	C	YTPG	C	S	C	GNYGF	C	YGTNGGTIFD	*Viola yedoensis*	Wang et al. (2008)
Cycloviolacin Y4	C	GES	C	VFIP	C	ITGVIG	C	S	C	SSNV	C	YLNGVP	*V. yedoensis*	Wang et al. (2008)
Cycloviolacin Y5	C	AES	C	VWIP	C	TVTALVG	C	S	C	SDKV	C	YNGIP	*V. yedoensis*	Wang et al. (2008)
Cycloviolin A	C	GES	C	VFIP	C	ISAAIG	C	S	C	KNKV	C	YRNGVIP	*Leonia cymosa*	Hallock et al. (2000)
Cycloviolin B	C	GES	C	YVLP	C	FTVG	C	T	C	TSSQ	C	FKNGTA	*L. cymosa*	Hallock et al. (2000)
Cycloviolin C	C	GES	C	VFIP	C	LTTVAG	C	S	C	KNKV	C	YRNGIP	*L. cymosa*	Hallock et al. (2000)
Cycloviolin D	C	GES	C	VFIP	C	ISAAIG	C	S	C	KNKV	C	YRNGFP	*L. cymosa*	Hallock et al. (2000)
Varv A	C	GET	C	VGGT	C	NTPG	C	S	C	SWPV	C	TRNGLPV	*L. cymosa*	Craik et al. (1999)
Varv D	C	GET	C	VGGS	C	NTPG	C	S	C	SWPV	C	TRNGLPI	*Viola arvensis*	Göransson et al. (1999)
Varv E	C	GET	C	VGGT	C	NTPG	C	S	C	SWPV	C	TRNLPI	*V. arvensis*	Svangård et al. (2004)
Varv F	C	GET	C	TLGT	C	YTAG	C	S	C	SWPV	C	TRNGVPI	*V. arvensis*	Svangård et al. (2004)
vhl-1	C	GES	C	AMISF	C	FTEVIG	C	S	C	KNKV	C	YLNSIS	*V. hederacea*	Chen et al. (2005)

Vaby A	**C**	GET	**C**	AGGT	**C**	NTPG	**C**	S	**C**	SWPI	**C**	TRNGLPV	*Viola abyssinica*	Yeshak et al. (2011)
Vaby D	**C**	GET	**C**	FGGT	**C**	NTPG	**C**	T	**C**	DPWPV	**C**	TRNGLPV	*V. abyssinica*	Yeshak et al. (2011)
Vila A	**C**	GES	**C**	VWIP	**C**	ISSAIG	**C**	S	**C**	KDKV	**C**	YRDGIP	*Viola labradorica*	Tang et al. (2010b)
Vila B	**C**	GES	**C**	VWIP	**C**	ISSAIG	**C**	S	**C**	RSKV	**C**	YRDGIP	*V. labradorica*	Tang et al. (2010b)
Vila D	**C**	GET	**C**	AFGT	**C**	YTGG	**C**	S	**C**	SWPV	**C**	TRNGIP	*V. labradorica*	Tang et al. (2010b)
Viphi A	**C**	GES	**C**	VFIP	**C**	ISSVIG	**C**	A	**C**	KSKV	**C**	YKNGSIP	*Viola philippica*	He et al. (2011)
Viphi D	**C**	GES	**C**	VFIP	**C**	ISSVIG	**C**	S	**C**	SSKV	**C**	YRNGIP	*V. philippica*	He et al. (2011)
Viphi E	**C**	GES	**C**	VFIP	**C**	ISAVIG	**C**	S	**C**	SNKV	**C**	YKNGSIP	*V. philippica*	He et al. (2011)
Viphi F	**C**	GES	**C**	VFIP	**C**	ISAIIG	**C**	S	**C**	SSKV	**C**	YKNGSIP	*V. philippica*	He et al. (2011)
Viphi G	**C**	GES	**C**	VFIP	**C**	ISAIIG	**C**	S	**C**	SNKV	**C**	YKNGSIP	*V. philippica*	He et al. (2011)
Viphi H	**C**	AES	**C**	VWIP	**C**	TVTAIVG	**C**	S	**C**	SWGV	**C**	YNGIP	*V. philippica*	He et al. (2011)
Vibi D	**C**	GET	**C**	FGGR	**C**	NTPG	**C**	T	**C**	SYPI	**C**	TRNGLPV	*Viola biflora*	Herrmann et al. (2008)
Vibi E	**C**	AES	**C**	VWIP	**C**	TVTALIG	**C**	G	**C**	SNKV	**C**	YNGIP	*V. biflora*	Herrmann et al. (2008)
Vibi G	**C**	GES	**C**	VFIP	**C**	LTSAIG	**C**	S	**C**	KSKV	**C**	YKNGTFP	*V. biflora*	Herrmann et al. (2008)
Vibi H	**C**	AES	**C**	VYIP	**C**	LTTVIG	**C**	S	**C**	KSKV	**C**	YKNGLLO	*V. biflora*	Herrmann et al. (2008)
Viba 15	**C**	GET	**C**	VGGT	**C**	NTPG	**C**	A	**C**	SWPV	**C**	TRNGLPV	*Viola baoshanensis, V. philippica*	Zhang et al. (2009)
Viba 17	**C**	GET	**C**	VGGT	**C**	NTPG	**C**	G	**C**	SWPV	**C**	TRNGLPV	*V. baoshanensis, V. philippica*	Zhang et al. (2009)
Hypa A	**C**	AES	**C**	VYIP	**C**	TITALLG	**C**	S	**C**	KNKV	**C**	YNGIP	*Hybanthus parviflorus*	Broussalis et al. (2001)
Cter 1	**C**	GES	**C**	VFIP	**C**	ITGAIG	**C**	S	**C**	KSKV	**C**	YRNGIP	*Clitoria ternatea*	Nguyen, Zhang, Ngan, et al. (2011)
Cter 2	**C**	GES	**C**	VQGE	**C**	YTPG	**C**	S	**C**	DWPI	**C**	KKNGEFLK	*C. ternatea*	Nguyen, Zhang, Ngan, et al. (2011)
Cter 3	**C**	GET	**C**	TLGT	**C**	YVPD	**C**	S	**C**	SWPI	**C**	MKNGLPT	*C. ternatea*	Nguyen, Zhang, Ngan, et al. (2011)
Cter 4	**C**	GES	**C**	VFIP	**C**	ITAAIG	**C**	S	**C**	KSKV	**C**	YRNGIP	*C. ternatea*	Nguyen, Zhang, Ngan, et al. (2011)
Cter 7	**C**	GES	**C**	VFIP	**C**	TVTALL	**C**	S	**C**	KDKV	**C**	YKNGIP	*C. ternatea*	Nguyen, Zhang, Ngan, et al. (2011)
Cter 10	**C**	GES	**C**	VYIP	**C**	TVTALL	**C**	S	**C**	KDKV	**C**	YKNGIP	*C. ternatea*	Nguyen, Zhang, Ngan, et al. (2011)
Cter 12	**C**	GES	**C**	VFIP	**C**	ITGAIG	**C**	S	**C**	KSKV	**C**	YRDGIP	*C. ternatea*	Nguyen, Zhang, Ngan, et al. (2011)
CterM	**C**	GET	**C**	TLGT	**C**	YVPD	**C**	S	**C**	SWPI	**C**	MKNGLPT	*C. ternatea*	Poth et al. (2011)

The six highly conserved cysteine residues are shown in bold. Backbone loops between successive cysteine residues are labelled 1–6.
*A full list of cyclotide sequences can be accessed from CyBase (Kaas & Craik, 2010; Wang, Kass, et al., 2008).

Plan, & Craik, 2006; Ding, Bai, & Qian, 2014; Dutton et al., 2004; Gerlach, Burman, Bohlin, Mondal, & Göransson, 2010; Göransson, Luijendijk, Johansson, Bohlin, & Claeson, 1999; Gustafson, Sowder, et al., 1994; Gustafson, Walter, 2000; Hallock et al., 2000; He et al., 2011; Herrmann et al., 2008; Ireland, Colgrave, & Craik, 2006; Jennings, West, Waine, Craik, & Anderson, 2001; Nguyen, Zhang, Ngan, et al., 2011, Pinto et al., 2012; Rosengren, Daly, Plan, Waine, & Craik, 2003; Saether et al., 1995; Svangård et al., 2004; Tang et al., 2010a, 2010b; Wang, Colgrave, et al., 2008; Wang, Colgrave, Ireland, Kaas, & Craik, 2009; Witherup et al., 1994; Wong, Taichi, Nishio, Nishiuchi, & Tam, 2011; Yeshak, Burman, Asres, & Göransson, 2011; Zhang et al., 2009). These cyclotides were originally discovered in a wide range of screening studies and have a diverse range of activities. A full list of all currently known cyclotide sequences is available on CyBase, http://www.cybase.org.au/ (Kaas & Craik, 2010; Wang, Kaas, Chiche, & Craik, 2008).

Overall it seems that cyclotides have evolved as a natural combinatorial template with many variations in the amino acid content of individual loops, ranging from highly hydrophilic such as in loop 1 to highly hydrophobic such as loop 5 of the Möbius cyclotides. The fact that natural cyclotides have such extensive variation has encouraged chemists to design their own modified cyclotides, principally those in which a bioactive epitope is grafted into a cyclotide framework. Such applications have been underpinned by a range of structure–activity studies, including point mutagenesis, scanning mutagenesis, truncation of cyclotide loops, acyclic permutation of the cyclic backbone, and selective removal or modification of disulfide bonds. Before discussing these engineered cyclotides, we begin with an overview of the "natural" activities of cyclotides, namely their host defense functions.

2. PESTICIDAL ACTIVITIES OF CYCLOTIDES

Several studies demonstrating their activities against various pests support the hypothesis that the evolution of cyclotides occurred primarily to deter plant predators. Table 2 lists the pesticidal cyclotides studied to date and the corresponding pest species against which the cyclotides have been tested against. Most research in this area has been on insecticidal activities but cyclotides also show promise as nematocidal and molluscicidal agents.

Table 2 Pesticidal Activities of Natural Cyclotides

Cyclotide	Pesticidal Activity			References
	Insecticidal	**Anthelmintic**	**Molluscicidal**	
kB1	*Helicoverpa punctigera* *Helicoverpa armigera*	*Haemonchus contortus, Trichostrongylus colubriformis, Necator americanus, Ancylostoma caninum*	*Pomacea canaliculata*	Colgrave, Kotze, Ireland, et al. (2008), Colgrave et al. (2009), Jennings et al. (2001), and Jennings et al. (2005)
kB2	*H. armigera*	*H. contortus, T. colubriformis, Schistosoma japonicum, Schistosoma mansoni*	*P. canaliculata*	Colgrave, Kotze, Ireland, et al. (2008), Jennings et al. (2005), Malagón et al. (2013), and Plan et al. (2008)
kB6		*H. contortus, T. colubriformis, A. caninum*		Colgrave, Kotze, Ireland, et al. (2008) and Colgrave et al. (2009)
kB7		*H. contortus, T. colubriformis*		Colgrave, Kotze, Ireland, et al. (2008)
Cycloviolacin O2		*H. contortus, T. colubriformis*	*P. canaliculata*	Colgrave, Kotze, Ireland, et al. (2008) and Plan et al. (2008)
Cycloviolacin O14		*H. contortus, T. colubriformis, N. americanus, A. caninum*		Colgrave, Kotze, Ireland, et al. (2008) and Colgrave et al. (2009)
Cycloviolacin O1, O3, O8, O13, O15, O16, O24, H3, Y4, Y5, varv A, varv E, vhl-1		*H. contortus, T. colubriformis*		Colgrave, Kotze, Ireland, et al. (2008)
Hypa A	*Ceratitis capitata*			Broussalis et al. (2010)
Parigidin-br1	*Diatraea saccharalis, Spodoptera frugiperda*			Pinto et al. (2012)
CterM	*H. armigera*			Poth et al. (2011)

2.1 Insecticidal Activity

The first account of the insecticidal activity of a cyclotide was in 2001 when kalata B1 (kB1) was shown to be detrimental to the major cotton pest *Helicoverpa punctigera*. When kB1 was added at 0.8 μmol per gram of artificial diet (a concentration similar to what is naturally found in leaves of *Oldenlandia affinis*) the *H. punctigera* mortality rate was 50% at 16 days. Furthermore, the survivors' development was arrested at the first instar, and on average weighed only 3 mg, compared to the control insects, which weighed of 284 mg (Jennings et al., 2001). The difference in morphology between control larvae and larvae fed with a diet containing kB1 is illustrated in Figure 1.

The insecticidal activity against the related lepidopteran pest *Helicoverpa armigera* appears to be somewhat lower. In a similar feeding trial the survivors weighed 112 mg or 135 mg after 14 days on an artificial diet supplemented with kB1 or kB2, respectively, and the mortality rate was 20–30%, respectively (Jennings et al., 2005). Similarly, when the *Clitoria ternatea* cyclotide, cterM, was supplemented in the diet of *H. armigera,* larval development was also stunted (Poth, Colgrave, Lyons, Daly, & Craik, 2011). Another study showed that the cyclotide paragidin-br1 from *Palicourea rigida* increased the mortality rate and decreased the viability of the lepidopteran species

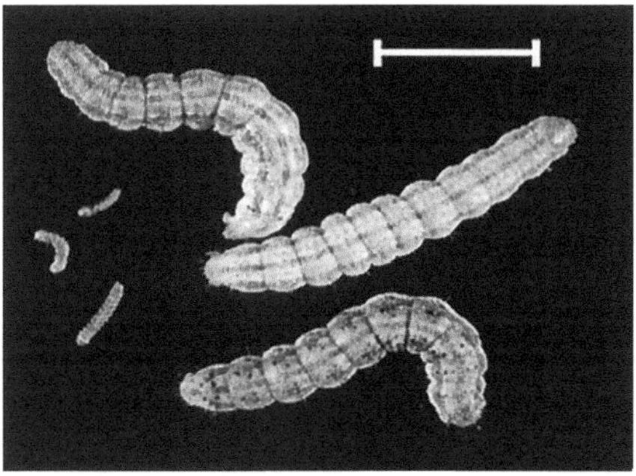

Figure 1 ***Effect of kalata B1 on the development of*** **Helicoverpa punctigera** ***larvae.*** The cyclotide dramatically decreased the growth and development of diet-fed larvae. After 16 days on an artificial diet supplemented with a kB1 concentration which is naturally found in leaves (0.8 μmol/g), larvae development was arrested at the first instar (left). Larvae fed with a control diet containing no cyclotides reached the fifth instar (right). *Reproduced with permission from Proceedings of the National Academy of Sciences USA, 98, 1064 (2001). Copyright 2015 National Academy of Sciences.*

Diatraea saccharalis and *Spodoptera frugiperda* (Pinto et al., 2012). Furthermore, development was delayed and mortality increased when the dipteran species *Ceratitis capitata* was fed with an artificial diet containing the cyclotide hypa A from *Hybanthus parviflorus* (Broussalis, Clemente, & Ferraro, 2010). Electron microscopic studies of kalata peptides have revealed that the insecticidal activity is due to their ability to cause cell lysis via epithelial cell plasma membrane disruption (Barbeta, Marshall, Gillon, Craik, & Anderson, 2008).

Cyclotides have also been shown to affect insects that are not necessarily detrimental to plants. For instance, a recent study showed that cyclotide-containing extracts from *Ixora coccinea* and *Allamanda violacea* inhibited the fifth larval instar development of the disease vectors *Aedes aegypti* and *Anopheles stephensi* (Werle, Kolmar, Albrecht, & Bernkop-Schnürch, 2008).

Overall, in our opinion cyclotides have promising potential as agricultural pest control agents and could be applied either topically to the host target plant or incorporated into transgenic plants. In the latter case they could, in principle, be utilized in combination with existing control agents such as Bt toxin to alleviate the development of resistance.

2.2 Anthelmintic Activity

Numerous cyclotides have been shown to be potent against parasitic helminths of sheep, cattle, dogs, and humans (Colgrave, Kotze, Huang, et al., 2008; Colgrave, Kotze, Ireland, Wang, & Craik, 2008; Colgrave et al., 2009). For example, one of the earliest studies showed that kB1, kB2, kB5, kB6, and kB7 inhibited larval development, decreased adult motility, and, to some extent, inhibited egg hatching of the sheep nematodes *Haemonchus contortus* and *Trichostrongylus colubriformis* (Colgrave, Kotze, Huang, et al., 2008). Figure 2 shows the percent larval inhibition of these two helminthic species in response to increasing concentrations of kB1 and kB2.

Among the cyclotides tested in this study, kB6 was shown to be the most potent against the larvae of both nematodes. Furthermore, when kB6 activity was compared to commercial anthelmintic drugs, its potency ($IC_{50} = 1.7\ \mu M$) was found to be comparable to levamisole ($IC_{50} = 8.9\ \mu M$) and naphthalophos ($IC_{50} = 7.5\ \mu M$), two of the most widely used anthelmintics (Colgrave, Kotze, Huang, et al., 2008). A similar study utilizing natural cyclotides from *Viola* species (varv A, varv E, cycloviolacin O1–O3, O8, O13–16, O24, H3, Y4, and Y5) demonstrated similar nematocidal effects (Colgrave, Kotze, Ireland, Wang, & Craik, 2008). Cycloviolacin O2, the most highly potent variant ($IC_{50} = 0.12\ \mu M$), was found to be

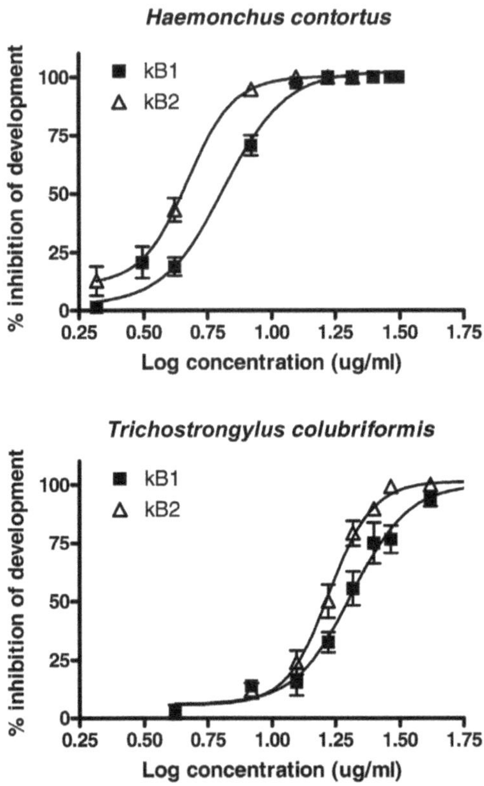

Figure 2 ***Effect of kalata B1 and kalata B2 on the development of egg to the third instar of the parasitic worms* Haemonchus contortus *and* Trichostrongylus colubriformis.** A dose-dependent inhibition of larvae development was observed in both species. Each data point is the mean ± SE of at least three replicates for every tested concentration. *Reproduced with permission from Biochemistry, 47, 5581 (2008). Copyright 2015 American Chemical Society.*

comparable in activity to the commercial anthelmintics thiabendazole (IC_{50} = 0.064 μM), levamisole (IC_{50} = 1.4 μM), and naphthalophos (IC_{50} = 0.73 μM). In another study, which determined cyclotide activity against parasitic nematodes of humans (*Necator americanus*) and dogs (*Ancylostoma caninum*), kB1, kB6, and cycloviolacin O14 showed significant anthelmintic activities (Colgrave et al., 2009).

Cyclotides are also effective in controlling trematodes. *Schistosoma japonica* and *Schistosoma mansoni* adults were killed within 5 min of addition of kB1 and kB2, albeit at relatively high concentrations (500–1000 μg/mL) (Malagón et al., 2013). That study showed that cyclotide potency was also likely to be dependent on trematode membrane composition, as exemplified by the varying effects of cyclotides against trematode samples of different sex, strain, and species.

When a series of alanine mutants of kB1 was assayed against nematode larvae, the activity was found to be notably reduced when any one of a particular set of residues clustered at the surface were mutated (Colgrave, Kotze, Huang, et al., 2008). Such a result might at first imply that the localized region of the cyclotide associated with these residues is a receptor binding face, but the study also showed that the naturally occurring L-isomer and the synthetic D-isomer of kB1 had similar activities, indicating that activities are chirality-independent. Furthermore, the study demonstrated that cyclotides caused a fluorescent dye to leak from vesicles used as model membrane mimetics. Together, these findings support the hypothesis that the anthelmintic activity of kB1 can be attributed to its ability to lyse membranes, rather than through interaction with stereospecific receptors. In support of this hypothesis, when a lysine scan was conducted to determine the effect of incorporating positively charged residues into kB1, it was found that six residues clustered on the surface-tolerated substitution and that nematocidal activity was enhanced when positively charged residues were introduced into this region (Huang, Colgrave, Clark, Kotze, & Craik, 2010). Because of the initial electrostatic interaction between the positively charged residues and the membrane, cyclotide adherence to the membrane was increased. This then increased the opportunity for the hydrophobic patch in the cyclotide to interact with the membrane, thus enabling peptide insertion. Once inserted, it has been suggested that cyclotides most likely self-associate and trigger pore formation (Huang et al., 2010). The types of mutagenesis studies that have been pivotal in reaching these conclusions are described in more detail in Section 5.2.

Much remains to be done to understand the mode of action of nematocidal activity of cyclotides but in addition to the studies described above, a range of new approaches, including those involving computational analysis, show great promise in assisting in this endeavor. For example, a recent study showed that when the hydrophobic patch and the positively charged residues on a cyclotide are asymmetrically distributed, the anthelmintic activity is proportional to the size of the surface that interacts with the membrane (Park, Strömstedt, & Göransson, 2014).

2.3 Molluscicidal Activity

An earlier study demonstrated the potential of cyclotides to control golden apple snail (*Pomacea canaliculata*), a pest of rice prevalent in the wetlands

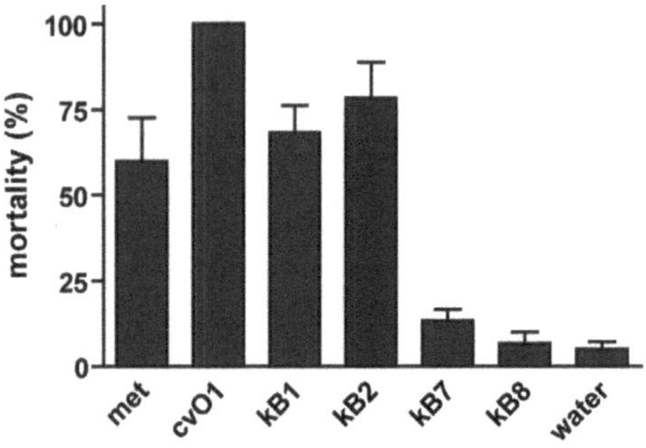

Figure 3 ***Molluscicidal activities of cyclotides.*** Golden apple snail (*Pomacea canaliculata*) was treated with either 65 μM of a cyclotide (cycloviolacin O1 (cvO1), kalata B1 (kB1), kB2, kB7, and kB8) or the commercially available molluscicide, metaldehyde. Each bar is the mean ± SE of six replicates except for cvO1 which had only four. Water was used as a control. *Reproduced with permission from Journal of Agricultural and Food Chemistry 56, 5237 (2008). Copyright 2015 American Chemical Society.*

of several Southeast Asian countries, but principally in the Philippines (Plan, Saska, Cagauan, & Craik, 2008). Figure 3 shows a comparison of the activities of the commercial molluscicide metaldehyde and the cyclotides cycloviolacin O1, kB1, kB2, kB7, and kB8.

At a tested concentration of 65 μM, cycloviolacin O1, kB1, and kB2 caused greater snail mortality (100%, 68%, and 78%, respectively) than metaldehyde (60%), whereas kB7 and kB8 did not. Encouragingly, when kB2 was tested against the nontarget organism Nile tilapia fish (*Oreochromis niloticus*), it was found to be less potent than the commonly used piscicidal agent, rotenone. Although the study was limited to golden apple snail, it demonstrated that certain cyclotides can both be highly potent and specific to a particular species of mollusc. As with insecticidal applications, and because of the ribosomal biosynthetic pathway of cyclotides, the potential exists to express molluscicidal cyclotides in transgenic rice as one approach to crop protection. With further studies the development of cyclotide-based molluscicidal agents or transgenic plants appears to be feasible.

3. PHARMACEUTICAL ACTIVITIES OF CYCLOTIDES

Numerous studies have demonstrated the potential pharmaceutical applications of natural cyclotides. Table 3 categorizes these activities into four groups, i.e., uterotonic, anti-HIV, antimicrobial, and immunosuppressive

Table 3 Pharmaceutical Activities of Natural Cyclotides

Cyclotide	Pharmaceutical Activity: Uterotonic	Anti-HIV	Antimicrobial	Immunosuppressive	References
kB1	Rat uteri	HIV-1 in CEM-SS	*Klebsiella oxytoca, Staphylococcus aureus, Micrococcus luteus, Candida kefyr, Escherichia coli*	Peripheral blood mononuclear cells	Daly et al. (2004), Gran (1973b), Gran et al. (2008), Grundemann et al. (2012), Gründemann et al. (2013), Tam et al. (1999), and Wang, Colgrave, et al. (2008)
kB2	Rat uteri		*Salmonella enterica, S. aureus, E. coli*		Dörnenburg (2010), Gran et al. (2008), and Pränting et al. (2010)
kB7	Rat uteri		*E. coli*		Gran et al. (2008)
kB8		HIV-1 in CEM-SS			Daly et al. (2006)
cirA		10 HIV strains in CEM-SS and MT-2	*Proteus vulgaris, S. aureus, Candida tropicalis, Candida kefyr*		Gustafson et al. (1994) and Tam et al. (1999)
cirB			*Pseudomonas aeruginosa, E. coli, P. vulgaris, K. oxytoca, S. aureus, C. kefyr*		Tam et al. (1999)
Circulin C–F		10 HIV strains in CEM-SS and MT2			Gustafson et al. (1994)
Cyclopsychotride A			*E. coli, P. aeruginosa, P. vulgaris, K. oxytoca, S. aureus, M. luteus, C. kefyr, C. tropicalis*		Tam et al. (1999)

(Continued)

Table 3 Pharmaceutical Activities of Natural Cyclotides—cont'd

	Pharmaceutical Activity				
Cyclotide	**Uterotonic**	**Anti-HIV**	**Antimicrobial**	**Immunosuppressive**	**References**
Hedyotide B1			*E. coli, Streptococcus salivarius*		Nguyen, Zhang, Wang, et al. (2011)
Clitoria ternatea CT1 and CT4			*E. coli, Klebsiella pneumonia, P. aeruginosa*		Nguyen, Zhang, Ngan, et al. (2011)
Cycloviolacin O2		HIV in U1	*S. enterica, S. aureus, E. coli, S. pyogenes, K. pneumonia, P. aeruginosa*		Dörnenburg (2010), Gerlach et al. (2013), and Pränting et al. (2010)
Vaby A and vaby D			*S. enterica, E. coli*		Pränting et al. (2010)
Cycloviolin A-D		HIV-1 in CEM-SS			Hallock et al. (2000)
Cycloviolacin O13, O14, O24		HIV-1 in CEM-SS			Ireland et al. (2008)
Cycloviolacin Y1, Y4, Y5, varv E		HIV (cell line unspecified)			Wang, Colgrave, et al. (2008)
Palicourein		HIV-1 in CEM-SS			Bokesch et al. (2001)
vhl-1		HIV strain unspecified			Chen et al. (2005)
Viola tricolor extract				Human peripheral lymphocytes	Hellinger et al. (2014)

activities. The cyclotides and their corresponding target cells/microbial organisms are listed in the table for convenience in the following discussion.

3.1 Uterotonic Activity

The earliest reported medically relevant activity of cyclotides was based on anecdotal reports of an African folk medicine made from the plant *O. affinis* (Gran, 1973a, 1973c). As noted in earlier chapters of this book, Congolese women were reported to accelerate labor and associated childbirth using a decoction known locally as "kalata kalata", made by boiling the above ground parts of the plant in water. Norwegian physician Lorents Gran (Gran, 1973b) discovered that the active uterotonic agent was a peptide, which he named kB1 and subsequently showed to have activity in isolated rat uteri and in placental strips (Gran, Sletten, & Skjeldal, 2008; Sletten & Gran, 1973). It should be noted that the use of this traditional African folk medicine was not a recommended practice, as it was associated with an increased incidence of obstetrical complications, in some cases leading to emergency cesarean sections (Gran, Sandberg, & Sletten, 2000). The existence of side effects associated with kB1 is not surprising given that its natural function is associated with its toxicity, albeit directed at plant pests. We thus do not suggest the use of natural cyclotides as medicinal agents, but believe their value is more as scaffolds in drug design. Nevertheless, natural cyclotides have been screened for a range of pharmaceutical applications, as described below.

3.2 Anti-HIV Activity

The anti-HIV activities of cyclotides were first reported in 1994, whereby circulin A and circulin B from *Chassalia parvifolia* prevented the development of HIV-related cytopathic effects (EC_{50} of 0.04–0.26 μM) and at the same time displayed strong cytotoxic activity against infected cells ($IC_{50} = 0.5$ μM) (Gustafson et al., 1994). Several years later novel circulins (C–F) were similarly demonstrated to inhibit the cytopathic effects of HIV in infected cell lines CEM-SS and MT-2 (EC_{50} of 0.05–0.27 μM) (Gustafson et al., 2000). Subsequent studies on various cyclotides from both the bracelet and Möbius subfamilies (cycloviolin A–D; cycloviolacin O2, O13, O14, O24, Y1, Y4, Y5; varv E; kB1; kB8; palicourein; and vhl-1) also demonstrated anti-HIV activities of cyclotides, with EC_{50} values ranging from 0.04 to 2.5 μM (Bokesch et al., 2001; Chen et al., 2005; Daly et al., 2006; Daly, Gustafson, & Craik, 2004; Gerlach et al., 2013; Hallock et al., 2000; Ireland, Wang, Wilson, Gustafson, & Craik, 2008; Wang, Colgrave, et al., 2008).

Many of the early studies on anti-HIV activities focused on bracelet cyclotides. However, in 2004, it was shown that kB1, a cyclotide from the Möbius subfamily (see Chapter 6 Structural Studies of Cyclotides for definitions of cyclotide subfamilies), exhibited anti-HIV activity comparable with the activities of previously reported bracelet cyclotides; the activity of kB1 was reported to be seemingly more therapeutically promising as it displayed reduced cytotoxic effects in normal cells (Daly et al., 2004). In the same study, it was observed that acyclic mutants of kB1 did not have anti-HIV activity, thus emphasizing the importance of an intact cyclic backbone for inhibiting HIV. Furthermore, the presence of a hydrophobic patch in loops 2 and 3 of bracelet cyclotides and in loops 5 and 6 of Möbius cyclotides were found to be necessary for anti-HIV activity and presumably interact with cell membranes (Ireland et al., 2008). The same study also showed that a negative charge in loop 5 of bracelet cyclotides reduces their cytotoxic activity. Designing sequences that enhance hydrophobicity in the regions noted above and modifying the charge of certain loops could aid in the development of cyclotide-based anti-HIV therapeutics (Ireland et al., 2008).

3.3 Antimicrobial Activity

In 1999, four synthetically produced cyclotides (kB1, circulin A, circulin B, and cyclopsychotride A) were tested for their antimicrobial properties and exhibited varying potencies (MIC of 0.19–56.5 μM) against several bacterial species (Tam, Lu, Yang, & Chiu, 1999), as summarized in Table 3. Subsequent studies similarly demonstrated varying effects of cyclotides on different species (Gran et al., 2008; Nguyen, Zhang, Ngan, et al., 2011; Nguyen, Zhang, Wang, et al., 2011; Pränting, Loov, Burman, Göransson, & Andersson, 2010). In addition to being specific to certain bacterial species, cyclotide antimicrobial activity might also be strain specific. In one study kB1 did not affect *Escherichia coli* ATCC 25922 (Tam et al., 1999), whereas another study showed inhibition of *E. coli* K12 at 5 μg/μL (Gran et al., 2008). Most of the tested cyclotides exhibited stronger antimicrobial activity in assays with low salt concentrations and were ineffective in high salt concentrations. This is probably because salts interfere with the electrostatic interactions that are necessary to facilitate initial contact between the cyclotides and bacterial membranes (Pränting et al., 2010). The general lack of antimicrobial activity at physiologically relevant salt concentrations suggests that natural cyclotides will probably not be practically useful as antimicrobial agents.

There is currently no reported study on the activity of cyclotides against plant pathogenic bacteria as almost all of the examined species so far have been human pathogens. However, one study looked into the potential toxic effects of cyclotides against soil bacteria (EC_{50} of 7–26 μM) and algae (EC_{50} of 12–140 μM). The same study also examined toxicity against two plant species, duckweed (EC_{50} of 9–40 μM) and lettuce (EC_{50} of 4–54 μM) (Ovesen et al., 2011). Further studies are required to assess the potential environmental toxicity of cyclotides.

3.4 Immunosuppressive Activity

A study in 2012 showed that an aqueous *O. affinis* extract inhibited the proliferation of activated peripheral blood mononuclear cells (PBMCs) in a dose-dependent manner (50–100 μg/mL) without a concomitant increase in the number of dead cells (Grundemann, Koehbach, Huber, & Gruber, 2012). An increase in cytotoxic effect was observed when the extract concentration was higher than 100 μg/mL. After demonstrating the antiproliferative property of the *O. affinis* crude extract, the study tested the effect of purified kB1, and it similarly showed a dose-dependent decrease in proliferation of PBMC over concentrations ranging from 1.8 to 14 μM, with an IC_{50} of 3.9 μM. Cytotoxicity was observed when the kB1 concentration was greater than 14 μM. The study further showed that the cyclotide-induced antiproliferation effect was neither due to apoptosis nor cell death, but rather due to inhibition of growth in a cytostatic manner. In essence, the study demonstrated that at a certain concentration range, cyclotides can act as immunosuppressants. They can be classified as such because, in contrast to cytotoxic agents that cause side effects, immunosuppressants simply maintain the cells in an inactive but viable state (Grundemann et al., 2012).

A subsequent study examined the immunosuppressive mode of action of kB1 (Gründemann et al., 2013). Like most biological activities of cyclotides, the immunosuppressive activity was lost after making mutations like [T8K], [V10A], or [V10K], which are in the hydrophilic region clustered around Glu-7. This region is thought to be responsible for pore formation in membranes and has been proposed to be driven via cyclotide self-association (Gründemann et al., 2013; Huang et al., 2010). The study further showed that the immunosuppressive activity of kB1 was mainly due to its effects on T-cell biology, as adding kB1 to activated T cells led to a decreased IL-2 release, a reduced expression of IL-2 surface receptor (CD25), and a reduced effector function as typified by an initial reduction of IFN-γ and TNF-α (Gründemann et al., 2013). A similar study showed that the cyclotides present in a *Viola tricolor* herbal extract exhibited immunosuppressive activity (Hellinger et al., 2014). Overall, there appears to be considerable interest in pursuing the immunosuppressive activities of cyclotides.

4. CYTOTOXIC ACTIVITIES OF CYCLOTIDES

The natural cyclotides that have been reported to be cytotoxic and the corresponding cell lines against which cyclotides have been tested are listed in Table 4. Although numerous cyclotides have been found to be cytotoxic to a range of human cancer cell lines, offering promise as anticancer agents, some of them also promote lysis of human red blood cells (RBCs), which is clearly an undesired activity from a pharmaceutical perspective. In the following sections we examine the relative toxicity of cyclotides to RBCs and cancer cells.

4.1 Hemolytic Activity

A study in 1999 that investigated the hemolytic activity of synthetically produced cyclotides showed that circulin A, circulin B, kB1, and cyclopsychotride A had the ability to lyse human RBCs but only at relatively high concentrations, with a HD_{50} range of 400−1500 μM (Tam et al., 1999). A more recent study suggested that the hemolytic activity of native kB1 is stronger than previously reported ($HD_{50} = 11.7$ μM) (Wang, Colgrave, et al., 2008). Subsequent studies have suggested that the hemolytic potency varies significantly from one cyclotide to another. For instance, the HD_{50} values of cyclotides from *Viola* sp. (cycloviolacin O2, O13, O14, O24, Y4, and Y5) are 8−36 μM (Ireland et al., 2006; Wang, Colgrave, et al., 2008) and the HD_{50} values of cyclotides from *C. ternatea* are 7−13 μM (Nguyen, Zhang, Ngan, et al., 2011).

A recent study has suggested that the hemolytic activity of cyclotides correlates with their nematocidal activity, thus implying that these biological activities potentially involve a similar membrane-based mechanism (Huang et al., 2010). Thus, one potential impediment to the development of cyclotides would be that mutation of residues implicated in hemolytic activity might simultaneously reduce the potencies of cyclotides against target cells if the latter employs the same membrane-based mechanism. Cyclotide-based therapeutics should ideally be devoid of hemolytic activity. Nevertheless, it should be noted that even the most highly hemolytic cyclotide, cter 1 ($HD_{50} = 7.1$ μM), is still significantly less potent than melittin, the "gold

Table 4 Cytotoxic Activities of Natural Cyclotides

Cyclotide	Cytotoxic Activity: Hemolytic Activity	Cytotoxic Activity: Cancer Cell Activity	References
cirA, cirB	Human red blood cells		Tam et al. (1999)
kB1	Human red blood cells		Tam et al. (1999)
kB2	Human red blood cells		Gran et al. (2008)
kB7	Human red blood cells		Gran et al. (2008)
Cyclopsychotride A (cpt A)	Human red blood cells		Tam et al. (1999)
Cycloviolacin O2	Human red blood cells	RPMI-8226, U-937, ACHN, CCRF-CEM, NCI-H69, MCF-7, U251, MDA-MB-231, A549, DU145, BEL-7402	Gerlach, Rathinakumar, et al. (2010), Ireland et al. (2006), Lindholm et al. (2002), and Tang et al. (2010a)
Cycloviolacin O13, O14, O24	Human red blood cells		Ireland et al. (2006)
Cter 1, cter 3, cter 4	Human red blood cells		Nguyen, Zhang, Ngan, et al. (2011)
Varv A		RPMI-8226, U-937, ACHN, CCRF-CEM, NCI-H69, HFF-1, MM96-L, HeLa, BGC-823	He et al. (2011), Lindholm et al. (2002), Svangård et al. (2004), and Tang et al. (2010a)
Varv E		RPMI-8226, U-937	Svangård et al. (2004)
Varv D		U251	Tang et al. (2010b)
Varv F		RPMI-8226, U-937, ACHN, CCRF-CEM, NCI-H69	Lindholm et al. (2002) and Svangård et al. (2004)
Vitri A		U-937 GTB, RPMI-8226, U251, MDA-MB-231, A549, DU145, BEL-7402	Svangård et al. (2004) and Tang et al. (2010a)

(Continued)

Table 4 Cytotoxic Activities of Natural Cyclotides—cont'd

	Cytotoxic Activity		
Cyclotide	**Hemolytic Activity**	**Cancer Cell Activity**	**References**
Vitri F		U251, MDA-MB-231, A549, DU145, BEL-7402	Tang et al. (2010a)
Vila A and vila B		U251, MDA-MB-231, DU145, BEL-7402	Tang et al. (2010b)
Vila D		U251	Tang et al. (2010b)
Vibi D, E, G, H		U-937	Herrmann et al. (2008)
Viphi A, D-E, F-G		HFF-1, MM96L, HeLa, BGC-823	He et al. (2011)
Viba 15, viba 17		HFF-1, MM96L, HeLa, BGC-823	He et al. (2011)
Psyles A, C, E		U-937, MCF-7	Gerlach, Burman, et al. (2010), Gerlach, Rathinakumar, et al. (2010)
Vaby A and D		U-937	Yeshak et al. (2011)
Hedyotide 5–9		Pancreatic cells (BxPC3, Capan-2, MOH1, PANC1)	Ding et al. (2014)
Cter 1 and 3		HeLa	Nguyen, Zhang, Ngan, et al. (2011)
Cter 2 and 4		HeLa, A549, A549/paclitaxel	Nguyen, Zhang, Ngan, et al. (2011) and Zhang et al. (2013)
Cter 7, 10, and 12		A549, A549/paclitaxel	Zhang et al. (2013)

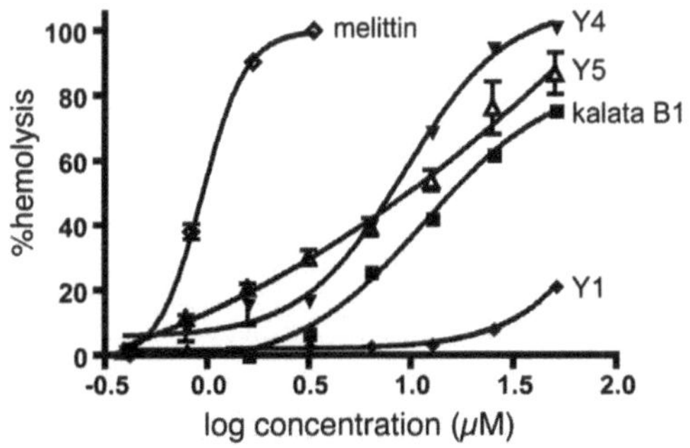

Figure 4 ***Hemolytic activities of selected cyclotides.*** Cyclotides kalata B1, cycloviolacin Y1, Y4, and Y5 from the Chinese herb *Viola yedoensis* in comparison to melittin, a highly hemolytic agent from bee venom. *Reproduced with permission from Journal of Natural Products, 71, 47 (2008). Copyright 2015 American Chemical Society.*

standard" hemolytic peptide derived from bee venom (Nguyen, Zhang, Ngan, et al., 2011; Wang, Colgrave, et al., 2008). The relative potency of selected cyclotides is illustrated in Figure 4, which compares the hemolytic activity of melittin and various cyclotides from the Chinese medicinal herb *Viola yedoensis*.

4.2 Cancer Cell Toxicity

There is a substantial literature on the anticancer properties of cyclotides. To date, numerous cyclotides have been tested against a range of human cancer cell lines, including breast cancer (MCF-7 and MDA-MB-231), cervical cancer (HeLa cells), gastric cancer (BGC-823), glioblastoma (U251), hepatocellular carcinoma (BEL-7402), leukemia (CCRF-CEM), lung carcinoma (A549), small cell lung cancer (NCI-H69), lymphoma (U-937-GTB/Vcr), melanoma (MM96-L), myeloma (RPMI-8226), pancreatic cancer (BxPC3, Capan-2, MOH1, and PANC1), prostate cancer (DU145), and renal adenocarcinoma (ACHN) (Ding et al., 2014; Gerlach, Burman, et al., 2010; Gerlach, Rathinakumar, et al., 2010; Herrmann et al., 2008; Huang et al., 2009; Lindholm et al., 2002; Svangård et al., 2004; Tang et al., 2010a, 2010b; Yeshak et al., 2011; Zhang, Xiao, Jin, Zhang, & Zhou, 2013). A number of drug-resistant cell lines were also found to be inhibited by various cyclotides (Gerlach, Burman, et al., 2010; Lindholm et al., 2002; Zhang et al., 2013). In all cases tested, the activity increased as the concentration increased, and the potency varied depending on the cyclotide used and the cancer cell line tested.

Although cyclotides are efficient at inhibiting the growth of numerous cancer cell lines, some cyclotides are just as detrimental to noncancerous cells (Henriques, Huang, Chaousis, Wang, & Craik, 2014). Thus, to be pharmaceutically relevant such cyclotides need to be tailored to preferentially recognize specific cancer cells. One way of doing this would be to target differences in cell surface composition. Figure 5 schematically illustrates some differences between the membrane composition of cancer cells and healthy cells and highlights the currently understood mechanism of cell disruption. A recent study suggested that because cancer cells have a greater amount of negatively charged phosphatidylserine phospholipid exposed on the surface compared to

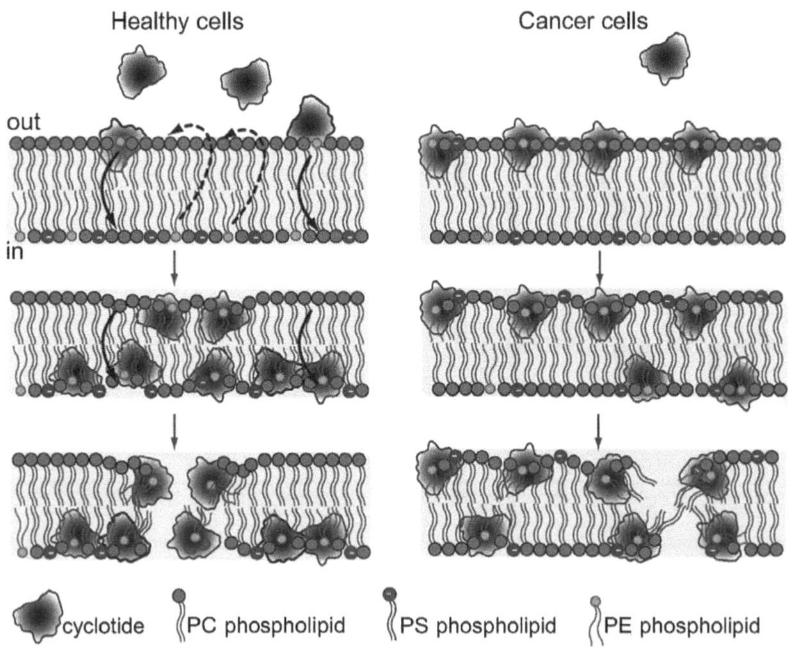

Figure 5 ***Mechanism of membrane binding of cyclotides.*** The mechanism in which cyclotides target and disrupt membranes of human healthy and cancer cells is represented in a model showing phosphatidylcholine (PC), phosphatidylethanolamine (PE), and phosphatidylserine (PS) phospholipids. Trans-bilayer movement of PE phospholipids is indicated in dark arrows. To bind to the membrane of both healthy and cancer cells, cyclotides target PE phospholipids. Insertion and membrane disruption follow. Because cancer cells have more exposed PE phospholipids, cyclotides can target cancer cells more effectively than healthy cells. However, PE phospholipids in healthy cells are more dynamic than cancer cells; hence, peptide insertion is more favorable in healthy cells. *Reproduced with permission from ChemBiochem, 15, 1956 (2014). Copyright 1999–2015 John Wiley & Sons, Inc.* (See color plate)

normal cells, increasing the overall positive charge of a cyclotide may be helpful (Henriques et al., 2014). Grafting an epitope into a cyclotide scaffold such that it preferentially binds target cancer cells is another possible approach to making cyclotides more effective anticancer agents (Poth, Chan, & Craik, 2013), as described in more detail in Section 5.5.

5. STRUCTURE–ACTIVITY RELATIONSHIPS

The development of solid-phase peptide chemistry methods for the synthesis of cyclotides was first reported in the late 1990s, with key papers from the groups of Tam (Tam & Yu, 1997, 1998; Tam et al., 1999) and Craik (Daly, Love, Alewood, & Craik, 1999). These studies were based on native chemical ligation approaches, which are described in more detail in Chapter 9 Chemical and Biological Production of Cyclotides. They demonstrated, using Boc-based solid-phase peptide chemistry, that native cyclotide sequences could be assembled and folded. Folding to form the correct disulfide bond connectivity to make the cystine knot structure was considered to be an important part of the synthetic strategy and two general approaches were used. Generally, Tam and coworkers used the approach of orthogonal protection of pairs of cysteine residues to selectively form individual disulfide bonds successively whereas Daly and colleagues opted for an undirected oxidation procedure (i.e., all three disulfide bonds simultaneously oxidized) followed by purification of the correct isomer from the mixture of correctly and incorrectly folded forms. Both approaches led to the production of sufficient quantities of cyclotides for structural and functional analyses.

Overall these synthetic studies laid the foundation for a range of mutagenesis studies on cyclotides that were useful in developing SARs and for engineering pharmaceutically active cyclotides. The broad categories of mutagenesis studies undertaken are summarized in Figure 6 and include the point mutation of individual residues, "scanning mutagenesis" in which a residue such as alanine is successively substituted at all positions within a sequence, as well as mutagenesis or selective deletion of individual disulfide bonds. Also indicated are studies involving the breaking of the peptide backbone in various different locations (acyclic permutation), and grafting studies where novel sequences are substituted into particular backbone loops of the cyclotide framework. We now examine each of these in turn, and although most of the SAR studies have been done using solid-phase peptide synthesis

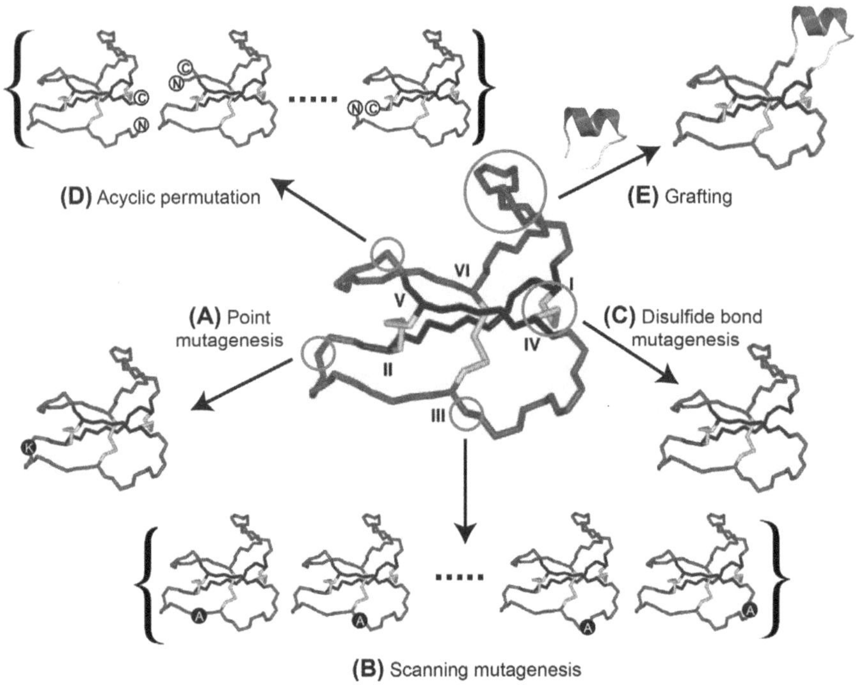

Figure 6 ***Overview of structure–activity relationship studies of cyclotides.*** The prototypical cyclotide kalata B1 is shown in the middle with the types of mutagenesis and structure–activity relationship studies that have been done illustrated schematically. (A) Point mutagenesis—individual residues are replaced with another residue (lysine, for example) to investigate the effect on activity. (B) Scanning mutagenesis—a residue (alanine, for example) is systematically substituted to all positions within the sequence. (C) Disulfide bond mutagenesis—each one of the three disulfide bonds were successively removed to investigate their role in structure and folding. (D) Acyclic permutation—acyclic derivatives are made by breaking single peptide bond in the backbone to investigate the effect of cyclization on structure and folding. (E) Grafting—bioactive epitopes were grafted into the cyclotide framework to investigate the plasticity of the framework being used for development of therapeutic and diagnostic agents. (See color plate)

it should be noted that SAR studies can also be achieved using biological approaches to cyclotide synthesis such as those described in Chapter 9 Chemical and Biological Production of Cyclotides.

5.1 Point Mutagenesis

Although this is the simplest approach to deriving SAR data for peptides there have been relatively few point mutagenesis studies on cyclotides. As illustrated in Figure 6, such a study involves replacing an individual amino acid residue in the sequence with a different residue. Typically this is done by replacing a residue that is presumed to play a functional role with a residue such as alanine to observe the effect on activity. Alanine is regarded as a "null" amino acid that is unlikely to introduce any new activity itself and thus loss of activity on Ala substitution is taken as evidence that the replaced residue was functionally active. The mutant peptide would typically be made from scratch either using solid-phase peptide chemistry or recombinant approaches, but it is also possible in certain cases to chemically modify a single residue in a native or preexisting peptide if there is only one residue of that type in the peptide and there is a chemical reaction that is specific for that amino acid. For example, Burman et al. used this approach to good effect by chemically modifying the Glu residue in loop 1 of cycloviolacin O2. By esterifying this residue they were able to establish a crucial role for the Glu residue in membrane interactions (Burman, Stromstedt, Malmsten, & Goransson, 2011). The two Lys residues in cycloviolacin O2 were likewise chemically modified via acetylation to neutralize their charge and further probe membrane binding. An example where solid-phase peptide chemistry was used to achieve the point mutagenesis was when Val10 in kB1 was replaced with Ala to help delineate factors responsible for immunosuppressive activity, as noted earlier in this chapter (Gründemann et al., 2013).

If conclusions about the role of a particular residue are to be drawn in any such mutagenesis studies it is important to check that any change in activity is due to a change in the nature of the residue (and hence its functional interaction with its molecular target) and not due to a more global conformational change that might be induced by its replacement. Indeed this is why Ala is preferred over Gly as a replacement residue, because Gly is typically more flexible and might lead to unexpected conformational changes when introduced into a peptide ligand. Typically we use NMR chemical shift analyses to check that any mutants made for SAR analysis retain the native peptide fold.

5.2 Scanning Mutagenesis

A logical extension of point mutagenesis is scanning mutagenesis. Most typically this is done in the form of Ala scanning mutagenesis, whereby all non-cysteine residues of a peptide are replaced one by one with alanine. This was applied to good effect by Simonsen and colleagues who made a series of 23 kB1 Ala mutants using this approach and showed that 22 of them folded correctly as judged by NMR (Simonsen et al., 2008). It is important to note that kB1 contains 29 amino acids but the six Cys residues were not replaced as they would affect the folding of the cystine knot. The authors measured the insecticidal activity of the 22 folded Ala scan mutants and found that activity was localized to nine residues, which are largely colocated in one patch on the surface of the peptide, as illustrated in Figure 7. These residues are similar to those implicated in hemolytic activity (Figure 7(B)) but are in a different location to the hydrophobic patch of kB1 (Figure 7(C)), which has been shown to be implicated in the binding of kB1 to model membranes (Shenkarev et al., 2006).

Scanning mutagenesis is not restricted to alanine residues and Huang and colleagues used lysine scanning mutagenesis to improve the nematocidal activity of cyclotides (Huang et al., 2010). Figure 8 shows dose–response curves for a range of lysine mutants of kB1 that showed that as well as a bioactive face and a membrane binding face, there was an "amendable" face on the cyclotide framework that could be used to improve activity (Figure 7(D)).

5.3 Disulfide Bond Mutants

kB1 contains three disulfide bonds and, in principle, there are three single disulfide-deficient mutants that can be made, in which each disulfide bond is successively removed, as well as several double disulfide-deficient mutants. The only one that has been extensively studied is that where the I–IV disulfide bond is removed, as illustrated in Figure 6. In that case, the removal of this disulfide bond was achieved by replacing the relevant pair of Cys residues with Ala residues. Interestingly, removal of the disulfide bond did not result in a markedly different structure from the parent cyclotide (Daly, Clark, & Craik, 2003), showing that to some extent this disulfide bond is dispensable and not essential for correct structure or folding. Nevertheless this disulfide-deficient variant was somewhat less stable than the native cyclotide, as would be expected. The advantage of a two-disulfide molecule over the native cystine knot would be its potentially lower cost of production but this is probably outweighed by the lower stability. Thus, at this time disulfide-deficient mutants of kB1 are mainly of academic rather than commercial interest. The disulfide mutants have been of particular interest in helping to delineate the folding pathways of kB1 and related cyclotides (Daly et al., 2003).

5.4 Acyclic Derivatives

Cyclotides are distinguished from most proteins by virtue of their circular backbone. In principle "acyclic permutants" of cyclotides can be made by

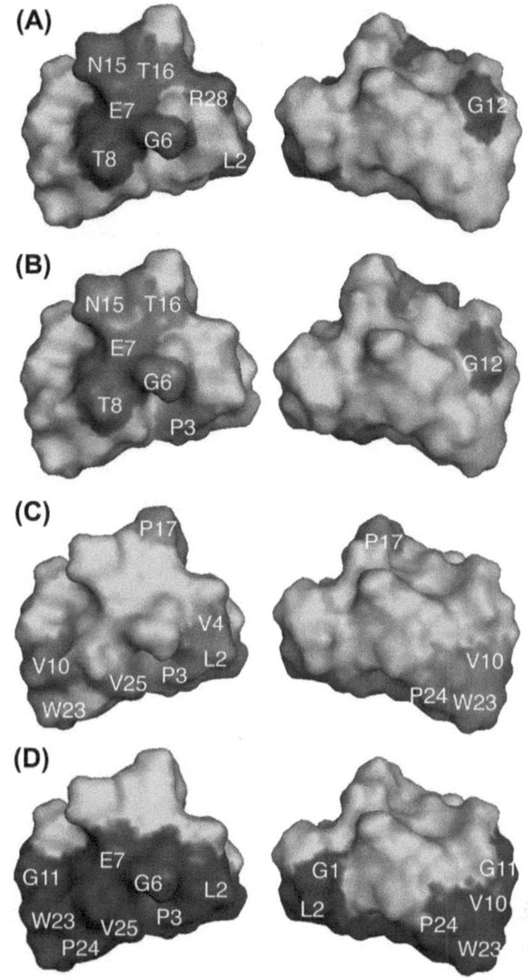

Figure 7 ***Localization of residues critical for hemolytic and insecticidal activity of kalata B1 and the hydrophobic patch of kalata B1.*** The figures on the right are rotated 180° relative to those on the left. (A) Mutated residues giving <20% hemolysis at 60 μM peptide concentration are shown in red. The wild-type peptide displays 69% hemolysis at this concentration. Functionally essential residues without concomitant structural perturbation are seen to cluster on one face of the molecule around E7. (B) The surface position of residues is critical for insecticidal activity. Mutated residues giving <70% insecticidal activity at 230 ppm are shown in orange, relative to the wild-type peptide normalized to 100% activity. (C) The hydrophobic residues of kalata B1 are highlighted in green, showing a dominant hydrophobic patch on the lower section of the framework. (D) Membrane-penetrating residues as measured by NMR are shown in dark blue (Shenkarev et al., 2006), showing a correspondence with the dominant hydrophobic surface patch. *Reproduced with permission from Journal of Biological Chemistry, 283, 9805 (2008). Copyright 2015 American Society for Biochemistry and Molecular Biology.* (See color plate)

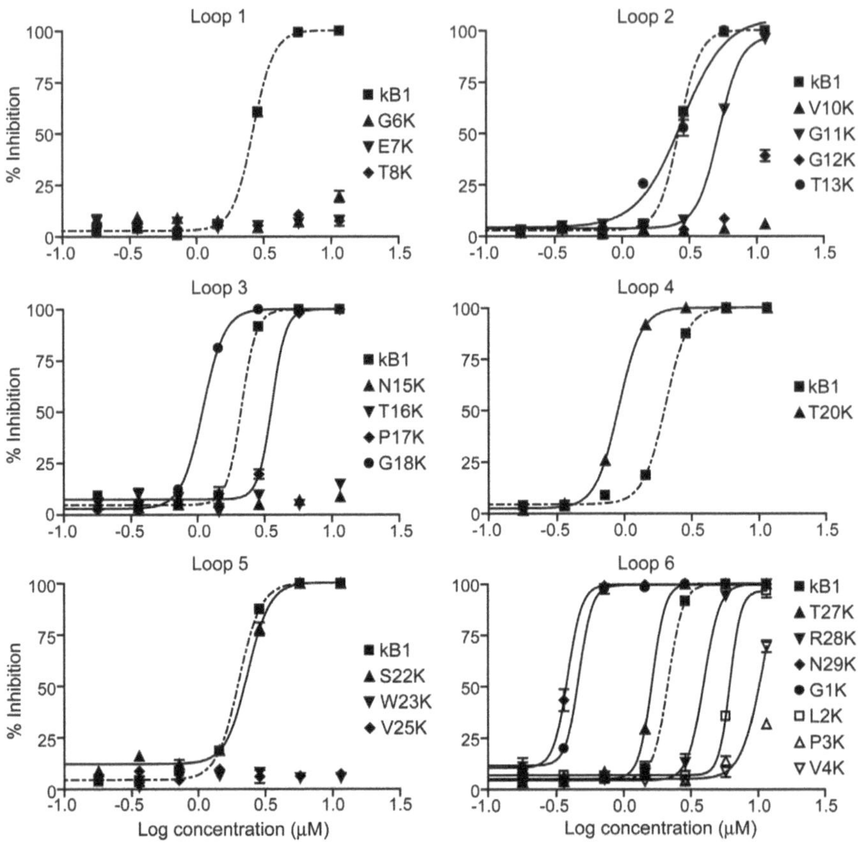

Figure 8 ***Dose–response curves for the inhibition of larval development.*** The graphs show the percentage inhibition of growth of eggs to third instar (L3) larvae of *Haemonchus contortus* of the 22 correctly folded lysine mutants of kalata B1. Each dose was tested in triplicate, and the *error bars* represent the S.D. per treatment. *Reproduced with permission from Journal of Biological Chemistry, 285, 10797 (2010). Copyright 2015 American Society for Biochemistry and Molecular Biology.*

breaking any individual peptide bond in the backbone, and so for kB1 there are 29 possible "acyclic permutants." A representative set of these acyclic permutants was studied by Daly and colleagues who showed that if the backbone is broken in the embedded ring of the cystine knot, i.e., between cysteines I and II (loop 1) or between cysteines IV and V (loop 4) then the peptide did not fold correctly, but breaking the backbone in any of the other four loops produced peptides that folded correctly (Daly & Craik, 2000).

In general, the acyclic permutants were somewhat less stable than the backbone cyclic parent cyclotide but typically adopted the same three-dimensional shape. Interestingly, none of the acyclic permutants displayed the hemolytic activity of the parent cyclotide (Barry, Daly, Clark, Sando, & Craik, 2003), nor the anti-HIV activity (Daly et al., 2004). Overall the results of studies of acyclic permutants suggest that the cyclic backbone does not itself define the three-dimensional structure of the cyclotides but is very important for stability. The fact that acyclic (i.e., linearized) variants of cyclotides display similar structures to the cyclic molecules is a valuable observation, as it suggests that during design and screening programs it should be possible to utilize simple acyclic derivatives to optimize activity and then transfer the design to a cyclic version to capture stability benefits, as would be required, for example, in a drug development program.

5.5 Loop Substitution and Grafting

Peptides generally bind to their targets with remarkably high affinity and specificity, which means that fewer off-target side effects are encountered relative to the case for small molecule drugs. However, the application of peptides as therapeutics has been restricted due to their low bioavailability and their susceptibility to proteolytic cleavage, thermal unfolding, and chemical degradation. These general limitations can potentially be overcome by exploiting the exceptionally stable cyclotide framework as a scaffold onto which bioactive peptide epitopes can be grafted. The idea of substituting one of the native loops of a cyclotide with a foreign bioactive epitope (referred to as grafting) is to combine the best of two worlds—the potent activity of a target epitope and the stability of the cyclotide framework (Craik, Čemažar, & Daly, 2006).

A complete list of grafting examples completed to date, with their pharmaceutically relevant applications is given in Table 5. Of the three subfamilies of cyclotides, most grafting studies so far have focused on cyclotide scaffolds from the Möbius (e.g., kB1) and trypsin inhibitor (e.g., MCoTI-I, or MCoTI-II) subfamilies as cyclotides from the bracelet subfamily are more difficult to fold.

The first proof-of-concept grafting study was reported in 2006, whereby two grafted analogues of the Möbius cyclotide kB1 were synthesized and structurally characterized (Clark, Daly, & Craik, 2006). To investigate the effect of the hydrophobic patch present in loop 5 on folding and biological activity, polar or charged residues were substituted into loop 5 and consequently the size of the hydrophobic patch was reduced. Both analogues, [W19K/P20N/V21K]kalata B1 and [P20D/V21K]kalata B1, retained the native fold of kB1. Furthermore, in contrast to the undesired hemolytic

Table 5 Summary of Published Grafting Studies Using Cyclic Peptide Framework as Scaffolds

Parent Cyclic Peptide	Activity	Potency	Application	References
Kalata B1	Proof-of-concept*	N/A	Confirm plasticity of scaffold	Clark et al. (2006)
	VEGF-A antagonist	$IC_{50} = 12\ \mu M^{\dagger}$	Cancer	Gunasekera et al. (2008)
	MC4R agonist	$K_i = 29\ nM^{\ddagger}$	Obesity	Eliasen et al. (2012)
	Bradykinin B_1 receptor antagonist	i.p. $= 1\ mg\ kg^{-1\S}$ p.o. $= 10\ mg\ kg^{-1¶}$	Chronic pain and inflammatory pain	Wong et al. (2012)
	Neuropilin-1 and -2 antagonist	$EC_{50} \approx 100\ nM^{\|}$	Angiogenesis and lymphangiogenesis	Getz, Cheneval, Craik, and Daugherty (2013)
	Immunomodulation	Clinical score: 28**	Multiple sclerosis	Wang et al. (2014)
MCoTI-I††	CXCR4 antagonist	$EC_{50} = 20\ nM$	Anti-HIV	Aboye et al. (2012)
	HIV-1 cell-entry blocker	$EC_{50} = 2\ nM$		
	Hdm2/HdmX antagonist	$K_d = 2.3\ nM^{\ddagger\ddagger}$	Antitumor	Ji et al. (2013)
MCoTI-II§§	FMDV 3C protease inhibitor	$K_i = 41\ \mu M$	Foot-and-mouth disease	Thongyoo, Roque-Rosell, Leatherbarrow, and Tate (2008)
	β-tryptase inhibitor	$K_i = 1–9\ nM$	Inflammation disorders	Sommerhoff et al. (2010) and Thongyoo, Bonomelli, Leatherbarrow, and Tate (2009)
	HLE inhibitor	$K_i = 40\ nM$		
	Angiogenic	$14\ \mu M^{¶¶}$	Cardiovascular/wound healing	Chan et al. (2011)
	BCR-ABL inhibitor	$IC_{50} = 1.3\ \mu M$	Chronic myeloid leukemia	Huang et al. (2015)

SFTI-1[‖‖]	Matriptase inhibitor	$K_i = 2.6, 6.4$ nM	Anti-tumor	Fittler, Avrutina, Empting, and Kolmar (2014) and Quimbar et al. (2013)
	20S proteasome inhibitor	$IC_{50} = 0.6–1.2$ μM	Anti-cancer	Dębowski et al. (2014)
	Kallikrein-related peptidase inhibitor	$IC_{50} = 170$ nM	Prostate cancer	de Veer et al. (2013)
	Angiogenic	3.6 μM[¶¶]	Cardiovascular/wound healing	Chan et al. (2011)
RTD-1[⋆⋆⋆]	Integrin receptor binding	$IC_{50} = 18$ nM	Anti-tumor and anti-cancer	Conibear et al. (2014)

⋆ Proof-of-concept study to show loops are tolerant to modification.
[†] IC_{50}, half maximal inhibitory concentration.
[‡] K_i, inhibition constant.
[§] i.p., intraperitoneal injection.
[¶] p.o., oral administration.
[‖] EC_{50}, half maximal effective concentration.
⋆⋆ Cumulative clinical score using experimental autoimmune encephalomyelitis mouse model was significantly lower than the cumulative score of control group.
[††] *Momordica cochinchinensis* trypsin inhibitor-I.
[‡‡] K_d, dissociation constant.
[§§] *Momordica cochinchinensis* trypsin inhibitor-II.
[¶¶] The single concentration in a chorioallantoic membrane assay that the grafted peptide displayed the highest angiogenic activity.
[‖‖] Sunflower trypsin inhibitor-1.
[⋆⋆⋆] Rhesus θ-defensin-1.

activity of kB1, both grafted analogues exhibited no hemolytic activity. The results of this study not only showed that cyclotides are tolerant to the grafting of foreign epitopes without disrupting the native cyclotide fold, but also enabled the possibility of enhancing the biopharmaceutical properties of cyclotides by increasing the polarity and charge on the surface of the grafted peptides. Thus, this study marked the first step toward grafting biologically active epitopes onto the exceptionally stable cyclotide framework for the development in peptide-based therapeutics.

Following this proof-of-concept work, a follow-on study that validated the grafting concept with a clinical target was reported in 2008 (Gunasekera et al., 2008). That study involved the grafting of epitopes with antiangiogenic activity into loops 2, 3, 5, or 6 of kB1 and included six grafted analogues, cpr1–6 (meaning **c**yclic **p**oly-**R**, as poly-R epitopes were grafted into different loops). As discussed in Section 5.4, loops 1 and 4 of kB1 are critical for the folding of the native conformation (Daly & Craik, 2000) and thus were not chosen for grafting to avoid complication in folding. The most active grafted analogue, cpr3, exhibited an IC_{50} of 12 μM against vascular endothelial growth factor-A in vitro. Overall, this study illustrated two positive outcomes of grafting—the in vitro stability of the linear epitope was increased significantly in the grafted cyclotide and the undesired cytotoxicity of kB1 was not observed in the grafted analogue.

In a more recent example that demonstrates the application of cyclotides against intracellular targets, Camarero and coworkers used the trypsin inhibitor cyclotide MCoTI-I as the framework and grafted a phage-selected α-helical peptide named PMI into loop 6 (Ji et al., 2013). They chose loop 6 because a previous study had suggested that this loop was amenable to sequence variation (Austin, Wang, Puttamadappa, Shekhtman, & Camarero, 2009). The most potent grafted cyclotide, MCo-PMI-K37R, bound to the p53-binding domain of the oncoprotein Hdm2 with a K_d of 2.3 nM and HdmX with a K_d of 9.7 nM.

The p53 pathway is a prime target for developing novel cancer therapeutics (Brown, Lain, Verma, Fersht, & Lane, 2009), and although a range of linear peptides has been explored over the years as Hdm2/HdmX antagonists (Bernal et al., 2010; Li et al., 2010; Li, Pazgier, Liu, Lu, & Lu, 2009; Pazgier et al., 2009), their unfavorable properties of low stability and bioavailability have limited their use as cancer therapeutics. The recent work has demonstrated that the cyclotide framework can potentially provide a solution to the low stability of linear peptides. The MCoTI-I framework was able to accommodate a 20-residue peptide in one of the backbone loops and the grafted cyclotide targeted an intracellular protein–protein interaction. The grafted cyclotide reduced the tumor burden in a mouse tumor xenograft model and the success of this approach opens the door to other intracellular targets involving protein–protein interactions. The work is described in more detail in Chapter 9 Chemical and Biological Production of Cyclotides and is described only briefly here to place it in context with other grafting studies.

The few examples described here are just some of the many grafting studies listed in Table 5, which illustrates the wide range of applications of cyclotides as frameworks for the development of therapeutic and diagnostic agents. The diverse peptide epitopes that have been grafted and are able to maintain relevant biological activities demonstrates the plasticity of the cyclotide framework. The promising in vitro and in vivo data associated with the grafting studies in Table 5 suggest that the cyclotide framework already offers a solution to the generally poor stability of peptide-based drugs. So far not much has been done to tackle the problem of low bioavailability of peptide-based drugs, but there is room for significant optimism. A recent grafting study using a kB1 scaffold for developing a peptidic bradykinin B_1 receptor antagonist as a treatment for inflammatory pain reported the oral activity of the grafted cyclotide (Wong et al., 2012). Although the results of the study were not conclusive in determining whether cyclization contributes to the oral bioavailability of the grafted cyclotide, the linear analogue exhibited significantly lower oral activity than the cyclic derivative. Another example that suggests that cyclization is important for oral activity was demonstrated by the cyclization of a bioactive 16-residue conotoxin using a linker sequence of six residues. The resulting cyclized conotoxin exhibited oral activity in an animal model of neuropathic pain, whereas the parent acyclic analogue did not (Clark et al., 2010). Although more validation is required, these two studies independently suggest that eliminating N and C termini may be one of the keys to overcome the low oral bioactivity of peptides.

6. CONCLUDING REMARKS

We hope that from this chapter readers will have gained an appreciation of the wide range of bioactivities of natural cyclotides and the substantial work done in delineating the relationships between structure and activity of both naturally occurring and synthetically designed cyclotides. Much of the mutagenesis work done so far in this field has been achieved using solid-phase peptide chemistry. This is a very powerful approach but can be expensive and there is an increasing interest in developing biological routes to cyclotide synthesis, including via bacterial, plant cell, and whole plant-based systems. In Chapter 8 Biosynthesis of cyclotides the underlying mechanisms of cyclotide biosynthesis are described, providing a foundation for describing a range of biologically oriented syntheses of cyclotides in Chapter 9 Chemical and Biological Production of Cyclotides.

ACKNOWLEDGMENTS

Work on cyclotides in our laboratory has been supported by the Australian Research Council (ARC) (DP150100443, DP0984390, LP130100550) and the National Health and Medical Research Council (NHMRC) (APP1047857, APP1009267, APP1028509). DJC is an NHMRC Professorial Fellow (APP1026501).

REFERENCES

Aboye, T. L., Ha, H., Majumder, S., Christ, F., Debyser, Z., Shekhtman, A., et al. (2012). Design of a novel cyclotide-based CXCR4 antagonist with anti-human immunodeficiency virus (HIV)-1 activity. *Journal of Medicinal Chemistry, 55*, 10729–10734.

Austin, J., Wang, W., Puttamadappa, S., Shekhtman, A., & Camarero, J. A. (2009). Biosynthesis and biological screening of a genetically encoded library based on the cyclotide MCoTI-I. *Chembiochem, 10*, 2663–2670.

Barbeta, B. L., Marshall, A. T., Gillon, A. D., Craik, D. J., & Anderson, M. A. (2008). Plant cyclotides disrupt epithelial cells in the midgut of lepidopteran larvae. *Proceedings of the National Academy of Sciences of the United States of America, 105*, 1221–1225.

Barry, D. G., Daly, N. L., Clark, R. J., Sando, L., & Craik, D. J. (2003). Linearization of a naturally occurring circular protein maintains structure but eliminates hemolytic activity. *Biochemistry, 42*, 6688–6695.

Bernal, F., Wade, M., Godes, M., Davis, T. N., Whitehead, D. G., Kung, A. L., et al. (2010). A stapled p53 helix overcomes HDMX-mediated suppression of p53. *Cancer Cell, 18*, 411–422.

Bokesch, H. R., Pannell, L. K., Cochran, P. K., Sowder, R. C., 2nd, McKee, T. C., & Boyd, M. R. (2001). A novel anti-HIV macrocyclic peptide from *Palicourea condensata*. *Journal of Natural Products, 64*, 249–250.

Broussalis, A. M., Clemente, S., & Ferraro, G. E. (2010). *Hybanthus parviflorus* (Violaceae): Insecticidal activity of a South American plant. *Crop Protection, 29*, 953–956.

Broussalis, A. M., Göransson, U., Coussio, J. D., Ferraro, G., Martino, V., & Claeson, P. (2001). First cyclotide from hybanthus (Violaceae). *Phytochemistry, 58*, 47–51.

Brown, C. J., Lain, S., Verma, C. S., Fersht, A. R., & Lane, D. P. (2009). Awakening guardian angels: drugging the p53 pathway. *Nature Reviews Cancer, 9*, 862–873.

Burman, R., Stromstedt, A. A., Malmsten, M., & Goransson, U. (2011). Cyclotide-membrane interactions: defining factors of membrane binding, depletion and disruption. *Biochimica et Biophysica Acta-Biomembranes, 1808*, 2665–2673.

Chan, L. Y., Gunasekera, S., Henriques, S. T., Worth, N. F., Le, S. J., Clark, R. J., et al. (2011). Engineering pro-angiogenic peptides using stable disulfide-rich cyclic scaffolds. *Blood, 118*, 6709–6717.

Chen, B., Colgrave, M. L., Daly, N. L., Rosengren, K. J., Gustafson, K. R., & Craik, D. J. (2005). Isolation and characterization of novel cyclotides from *Viola hederacea*: solution structure and anti-HIV activity of vhl-1, a leaf-specific expressed cyclotide. *Journal of Biological Chemistry, 280*, 22395–22405.

Clark, R. J., Daly, N. L., & Craik, D. J. (2006). Structural plasticity of the cyclic-cystine-knot framework: implications for biological activity and drug design. *Biochemistry Journal, 394*, 85–93.

Clark, R. J., Jensen, J., Nevin, S. T., Callaghan, B. P., Adams, D. J., & Craik, D. J. (2010). The engineering of an orally active conotoxin for the treatment of neuropathic pain. *Angewandte Chemie International Edition, 49*, 6545–6548.

Colgrave, M. L., Kotze, A. C., Huang, Y. H., Grady, J. O., Simonsen, S. M., & Craik, D. J. (2008). Cyclotides: natural, circular plant peptides that possess significant activity against gastrointestinal nematode parasites of sheep. *Biochemical Journal, 47*, 5581–5589.

Colgrave, M. L., Kotze, A. C., Ireland, D. C., Wang, C. K., & Craik, D. J. (2008). The anthelmintic activity of the cyclotides: natural variants with enhanced activity. *Chembiochem, 9*, 1939–1945.

Colgrave, M. L., Kotze, A. C., Kopp, S., McCarthy, J. S., Coleman, G. T., & Craik, D. J. (2009). Anthelmintic activity of cyclotides: in vitro studies with canine and human hookworms. *Acta Tropica, 109*, 163–166.

Conibear, A. C., Bochen, A., Rosengren, K. J., Stupar, P., Wang, C., Kessler, H., et al. (2014). The cyclic cystine ladder of theta-defensins as a stable, bifunctional scaffold: a proof-of-concept study using the integrin-binding RGD motif. *Chembiobhem, 15*, 451–459.

Craik, D. J., Cĕmažar, M., & Daly, N. L. (2006). The cyclotides and related macrocyclic peptides as scaffolds in drug design. *Current Opinion in Drug Discovevery and Development, 9*, 251–260.

Craik, D. J., Daly, N. L., Bond, T., & Waine, C. (1999). Plant cyclotides: a unique family of cyclic and knotted proteins that defines the cyclic cystine knot structural motif. *Journal of Molecular Biology, 294*, 1327–1336.

Daly, N. L., & Craik, D. J. (2000). Acyclic permutants of naturally occurring cyclic proteins. Characterization of cystine knot and β-sheet formation in the macrocyclic polypeptide kalata B1. *Journal of Biological Chemistry, 275*, 19068–19075.

Daly, N. L., Clark, R. J., & Craik, D. J. (2003). Disulfide folding pathways of cystine knot proteins. Tying the knot within the circular backbone of the cyclotides. *Journal of Biological Chemistry, 278*, 6314–6322.

Daly, N. L., Clark, R. J., Plan, M. R., & Craik, D. J. (2006). Kalata B8, a novel antiviral circular protein, exhibits conformational flexibility in the cystine knot motif. *Biochemical Journal, 393*, 619–626.

Daly, N. L., Gustafson, K. R., & Craik, D. J. (2004). The role of the cyclic peptide backbone in the anti-HIV activity of the cyclotide kalata B1. *Federation of European Biochemical Societies Letters, 574*, 69–72.

Daly, N. L., Love, S., Alewood, P. F., & Craik, D. J. (1999). Chemical synthesis and folding pathways of large cyclic polypeptides: studies of the cystine knot polypeptide kalata B1. *Biochemistry, 38*, 10606–10614.

Dębowski, D., Pikuła, M., Lubos, M., Langa, P., Trzonkowski, P., Lesner, A., et al. (2014). Inhibition of human and yeast 20S proteasome by analogues of trypsin inhibitor SFTI-1. *PLoS One, 9*, e89465.

Ding, X., Bai, D., & Qian, J. (2014). Novel cyclotides from *Hedyotis biflora* inhibit proliferation and migration of pancreatic cancer cell in vitro and in vivo. *Medicinal Chemical Research, 23*, 1406–1413.

Dörnenburg, H. (2010). Cyclotide synthesis and supply: from plant to bioprocess. *Biopolymers: Peptide Science, 94*, 602–610.

Dutton, J. L., Renda, R. F., Waine, C., Clark, R. J., Daly, N. L., Jennings, C. V., et al. (2004). Conserved structural and sequence elements implicated in the processing of gene-encoded circular proteins. *Journal of Biological Chemistry, 279*, 46858–46867.

Eliasen, R., Daly, N. L., Wulff, B. S., Andresen, T. L., Conde-Frieboes, K. W., & Craik, D. J. (2012). Design, synthesis, structural and functional characterization of novel melanocortin agonists based on the cyclotide kalata B1. *Journal of Biological Chemistry, 287*, 40493–40501.

Fittler, H., Avrutina, O., Empting, M., & Kolmar, H. (2014). Potent inhibitors of human matriptase-1 based on the scaffold of sunflower trypsin inhibitor. *Journal of Peptide Science, 20*, 415–420.

Gerlach, S. L., Burman, R., Bohlin, L., Mondal, D., & Göransson, U. (2010). Isolation, characterization, and bioactivity of cyclotides from the Micronesian plant *Psychotria leptothyrsa*. *Journal of Natural Products, 73*, 1207–1213.

Gerlach, S. L., Rathinakumar, R., Chakravarty, G., Göransson, U., Wimley, W. C., Darwin, S. P., et al. (2010). Anticancer and chemosensitizing abilities of cycloviolacin O_2 from *Viola odorata* and psyle cyclotides from *Psychotria leptothyrsa*. *Biopolymers, 94*, 617–625.

Gerlach, S. L., Yeshak, M., Göransson, U., Roy, U., Izadpanah, R., & Mondal, D. (2013). Cycloviolacin O_2 (CyO_2) suppresses productive infection and augments the antiviral efficacy of nelfinavir in HIV-1 infected monocytic cells. *Biopolymers, 100*, 471–479.

Getz, J. A., Cheneval, O., Craik, D. J., & Daugherty, P. S. (2013). Design of a cyclotide antagonist of neuropilin-1 and -2 that potently inhibits endothelial cell migration. *ACS Chemical Biology, 8*, 1147–1154.

Göransson, U., Luijendijk, T., Johansson, S., Bohlin, L., & Claeson, P. (1999). Seven novel macrocyclic polypeptides from *Viola arvensis*. *Journal of Natural Products, 62*, 283–286.

Gran, L. (1973a). Isolation of oxytocic peptides from *Oldenlandia affinis* by solvent extraction of tetraphenylborate complexes and chromatography on sephadex LH-20. *Lloydia, 36*, 207–208.

Gran, L. (1973b). On the effect of a polypeptide isolated from "Kalata-Kalata" (*Oldenlandia affinis* DC) on the oestrogen dominated uterus. *Acta Pharmacologica et Toxicologica, 33*, 400–408.

Gran, L. (1973c). Oxytocic principles of *Oldenlandia affinis*. *Lloydia, 36*, 174–178.

Gran, L., Sandberg, F., & Sletten, K. (2000). *Oldenlandia affinis* (R&S) DC. A plant containing uteroactive peptides used in African traditional medicine. *Journal of Ethnopharmacology, 70*, 197–203.

Gran, L., Sletten, K., & Skjeldal, L. (2008). Cyclic peptides from *Oldenlandia affinis* DC. Molecular and biological properties. *Chemistry and Biodiversity, 5*, 2014–2022.

Gründemann, C., Koehbach, J., Huber, R., & Gruber, C. W. (2012). Do plant cyclotides have potential as immunosuppressant peptides? *Journal of Natural Products, 75*, 167–174.

Gründemann, C., Thell, K., Lengen, K., Garcia-Käufer, M., Huang, Y.-H., Huber, R., et al. (2013). Cyclotides suppress human T-lymphocyte proliferation by an interleukin 2-dependent mechanism. *PLoS One, 8*, e68016.

Gunasekera, S., Foley, F. M., Clark, R. J., Sando, L., Fabri, L. J., Craik, D. J., et al. (2008). Engineering stabilized vascular endothelial growth factor-A antagonists: synthesis, structural characterization, and bioactivity of grafted analogues of cyclotides. *Journal of Medicinal Chemistry, 51*, 7697–7704.

Gustafson, K. R., Sowder, R. C. I., Henderson, L. E., Parsons, I. C., Kashman, Y., Cardellina, J. H. I., et al. (1994). Circulins A and B: novel HIV-inhibitory macrocyclic peptides from the tropical tree *Chassalia parvifolia*. *Journal of the American Chemical Society, 116*, 9337–9338.

Gustafson, K. R., Walton, L. K., Sowder, R. C. I., Johnson, D. G., Pannell, L. K., Cardellina, J. H. I., et al. (2000). New circulin macrocyclic polypeptides from *Chassalia parvifolia*. *Journal of Natural Products, 63*, 176–178.

Hallock, Y. F., Sowder, R. C. I., Pannell, L. K., Hughes, C. B., Johnson, D. G., Gulakowski, R., et al. (2000). Cycloviolins A-D, anti-HIV macrocyclic peptides from Leonia cymosa. *Journal of Organic Chemistry, 65*, 124–128.

He, W., Chan, L. Y., Zeng, G., Daly, N. L., Craik, D. J., & Tan, N. (2011). Isolation and characterization of cytotoxic cyclotides from *Viola philippica*. *Peptides, 32*, 1719–1723.

Hellinger, R., Koehbach, J., Fedchuk, H., Sauer, B., Huber, R., Gruber, C. W., et al. (2014). Immunosuppressive activity of an aqueous *Viola tricolor* herbal extract. *Journal of Ethnopharmacology, 151*, 299–306.

Henriques, S. T., Huang, Y. H., Chaousis, S., Wang, C. K., & Craik, D. J. (2014). Anticancer and toxic properties of cyclotides are dependent on phosphatidylethanolamine phospholipid targeting. *Chembiochem, 15*, 1956–1965.

Herrmann, A., Burman, R., Mylne, J. S., Karlsson, G., Gullbo, J., Craik, D. J., et al. (2008). The alpine violet, *Viola biflora*, is a rich source of cyclotides with potent cytotoxicity. *Phytochemistry, 69*, 939–952.

Huang, Y.-H., Colgrave, M. L., Daly, N. L., Keleshian, A., Martinac, B., & Craik, D. J. (2009). The biological activity of the prototypic cyclotide kalata B1 is modulated by the formation of multimeric pores. *Journal of Biological Chemistry, 284*, 20699–20707.

Huang, Y.-H., Colgrave, M. L., Clark, R. J., Kotze, A. C., & Craik, D. J. (2010). Lysine-scanning mutagenesis reveals a previously unidentified amendable face of the cyclotide kalata B1 for the optimisation of nematocidal activity. *Journal of Biological Chemistry, 285*, 10797–10805.

Huang, Y.-H., Henriques, S. T., Wang, C. K., Thorstholm, L., Daly, N. L., Kaas, Q., et al. (2015). Design of substrate-based BCR-ABL kinase inhibitors using the cyclotide scaffold. *Scientific Reports, 5*, 12974.

Ireland, D. C., Colgrave, M. L., & Craik, D. J. (2006). A novel suite of cyclotides from *Viola odorata*: sequence variation and the implications for structure, function and stability. *Biochemical Journal, 400*, 1–12.

Ireland, D. C., Wang, C. K., Wilson, J. A., Gustafson, K. R., & Craik, D. J. (2008). Cyclotides as natural anti-HIV agents. *Biopolymers, 90*, 51–60.

Jennings, C., West, J., Waine, C., Craik, D., & Anderson, M. (2001). Biosynthesis and insecticidal properties of plant cyclotides: the cyclic knotted proteins from *Oldenlandia affinis*. *Proceedings of the National Academy of Sciences of the United States of America, 98*, 10614–10619.

Jennings, C. V., Rosengren, K. J., Daly, N. L., Plan, M., Stevens, J., Scanlon, M. J., et al. (2005). Isolation, solution structure, and insecticidal activity of kalata B2, a circular protein with a twist: do Möbius strips exist in nature? *Biochemistry, 44*, 851–860.

Ji, Y., Majumder, S., Millard, M., Borra, R., Bi, T., Elnagar, A. Y., et al. (2013). In vivo activation of the p53 tumor suppressor pathway by an engineered cyclotide. *Journal of the American Chemical Society, 135*, 11623–11633.

Kaas, Q., & Craik, D. J. (2010). Analysis and classification of circular proteins in CyBase. *Biopolymers: Peptide Science, 94*, 584–591.

Li, C., Pazgier, M., Liu, M., Lu, W.-Y., & Lu, W. (2009). Apamin as a template for structure-based rational design of potent peptide activators of p53. *Angewandte Chemie International Edition, 48*, 8712–8715.

Li, C., Pazgier, M., Li, C., Yuan, W., Liu, M., Wei, G., et al. (2010). Systematic mutational analysis of peptide inhibition of the p53-MDM2/MDMX interactions. *Journal of Molecular Biology, 398*, 200–213.

Lindholm, P., Göransson, U., Johansson, S., Claeson, P., Gulbo, J., Larsson, R., et al. (2002). Cyclotides: a novel type of cytotoxic agents. *Molecular Cancer Therapeutics, 1*, 365–369.

Malagón, D., Botterill, B., Gray, D. J., Lovas, E., Duke, M., Gray, C., et al. (2013). Anthelminthic activity of the cyclotides (kalata B1 and B2) against schistosome parasites. *Biopolymers, 100*, 461–470.

Nguyen, G. K. T., Zhang, S., Ngan, T. K. N., Phuong, Q. T. N., Chiu, M. S., Hardjojo, A., et al. (2011). Discovery and characterization of novel cyclotides originated from chimeric precursors consisting of albumin-1 chain a and cyclotide domains in the Fabaceae family. *Journal of Biological Chemistry, 286*, 24275–24287.

Nguyen, G. K. T., Zhang, S., Wang, W., Wong, C. T. T., Ngan, T. K. N., & Tam, J. P. (2011). Discovery of a linear cyclotide from the bracelet subfamily and its disulfide mapping by top-down mass spectrometry. *Journal of Biological Chemistry, 286*, 44833–44844.

Ovesen, R. G., Brandt, K. K., Göransson, U., Nielsen, J., Hansen, H. C., & Cedergreen, N. (2011). Biomedicine in the environment: cyclotides constitute potent natural toxins in plants and soil bacteria. *Environmental Toxicology and Chemistry, 30*, 1190–1196.

Park, S., Strömstedt, A. A., & Göransson, U. (2014). Cyclotide structure–activity relationships: qualitative and quantitative approaches linking cytotoxic and anthelmintic activity to the clustering of physicochemical forces. *PLoS One, 9*, e91430.

Pazgier, M., Liu, M., Zou, G., Yuan, W., Li, C., Li, C., et al. (2009). Structural basis for high-affinity peptide inhibition of p53 interactions with MDM2 and MDMX. *Proceedings of the National Academy of Sciences United States of America, 106*, 4665–4670.

Pinto, M. F. S., Fensterseifer, I. C. M., Migliolo, L., Sousa, D. A., de Capdville, G., Arboleda-Valencia, J. W., et al. (2012). Identification and structural characterization of novel cyclotide with activity against an insect pest of sugar cane. *Journal of Biological Chemistry, 287*, 134–147.

Plan, M. R. R., Saska, I., Cagauan, A. G., & Craik, D. J. (2008). Backbone cyclised peptides from plants show molluscicidal activity against the rice pest *Pomacea canaliculata* (Golden Apple Snail). *Journal of Agricultural and Food Chemistry, 56*, 5237–5241.

Poth, A. G., Colgrave, M. L., Lyons, R. E., Daly, N. L., & Craik, D. J. (2011). Discovery of an unusual biosynthetic origin for circular proteins in legumes. *Proceedings of the National Academy of Sciences of the United States of America, 108*, 10127–10132.

Poth, A. G., Chan, L. Y., & Craik, D. J. (2013). Cyclotides as grafting frameworks for protein engineering and drug design applications. *Biopolymers, 100*, 480–491.

Pränting, M., Loov, C., Burman, R., Göransson, U., & Andersson, D. I. (2010). The cyclotide cycloviolacin O_2 from *Viola odorata* has potent bactericidal activity against gram-negative bacteria. *Journal of Antimicrobial Chemotherapy, 65*, 1964–1971.

Quimbar, P., Malik, U., Sommerhoff, C. P., Kaas, Q., Chan, L. Y., Huang, Y.-H., et al. (2013). High-affinity cyclic peptide matriptase inhibitors. *Journal of Biological Chemistry, 288*, 13885–13896.

Rosengren, K. J., Daly, N. L., Plan, M. R., Waine, C., & Craik, D. J. (2003). Twists, knots, and rings in proteins. Structural definition of the cyclotide framework. *Journal of Biological Chemistry, 278*, 8606–8616.

Saether, O., Craik, D. J., Campbell, I. D., Sletten, K., Juul, J., & Normano, D. G. (1995). Elucidation of the primary and three-dimensional structure of the uterotonic polypeptide kalata B1. *Biochemistry, 34*.

Shenkarev, Z. O., Nadezhdin, K. D., Sobol, V. A., Sobol, A. G., Skjeldal, L., & Arseniev, A. S. (2006). Conformation and mode of membrane interaction in cyclotides – spatial structure of kalata B1 bound to a dodecylphosphocholine micelle. *Federation of European Biochemical Societies Journal, 273*, 2658–2672.

Simonsen, S. M., Sando, L., Rosengren, K. J., Wang, C. K., Colgrave, M. L., Daly, N. L., et al. (2008). Alanine scanning mutagenesis of the prototypic cyclotide reveals a cluster of residues essential for bioactivity. *Journal of Biological Chemistry, 283*, 9805–9813.

Sletten, K., & Gran, L. (1973). Some molecular properties of kalatapeptide B-1 A uterotonic polypeptide isolated from *Oldenlandia affinis* DC. *Meddelelser fra Norsk Farmaceutisk Selskap, 7-8*, 69–82.

Sommerhoff, C. P., Avrutina, O., Schmoldt, H. U., Gabrijelcic-Geiger, D., Diederichsen, U., & Kolmar, H. (2010). Engineered cystine knot miniproteins as potent inhibitors of human mast cell tryptase β. *Journal of Molecular Biology, 395*, 167–175.

Svangård, E., Göransson, U., Hocaoglu, Z., Gullbo, J., Larsson, R., Claeson, P., et al. (2004). Cytotoxic cyclotides from *Viola tricolor*. *Journal of Natural Products, 67*, 144–147.

Tam, J. P., & Lu, Y.-A. (1997). Synthesis of large cyclic cystine-knot peptide by orthogonal coupling strategy using unprotected peptide precursors. *Tetrahedron Letters, 38*, 5599–5602.

Tam, J. P., & Lu, Y.-A. (1998). A biomimetic strategy in the synthesis and fragmentation of cyclic protein. *Protein Science, 7*, 1583–1592.

Tam, J. P., Lu, Y. A., Yang, J. L., & Chiu, K. W. (1999). An unusual structural motif of antimicrobial peptides containing end-to-end macrocycle and cystine-knot disulfides. *Proceedings of the National Academy of Sciences of the United States of America, 96*, 8913–8918.

Tang, J., Wang, C. K., Pan, X., Yan, H., Zeng, G., Xu, W., et al. (2010a). Isolation and characterization of cytotoxic cyclotides from *Viola tricolor*. *Peptides, 31*, 1434–1440.

Tang, J., Wang, C. K., Pan, X., Yan, H., Zeng, G., Xu, W., et al. (2010b). Isolation and characterization of bioactive cyclotides from *Viola labradorica*. *Helvetica Chimica Acta, 93*, 2287–2295.

Thongyoo, P., Roque-Rosell, N., Leatherbarrow, R. J., & Tate, E. W. (2008). Chemical and biomimetic total syntheses of natural and engineered MCoTI cyclotides. *Organic and Biomolecular Chemistry, 6*, 1462–1470.

Thongyoo, P., Bonomelli, C., Leatherbarrow, R. J., & Tate, E. W. (2009). Potent inhibitors of β-tryptase and human leukocyte elastase based on the MCoTI-II scaffold. *Journal of Medicinal Chemistry, 52*, 6197–6200.

de Veer, S. J., Ukolova, S. S., Munro, C. A., Swedberg, J. E., Buckle, A. M., & Harris, J. M. (2013). Mechanism-based selection of a potent kallikrein-related peptidase 7 inhibitor from a versatile library based on the sunflower trypsin inhibitor SFTI-1. *Peptide Science, 100*, 510–518.

Wang, C. K., Colgrave, M. L., Gustafson, K. R., Ireland, D. C., Göransson, U., & Craik, D. J. (2008). Anti-HIV cyclotides from the Chinese medicinal herb *Viola yedoensis*. *Journal of Natural Products, 71*, 47–52.

Wang, C. K., Kaas, Q., Chiche, L., & Craik, D. J. (2008). CyBase: a database of cyclic protein sequences and structures, with applications in protein discovery and engineering. *Nucleic Acids Research, 36*, D206–D210.

Wang, C. K., Colgrave, M. L., Ireland, D. C., Kaas, Q., & Craik, D. J. (2009). Despite a conserved cystine knot motif, different cyclotides have different membrane binding modes. *Biophysical Journal, 97*, 1471–1481.

Wang, C. K., Gruber, C. W., Cemazar, M., Siatskas, C., Tagore, P., Payne, N., et al. (2014). Molecular grafting onto a stable framework yields novel cyclic peptides for the treatment of multiple sclerosis. *ACS Chemical Biology, 9*, 156–163.

Werle, M., Kolmar, H., Albrecht, R., & Bernkop-Schnürch, A. (2008). Characterisation of the barrier caused by luminally secreted gastro-intestinal proteolytic enzymes for two novel cystine-knot microproteins. *Amino Acids, 35*, 195–200.

Witherup, K. M., Bogusky, M. J., Anderson, P. S., Ramjit, H., Ransom, R. W., Wood, T., et al. (1994). Cyclopsychotride A, A biologically active, 31-residue cyclic peptide isolated from *Psychotria longipes*. *Journal of Natural Products, 57*, 1619–1625.

Wong, C. T. T., Taichi, M., Nishio, H., Nishiuchi, Y., & Tam, J. P. (2011). Optimal oxidative folding of the novel antimicrobial cyclotide from *Hedyotis biflora* requires high alcohol concentrations. *Biochemistry, 50*(33).

Wong, C. T. T., Rowlands, D. K., Wong, C. H., Lo, T. W. C., Nguyen, G. K. T., Li, H. Y., et al. (2012). Orally active peptidic bradykinin B-1 receptor antagonists engineered from a cyclotide scaffold for inflammatory pain treatment. *Angewandte Chemie International Edition in English, 51*, 5620–5624.

Yeshak, M. Y., Burman, R., Asres, K., & Göransson, U. (2011). Cyclotides from an extreme habitat: characterization of cyclic peptides from *Viola abyssinica* of the Ethiopian highlands. *Journal of Natural Products, 74*, 727–731.

Zhang, J., Liao, B., Craik, D. J., Li, J. T., Hu, M., & Shu, W. S. (2009). Identification of two suites of cyclotide precursor genes from metallophyte *Viola baoshanensis*: cDNA sequence variation, alternative RNA splicing and potential cyclotide diversity. *Gene, 431*, 23–32.

Zhang, S., Xiao, K. Z., Jin, J., Zhang, Y., & Zhou, W. (2013). Chemosensitizing activities of cyclotides from *Clitoria ternatea* in paclitaxel-resistant lung cancer cells. *Oncology Letters, 5*, 641–644.

CHAPTER EIGHT

Biosynthesis of Cyclotides

Thomas Shafee, Karen Harris, Marilyn Anderson[1]
Department of Biochemistry and Genetics, La Trobe Institute for Molecular Science, La Trobe University, Melbourne, Vic, Australia
[1]Corresponding author: E-mail: M.Anderson@latrobe.edu.au

Contents

Abstract

Cyclotides are gene-encoded, ribosomally produced cyclic proteins. Their biosynthesis is a multistep process involving several enzymes for translation, disulphide isomerization, transport, cleavage and, finally, cyclization. Here, we describe this process, focusing on the cyclization step and the vacuolar processing enzyme (VPE) that performs it.

Cyclotide precursor proteins contain prodomains that direct organelle targeting and contain sequences necessary for cyclization in addition to the one or more

Advances in Botanical Research, Volume 76
ISSN 0065-2296
http://dx.doi.org/10.1016/bs.abr.2015.08.005

domains that make up the final cyclic product. The VPEs are a family of cysteine proteases that perform cyclization by an adaptation of their more commonly described proteolysis mechanism. VPE-mediated cyclization is dependent on the presence of short recognition sequences (within and flanking the mature cyclotide domain) which have been characterized in vivo and in vitro for the prototypical cyclotide kalata B1. Finally, we compare cyclization by VPEs to other methods of protein cyclization found in nature or developed for biotechnology. Understanding cyclotide biosynthesis is important for both fundamental science, and for broader protein cyclization applications.

1. INTRODUCTION

Cyclotides are small, disulphide-rich plant proteins with the unusual property of having a cyclic peptide backbone (Conlan & Anderson, 2011; Craik, Daly, Bond, & Waine, 1999). They are highly thermally and proteolytically stable due to the lack of free peptide termini, the knotted topology of their disulphides and their compact size (around 30 amino acids) (Colgrave & Craik, 2004).

Cyclotides perform a variety of pathogen defense roles in the plants that produce them and have been described variously as insecticidal (Gruber, Cemazar, Anderson, & Craik, 2007; Jennings, West, Waine, Craik, & Anderson, 2001; Mylne, Wang, van der Weerden, & Craik, 2010), nematocidal (Colgrave, Huang, et al., 2010; Colgrave et al., 2008; Simonsen et al., 2008), molluscicidal (Plan, Saska, Cagauan, & Craik, 2008), bactericidal (Pränting, Lööv, Burman, Göransson, & Andersson, 2010), antiviral (Wang et al., 2008), hemolytic (Henriques et al., 2011) and phytotoxic (Ovesen et al., 2011). Their activity marks them as interesting agricultural and pharmaceutical engineering targets (Craik, Mylne, & Daly, 2010; Craik, Swedberg, Mylne, & Cemazar, 2012; Dörnenburg, 2010), and their stability allows them to be used as scaffolds for displaying engineered peptides (Ji et al., 2013).

Investigations into the biosynthesis of cyclic peptides are important to both fundamental and applied science. Firstly, cyclotides are circularized by a modified protease mechanism that appears to have convergently evolved in multiple other cyclic protein biosynthesis pathways. Additionally, understanding cyclization is necessary for recombinant cyclotide production and as a biotechnological tool for cyclizing other proteins of interest. This may also provide the means to overcome the poor expression of correctly folded, mature cyclotides in transgenic plants, enhancing the agricultural value of this group of peptides.

In this chapter we review the current evidence on cyclotide biosynthesis, focusing on the processing events that occur, and the enzymes that facilitate them.

2. CYCLOTIDES ARE GENE PRODUCTS

Plant cyclotides are gene-encoded, unlike many small cyclic peptides from bacteria and fungi that are synthesized by multienzyme complexes (Conlan, Gillon, Barbeta, & Anderson, 2011). The cyclotides are produced by cleavage and cyclization of cyclotide domains (30±4 residues in length) embedded in precursor proteins. They have been described for the Rubiaceae (Jennings et al., 2001), Violaceae (Dutton et al., 2004), Cucurbitaceae (Mylne et al., 2012), Solanaceae (Poth et al., 2012) and Fabaceae (Poth, Colgrave, Philip, et al., 2011).

The first cyclotide transcripts were obtained from *Oldenlandia affinis* (Rubiaceae) which produces the prototypical kalata B1 (kB1) and B2 (kB2) cyclotides of the Möbius subclass (Jennings et al., 2001). Transcripts encoding cycloviolacin cyclotides from both the Möbius and Bracelet subclasses were then obtained from *Viola odorata* (Violaceae) (Dutton et al., 2004). Genes from the Rubiaceae and the Violaceae encode similar precursors comprised of an endoplasmic reticulum (ER) signal sequence and a propeptide, followed by up to three cyclotide domains each flanked by conserved N-terminal and C-terminal prodomains (Figure 1). The cyclotide domains in the Möbius group shared greater than 79% sequence identity at the amino acid level irrespective of whether they were from the Rubiaceae or Violaceae. However when the Möbius and Bracelet mature cyclotide domains were compared they shared only 40% identity (Dutton et al., 2004) suggesting that these subclasses were inherited from an earlier ancestor in the Eudicot lineage and have evolved separately. Conservation of the 16 amino acid N-terminal repeat (Ntr) sequences in the *O. affinis* precursors led to the original hypothesis that the Ntr had a role in cyclotide processing. However there is a lack of sequence conservation between the Ntrs from *O. affinis* and *V. odorata* precursors and thus a role in processing was considered unlikely. The observation that the Ntrs from both species form amphipathic helices led to the new hypothesis that they may function in protein folding or subcellular targeting (Dutton et al., 2004).

Relatively little is known about the regulation of cyclotide gene expression. Some cyclotides are expressed constitutively, whilst expression of others varies during the year (Trabi et al., 2004), presumably induced by

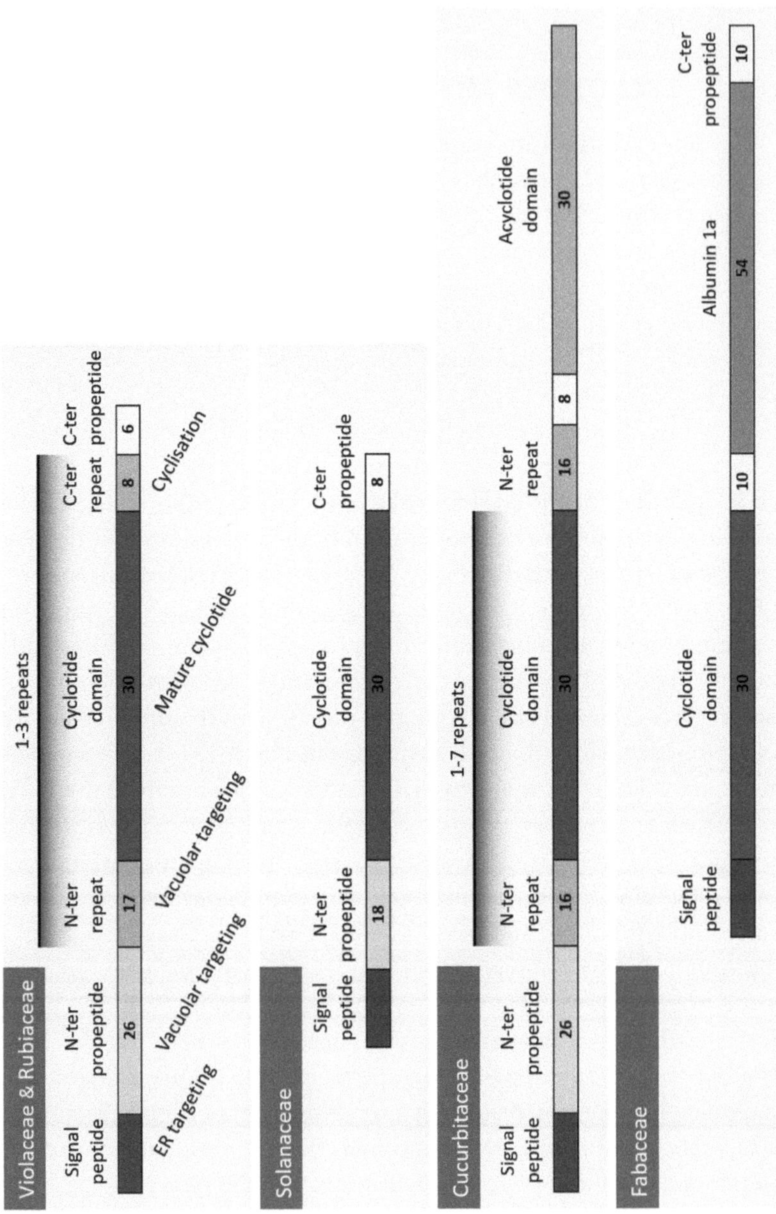
Violaceae & Rubiaceae
1-3 repeats
Signal peptide
N-ter propeptide
N-ter repeat
Cyclotide domain
C-ter repeat
C-ter propeptide
26
17
30
8
6
ER targeting
Vacuolar targeting
Vacuolar targeting
Mature cyclotide
Cyclisation
Solanaceae
Signal peptide
N-ter propeptide
Cyclotide domain
C-ter propeptide
18
30
8
Cucurbitaceae
1-7 repeats
Signal peptide
N-ter propeptide
N-ter repeat
Cyclotide domain
N-ter repeat
Acyclotide domain
26
16
30
16
8
30
Fabaceae
Signal peptide
Cyclotide domain
Albumin 1a
C-ter propeptide
30
10
54
10

Figure 1 ***Schematic of cyclotide precursors. Violaceae and Rubiaceae***: Cyclotide precursors are composed of an endoplasmic reticulum (ER) signal peptide, an N-terminal propeptide and a region composed of an N-terminal repeat, a mature cyclotide domain and a C-terminal repeat. This region can be repeated up to three times and is followed by an additional short sequence at the C-terminus. ***Solanaceae***: Cyclotide transcripts from *Petunia hybrida* encode a precursor with an ER sequence and one N-terminal propeptide of 18 amino acids followed by a mature cyclotide domain and a short C-terminal propeptide. ***Cucurbitaceae***: Cyclotide transcripts from *Momordica cochinchinensis* in the Cucurbitaceae family encode precursors with an ER signal sequence followed by an N-terminal propeptide of 26 amino acids and a region composed of an N-terminal repeat sequence of 16 amino acids and a mature cyclotide domain. This region is repeated up to seven times and is separated by 8 amino acids from a terminal acyclotide domain which is a typical noncyclic knottin. ***Fabaceae***: Cyclotide transcripts from the butterfly pea, *Clitoria ternatea*, from the Fabaceae family encode a precursor with an ER signal peptide followed immediately by a mature cyclotide domain that is flanked at the C-terminus by a linking peptide and the albumin-1a chain. The cyclotide domain replaces the PA1b subunit that is present in typical albumin-1 genes. (See color plate)

seasonal changes in biotic and abiotic stress (Basse, 2005; Seydel, Gruber, Craik, & Dörnenburg, 2007). Similarly cyclotides appear to be localized to different plant tissues (Mylne et al., 2010; Trabi & Craik, 2004).

The cyclotide precursors from the Solanaceae, the Cucubitaceae and the Fabaceae all have substantial differences to those from the Rubiaceae and the Violaceae (Figure 1). Precursors from the Solanaceae have only one mature cyclotide domain and do not have an N-terminal propeptide. In contrast, precursors from *Momordica cochinchinensis* from the Cucurbitaceae have up to seven cyclotide domains as well as one acyclotide domain at the C-terminus and are potent trypsin inhibitors (Mylne et al., 2012). Finally, the cyclotides from the Fabaceae are embedded within an albumin seed storage protein (Poth, Colgrave, Philip, et al., 2011). Thus, there has either been more than one independent origin for the cyclotide genes in the Eudicots, or they were present in a very early ancestor. If the latter is true they will be found in many more plant families. This also has implications for the level of conservation of the maturation process, which will be discussed in Section 4.4.

3. THE CYCLOTIDE PROCESSING PATHWAY

All the cyclotide precursors from the five Eudicot families illustrated in Figure 1 have typical ER targeting sequences and thus they enter the secretory pathway where folding and disulphide bond formation, aided by protein disulphide isomerase occurs (Gruber, Čemažar, et al., 2007) (Figure 2). After folding, the precursor leaves the ER and transits the Golgi apparatus. In the absence of specific targeting signals, proteins would be secreted via the default pathway. However, the cyclotide precursors are targeted to the prevacuolar compartment and the vacuole for proteolytic processing and cyclization (Figure 2).

Subcellular targeting to the vacuole was demonstrated using a series of green fluorescent protein (GFP) constructs and transient expression in *Nicotiana benthamiana* (Conlan et al., 2011). The full length cyclotide precursor from *O. affinis*, *Oak1*, directed GFP to the vacuole (Figure 3) demonstrating that excision and cyclization of the kB1 cyclotide occurred in the vacuole and that cyclotides were stored in this compartment. Further GFP constructs were used to identify the vacuolar targeting signals within the precursor. The N-terminal propeptide and the N-terminal repeat sequence from *Oak1* were both sufficient to target the precursor to the vacuole. The C-terminal

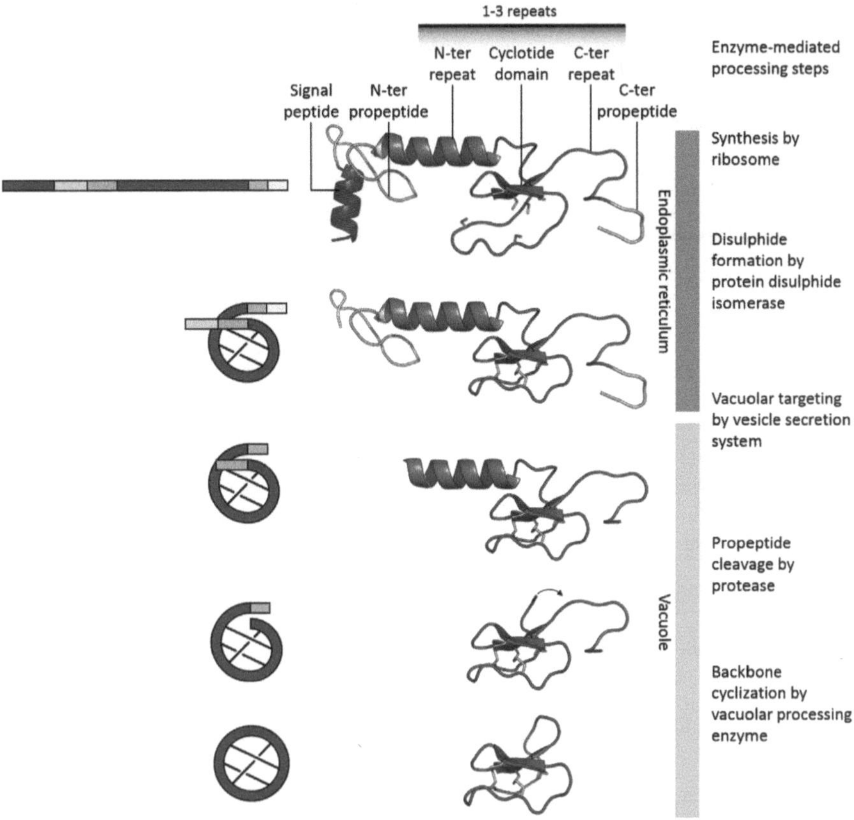

Figure 2 ***Cyclotide processing pathway.*** The cyclotide precursor is comprised of several parts: Firstly a hydrophobic signal peptide (thought to be α-helical (Nakamura & Matsuoka, 1993)) targets the precursor to the ER. The N-terminal propeptide and the helical N-terminal repeat contain vacuolar targeting sequences (Conlan et al., 2011) (PDB:1WN8). The cyclotide domain contains the cystine knot motif (Rosengren, Daly, Plan, Waine, & Craik, 2003) (PDB:1NB1). A final C-terminal propeptide of unknown structure is necessary for cyclization. The cyclotide domain folds in the ER. Disulphide bond formation is mediated by the oxidizing environment and protein disulphide isomerase. The first signal peptide is cleaved and the remaining protein is secreted to the vacuole, whereupon the N-terminal propeptide is also cleaved by an unknown protease. Finally, the protease, vacuolar processing enzyme performs backbone cyclization by transpeptidation to yield the final cyclic product. (See color plate)

propeptide, with the C-terminal repeat sequence which is essential for cyclization, was not a targeting signal. Targeting of the precursor to the acidic vacuolar compartment supported the early hypothesis that a vacuolar processing enzyme (VPE) had a pivotal role in excision and cyclization of the cyclotide domains from the cyclotide precursors (Gillon et al., 2008).

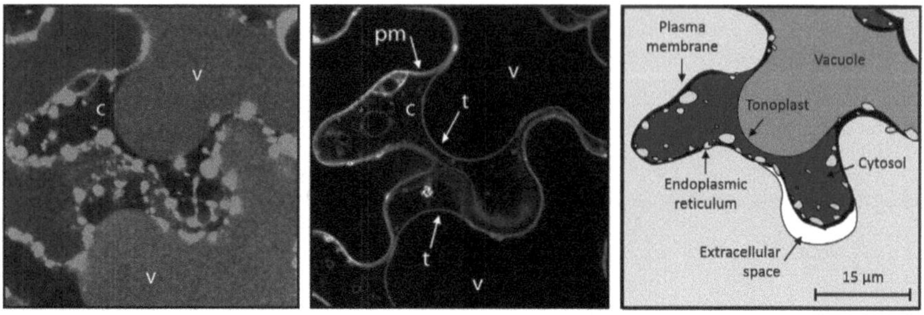

Figure 3 ***Vacuolar targeting by the signal peptide and*** **Oak1** ***precursor.*** Transient expression of GFP linked to the *Oak1* cyclotide precursor from *Oldenlandia affinis* in leaf epidermal cells from *Nicotiana benthamiana*. Left panel: GFP fluorescence from GFP-Oak1 in the ER and Golgi on the periphery of the cell and GFP in the vacuole (v) after proteolytic processing and production of the kB1 cyclotide. Middle panel: Fluorescence from the membrane dye FM 5−95 highlighting the plasma membrane (pm), the tonoplast (t, vacuolar membrane) and other small membranous components. Right panel: Diagram illustrating the relevant subcellular compartments of one of the cells shown in the fluorescent micrographs. *Modified from Conlan et al. (2011).* (See color plate)

4. SEQUENCE REQUIREMENTS FOR CYCLIZATION

The maturation of cyclotides from their precursor proteins to the final, mature products requires several processing steps (Figure 2). The penultimate processing event removes the N-terminal region, producing a linear precursor containing the C-terminal propeptide. In the final processing step, the C-terminal propeptide is removed and the free N- and C-termini are ligated to yield the mature, cyclic peptide. This is thought to be achieved enzymatically by intramolecular transpeptidation; an event that requires the N- and C-termini to be brought into close proximity (Figure 4). The active site must therefore accommodate residues from both the N- and C-termini of the nascent cyclotide, suggesting that the propensity of a given substrate to be cyclized may be driven by the amino acid composition at both termini. In this section, the sequence requirements around these processing sites will be discussed, particularly in relation to the prototypic cyclotide kB1, drawing on evidence from precursor sequence conservation, transgenic cyclotide production and substrate preference of putative cyclases.

4.1 Sequence Conservation

Sequence analysis of the N- and C-terminal processing sites of 36 cyclotide precursors from plants of the Rubiaceae and Violaceae families led to the identification of a conserved C-terminal tripeptide motif at each cleavage

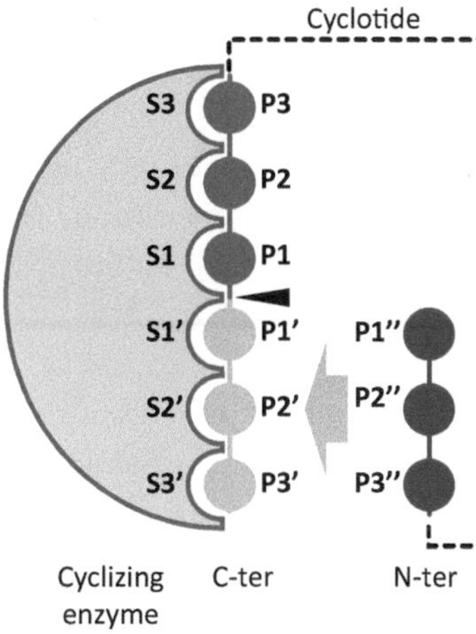

Figure 4 ***Enzyme binding sites.*** The cyclizing enzyme binds firstly to the C-terminus of the cyclotide precursor and subsequently to the N-terminus using the same binding site. The binding sites are labeled S1 to S3 before the cleavage site, and S1′ to S3′ after the cleavage site (indicated by a black triangle). The peptide residues that bind to these sites are similarly labeled P1 to P3 before and P1′ to P3′ after the cleavage site. Since the cyclotide N-terminus binds to the enzyme as a second substrate after cleavage, its residues are labeled P1″ to P3″. (See color plate)

site (Gillon et al., 2008). Extending this to include a total of 113 cyclotide sequences supports this early conclusion (Figure 5). The first position of this motif prefers a residue with a small side chain (typically Gly or Ser); the second position is represented by a conserved hydrophobic residue (typically Leu or Ile) and the third site is most often occupied by Pro, although this is variable at the C-terminal processing site. Gly_1 and Leu_{31} are particularly highly favored and are present in >90% of sequences analyzed here

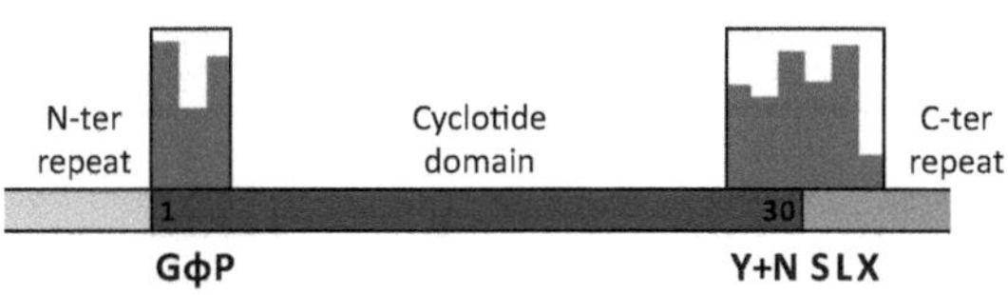

Figure 5 ***Cyclotide terminal processing sequences.*** The N-terminal sequence (P1″ to P3″) and C-terminal processing sequence (P3 to P1 and P1′ to P3′) based on an alignment of 113 cyclotide precursor sequences from the Rubiceae and Violaceae. The consensus sequence is shown below, where "ϕ" is Leu/Ile, "+" is Arg/Lys and "X" is any residue. The bar-chart shows the percentage conservation at each residue position. Note the similarity of GϕP to SLX that have to bind to the same active site pockets of the cyclizing enzyme.

indicating they are under strong selective pressure. However, whilst Gly_1 itself is highly conserved, some cyclotide sequences include an extra residue on its C-terminal side suggesting that sequence conservation of subsequent residues in the N-terminal tripeptide motif may not be as crucial for cyclotide biosynthesis. Strong sequence conservation of Asn/Asp at the C-terminal P1 site has also been observed and of the 282 cyclotide sequences currently available in CyBase, 78% contain Asn_{29} and 15% contain Asp at this position (Mulvenna, Wang, & Craik, 2006). This, in combination with other lines of evidence, led to the prediction that VPEs, an Asx-specific group of vacuolar proteases introduced in Section 3, were responsible for cleavage of the C-terminal propeptide and head—tail ligation (discussed in detail in Section 5).

4.2 Mutagenesis in Transgenic Plants

Analysis of the sequence conservation around the N- and C-terminal processing sites highlighted residues that were likely to be important for cyclotide maturation and guided *in planta* mutagenesis of precursors of the prototypic cyclotide kB1 (encoded by the *Oak1* gene) (summarized in Figure 6) (Gillon et al., 2008). In this cyclotide, the tripeptide motif is represented by Gly—Leu—Pro at both the C- and N-terminal processing sites. Both *Nicotiana tabacum* and *Arabidopsis thaliana* transformed with the wt *Oak1* gene could stably express correctly processed cyclic kB1, although a number of linear processing variants were also evident. Following mutation of the conserved Asn_{29} to either Asp or Ala, no cyclic product was detected. To further investigate sequence requirements, transient expressions of additional variants were carried out in *N. benthamiana*. This series of experiments showed that truncation of the C-terminal propeptide from GLPSLAA to GLP did not affect cyclization, however no cyclization was evident when *Oak1* was terminated after Asn_{29}. Similarly, no cyclization was observed when Gly_1 or Leu_2 within the N-terminal Gly—Leu—Pro motif were substituted with Ala.

A further suite of *Oak1* mutants were analyzed in a more recent report (summarized in Figure 6) (Conlan et al., 2012). Here, transient expression of *O. affinis Oak1* incorporating mutations around the C- and N-terminal processing sites was carried out in *N. benthamiana*. At the N-terminus, neither of the two residues targeted (Lys_{-1} and Leu_{-2}) were required for cyclization. However, substitution of Leu_{-2} did result in the appearance of additional residues at the N-terminus suggesting that this mutation hinders clean cleavage of the Ntr from the cyclotide domain in *N. benthamiana*. The lack of

... K_{-5} L_{-4} Q_{-3} L_{-2} K_{-1} G_1 L_2 P_3 .... T_{27} R_{28} N_{29} G_{30} L_{31} P_{32} S_{33} L_{34} ...

P1″ P2″ P3″ P3 P2 P1 | P1′ P2′ P3′

N-ter propeptide

Mutation	Cyclic product
$L_{-2}A$	✓
$K_{-1}A$	✓

kB1 cyclotide domain

Mutation	Cyclic product
G_1A	✗
L_2A	✗
$R_{28}A$	✓
$N_{29}A$	✗
$N_{29}D$	✗
$N_{29}Q$	✗

C-ter propeptide

Mutation	Cyclic product	Mutation	Cyclic product
$G_{30}A$	✓	[GAPSL*]	✗
$G_{30}F$	✗	[GAPS*]	✗
$L_{31}A$	✗	[GAP*]	✗
$L_{31}I$	✓	[GIP*]	✓
$L_{31}A+P_{32}A$	✗	[ALA*]	✓
$L_{31}A+P_{32}A$	✗	[LGP*]	✗
N_{29}*	✗	[GGL*]	✗
G_{30}*	✗	[SLP*]	✓
L_{31}*	✓	[GLA*]	✓
P_{32}*	✓		

Figure 6 ***Processing efficiency of* Oak1 *gene mutants.*** The *Oak1* gene was mutated to encode amino acid substitutions and/or truncations in the N-terminal propeptide, mature cyclotide domain (29 residues long) and the C-terminal propeptide and was then introduced into tobacco plants. The presence or absence of cyclic product was determined by mass spectrometry and is indicated by ✓ or ✗ respectively. * denotes C-terminal truncation after the indicated residue. Brackets denote combined truncation and mutation of the C-terminal propeptide. *Adapted from references Conlan et al. (2012) and Gillon et al. (2008).*

apparent sequence specificity around this cleavage site suggests that the residue immediately upstream of the cyclotide domain may not be important, at least in this system. Cleavage may instead be driven by surface-exposure and accessibility leading to nonspecific targeting by various proteases, or by a single enzyme with limited sequence specificity.

At the C-terminal cleavage site of kB1, a conservative $Asn_{29}Gln$ substitution abolished the production of cyclic peptide in *N. benthamiana*, further supporting a role for VPEs as the mediators of this processing event. At the P2 position, substitution of Arg_{28} with Ala or Lys did not preclude production of cyclic protein suggesting this position is not important for cyclization. However, overall expression of these variants was reduced compared to native *Oak1*, perhaps as a result of misfolded precursor being unable to exit the ER, demonstrating that sequence variation around these processing sites is also limited by factors other than specificity of the processing enzymes. At the P1′ position, substitution of the native Gly with the correspondingly small residues Ala or Ser was well tolerated, whilst mutation to the larger, bulky residue, Phe, resulted in a lack of correctly processed cyclic product. This suggested that the conservation of small residues at this position was attributable to the constraints imposed by enzyme specificity. The presence of Pro at the P3′ position was deemed nonessential since cyclic product was still detected following substitution with Ala. Consistent with this, a series of truncations of the C-terminal propeptide identified the dipeptide Gly—Leu as the minimal C-terminal sequence required for cyclization. Interestingly, whilst no cyclic product was detected in the P2′ $Leu_{31}Ala$ mutant, an unusual posttranslational modification was observed: hydroxylation of the neighboring proline. When a more conservative substitution of $Leu_{31}Ile$ was introduced, both cyclic kB1 and a small amount of the hydroxylated product were produced. Whilst not directly tested, it is possible that hydroxylation of this Pro could also directly affect cyclization and a requirement to limit this event could be contributing to sequence conservation in this region.

A limitation of studying processing of cyclotide precursors in transgenic plants is that they are necessarily carried out with species that do not natively produce cyclotides and so are unlikely to provide optimal conditions for processing cyclic peptides. This is exemplified by the relatively small proportion of the total *Oak1* gene product that is correctly processed to the mature, cyclic peptide in transgenic tobacco (Conlan et al., 2012; Gillon et al., 2008). Small changes in the level of cyclic product may therefore render the product below the detection limit of the assay and drawing absolute conclusions about essential residues for cyclization should be approached with caution. It remains to be determined if any of the aberrantly processed forms are relevant in native cyclotide producing plants where the high level of cyclotide expression and lack of detection of any linear forms imply a highly efficient maturation process. However, native cyclotide producers are currently poor models for probing cyclization requirements since they are not readily amenable to transformation and production of transgenic cyclotides is obscured by native cyclotides. Co-expression of cyclotides and VPEs coupled with advances in cyclotide detection could facilitate improvements in this area. Additionally, the recent discovery of cyclotides in the Solanaceae has revealed subtle sequence variation in the regions flanking the mature cyclotide, perhaps reflecting the specificity of processing enzymes within this plant family (Poth et al., 2012). Tailoring cyclotide precursor sequences within model plant systems to reflect the substrate specificity of endogenous VPEs may therefore improve the utility of this technique.

4.3 Atypical Cyclotide Maturation

Cyclotides have thus far been discovered in the Rubiaceae, Violaceae, Cucurbitaceae, Fabaceae and Solanaceae plant families and are predicted to be present in many more (Craik & Malik, 2013; Gruber et al., 2008). Most known sequences belong to the Rubiaceae and Violaceae and are reported to be processed in a similar manner from precursor proteins containing one, two or three cyclotide domains (Dutton et al., 2004). However, the biosynthesis of at least some cyclotides from other plant families does not follow this established maturation process and mature cyclotides are produced from precursors with varying gene structures (see Figure 1). Furthermore, the *M. cochinchinensis* trypsin inhibitor II (MCoTI-II) precursor protein contains the less commonly observed Asn at the N-terminal P1 site, implicating a VPE in processing of both the C- and N-terminal propeptides (Mylne et al., 2012). In such cases it will be of interest to determine if a single VPE has evolved to act as both a protease and cyclase, or if different allelic paralogs carry out the two different functions.

Cyclotides recently discovered in *Clitoria ternatea* (butterfly pea) of the Fabaceae family are also produced from genes embedded within an albumin precursor protein (Poth, Colgrave, Philip, et al., 2011). The *C. ternatea* cyclotides do not contain an N-terminal prodomain or an Ntr region as with typically reported cyclotide precursors; rather, their ER signal sequence is immediately followed by the cyclotide domain. This suggests that, as well as gene structure, the processing events leading to mature cyclotide can vary across plant families. Interestingly, the sequences of cyclotides from *C. ternatea* differ in that the consensus C-terminal processing motif is His—Val—Ile (compared to Gly—Leu—Pro in kB1) (Figure 7). In particular, His is not observed in the P1′ position in the C-terminal propeptide of any other cyclotide precursors and reflects a divergence from the small residues that are typically accommodated here (Poth, Colgrave, Lyons, Daly, & Craik, 2011). Similarly, Glu is found at this site in the recently reported cyclotides from the Solanaceae (Poth et al., 2012). This raises the possibility that the specificity of the enzyme responsible for cyclization may also differ between plant families, providing support for a more tailored approach to cyclotide production in transgenic plants. Alternatively, the sequence requirements for enzymatic cyclization may be less stringent than initially suggested by the overall sequence conservation and transgenic data.

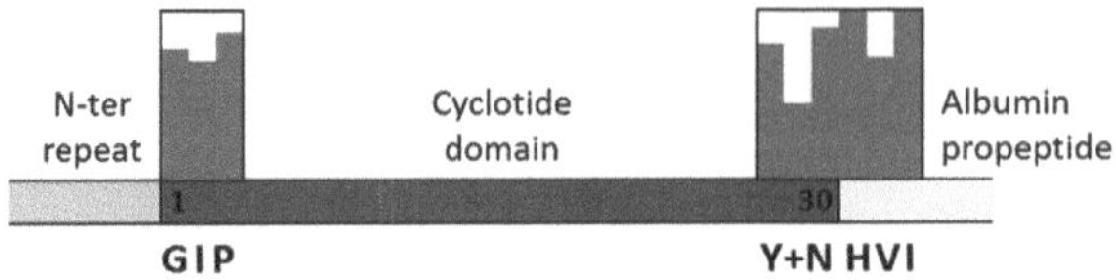

Figure 7 Clitoria ternatea *cyclotide terminal processing sequences.* The cyclotide precursor from *Clitoria ternatea* differs from the standard form. The cyclotide domain is located at the beginning of the precursor, before an albumin domain that is also processed into a mature protein. The N-terminal sequence (P1″ to P3″) and C-terminal processing sequence (P3 to P1 and P1′ to P3′) is based on an alignment of 11 *Clitoria ternatea* cyclotide precursor sequences. The consensus sequence is shown below, where "+" is Arg/Lys. The bar-chart shows the percentage conservation at each residue position.

Acyclic, linear peptides that share high sequence homology with the cyclotides—including the characteristic cystine knot motif—have also been identified in an increasing number of plant families (Gerlach, Burman, Bohlin, Mondal, & Go, 2010; Ireland, Colgrave, Nguyencong, Daly, & Craik, 2006; Nguyen et al., 2013; Nguyen, Lim, Nguyen, & Tam, 2012; Nguyen et al., 2011; Poth et al., 2012). These peptides, termed "acyclotides" (Poth et al., 2012), can share structural similarity with their cyclic counterparts (Ireland et al., 2006) and retain some native cyclotide bioactivities (Gerlach et al., 2010; Nguyen et al., 2013, 2012). A common feature of the acyclotides is the absence of Asx at the C-terminus further supporting the importance of Asx-specific processing in the enzymatic maturation of backbone-cyclized peptides.

4.4 Substrate Sequence Requirements for VPE-Mediated Transpeptidation

Recently it has been reported that a VPE, named butelase 1, isolated from *C. ternatea* is able to cyclize a number of peptides (including kB1) containing an appropriate recognition sequence (Nguyen et al., 2014). The catalytic mechanism of VPEs and their role in cyclization are discussed in more detail in Section 5, however it is worth noting here that this cyclase required the addition of only a small recognition sequence to unrelated, acyclic peptides to render these peptides cyclizable. For example, thanatin, an antimicrobial peptide from insects (Fehlbaum et al., 1996) only required the addition of a C-terminal Asn and the dipeptide His—Val (representing the C-terminal prodomain of the native substrate) for efficient butelase 1-mediated cyclization. Therefore the addition of three residues (Asn—His—Val) can provide sufficient affinity to the enzyme—substrate interaction for cyclization to occur, provided the N-terminus can also be accommodated by the active site. This is consistent with the *in planta* studies showing that only two residues C-terminal to the P1 Asn were required for cyclization of the prototypic cyclotide kB1.

At the N-terminus of the cyclotides identified in *C. tenatea* the established tripeptide motif is retained and most commonly exemplified by Gly—Ile—Pro, although His has also been observed in place of Gly (Figure 7) (Poth, Colgrave, Lyons, et al., 2011). Accordingly, the N-terminus of most of the synthetic peptides cyclized by butelase 1 contained a similar sequence (Gly—Val or Gly—Leu), either natively or engineered. However, when the N-terminal specificity of butelase 1 was assessed in the context of model peptides, most amino acids could be accommodated in place of Gly with similar efficiency (Figure 8). This apparent N-terminal flexibility is incongruent with predictions based on sequence conservation and data from transgenic plants. For example, an N-terminal Phe was readily accepted by butelase 1 whereas, in transgenic plants, the same substitution prevented cyclization of kB1 precursors (Conlan et al., 2012). This may be due to a greater stringency of *in planta* over in vitro experiments or indicate a real difference in substrate specificity of VPEs from different sources.

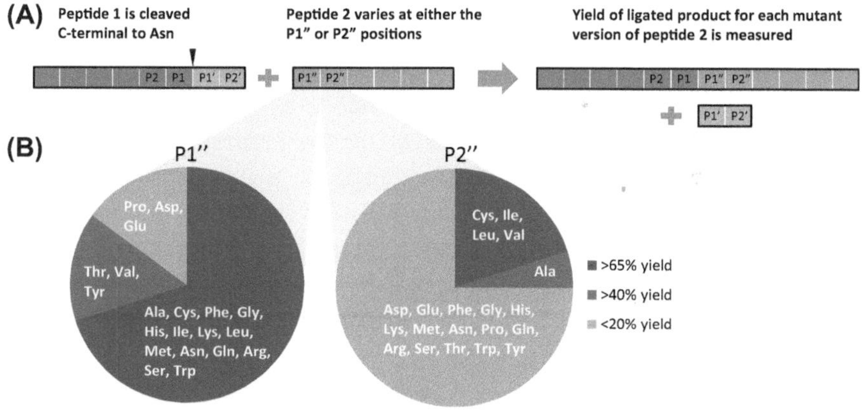

Figure 8 ***Substrate specificity of butelase 1 for the incoming N-terminus during transpeptidation.*** (A) Butelase 1 cleaves after a P1 Asn (black triangle) in an 8-residue model peptide and ligates the newly exposed C-terminus to the N-terminus of a second peptide. In the second peptide, either the P1″ or P2″ residue is substituted with each of the 20 amino acids and the effect on yield examined. (B) Yields of ligated product from model peptides with variations at the P1″ and P2″ positions. *Adapted from reference Nguyen et al.(2014).* (See color plate)

The reported P2″ preference of butelase 1 for the hydrophobic amino acids Leu, Ile and Val in the context of model peptides is more consistent with previous reports (Figure 8(B)). For example, the presence of Ala at this position resulted in a dramatic reduction in the ligation efficiency of model peptides by butelase 1, as well as precluding the production of cyclic kB1 from the mutagenized *Oak1* gene in transgenic plants (Gillon et al., 2008; Nguyen et al., 2014). However, hydrophobicity at this site was not an absolute requirement for butelase 1 activity since cyclization of thanitin, which contains Ser at this position, was still achieved; albeit at a slower rate.

Whilst it remains to be determined if the broad specificity reported for butelase 1 is applicable to a full-length cyclotide, and to cyclization *in planta*, the apparent lack of preference for an N-terminal Gly raises the possibility that VPE specificity is not be driving the high level of conservation at this position. This is unlikely to be driven by functional constraints since Ala and Lys scanning of the prototypic cyclotide kB1 demonstrated that substitution at position 1 did not impede its insecticidal or nematocidal activity (Huang, Colgrave, Clark, Kotze, & Craik, 2010; Simonsen et al., 2008). It may, however, be dictated by the specificity of the as yet unidentified protease that mediates release of the Ntr. Alternatively, folding efficiency may be impaired following substitution at this position. Indeed, the yield of chemically synthesized Gly_1Ala was only 50% that of wild type providing support for this hypothesis (Simonsen et al., 2008).

4.5 Summary

Sequence conservation around the C- and N-terminal processing sites provided early insight into the requirements for cyclization and this was supported by targeted mutagenesis in transgenic plants. However, these motifs may not be applicable to all families of cyclotide producing plants. The recent isolation of butelase 1 demonstrated that cyclization of unrelated peptides could be achieved following the addition of a small recognition motif and that sequence requirements outside of this appear minimal. It is unknown if the substrate promiscuity displayed by butelase 1 is representative of all VPEs that can achieve cyclization, or if it uniquely accommodates the sequences expressed in *C. ternatea*. Isolation of VPEs from other plant species, as well as combining cyclotide and VPE expression to improve the utility of transgenic studies will aid in answering this question and further elucidating the mechanisms of cyclotide maturation.

5. CIRCULARIZATION OF CYCLOTIDES IS MEDIATED BY A VACUOLAR PROCESSING ENZYME (VPE)

VPEs (also known as asparaginyl endopeptidases (AEPs) or legumains) are proposed to mediate peptide cyclization by a promiscuous transpeptidase activity—a modification of their native hydrolysis mechanism. In this section, we will outline the evidence that led to the identification of VPEs as key players in cyclotide maturation and discuss our current understanding of the mechanisms behind cyclization.

5.1 VPE is the Likey Cyclotide Cyclase

Initial attempts to identify potential enzymes that might be responsible for cyclotide cyclization identified VPEs as likely candidates. VPEs exclusively cleave C-terminal to Asn/Asp (Hara-Nishimura, Inoue, & Nishimura, 1991; Hiraiwa, Nishimura, & Hara-Nishimura, 1999) and this substrate specificity matches the high sequence conservation of Asn/Asp at the cyclotide C-terminus (Gillon et al., 2008). Additionally, their vacuolar localization is consistent with the reported production and storage of cyclotides within this compartment (Conlan et al., 2011). Strengthening this early conclusion was the report that the lectin, concanavalin A (Con A), from *Canavalia ensiformis* (jackbean) is matured by a VPE (Sheldon, Keen, & Bowles, 1996). Con A maturation involves VPE-mediated proteolysis and transpeptidation (though not backbone cyclization), demonstrating the capability of this group of enzymes (Figure 9). It was speculated that the ligation function of VPE was driven by the close proximity of the N- and C-termini of the intermediate Con A processing products to preference transpeptidation (Sheldon et al., 1996). More recently a VPE from jackbean was demonstrated to be capable of producing cyclic peptide from a linear precursor, albeit inefficiently (Bernath-Levin et al., 2015).

Evidence *in planta* that cyclotide transpeptidation is mediated by VPEs came from studies where suppression of VPE activity resulted in a decreased amount of cyclic product in transgenic plants (Mylne et al., 2012; Saska et al., 2007). A limitation of this approach (as with the *in planta* mutagenesis studies discussed in Section 4) is that the plant species used in these studies do not natively produce cyclotides and consequently the amount of cyclic product produced is relatively low, making it difficult to draw absolute conclusions. However, the finding that butelase 1 from *C. ternatea* could cyclize the *O. affinis* cyclotide kB1 provided the first direct in vitro evidence for the role of VPEs in cyclotide maturation.

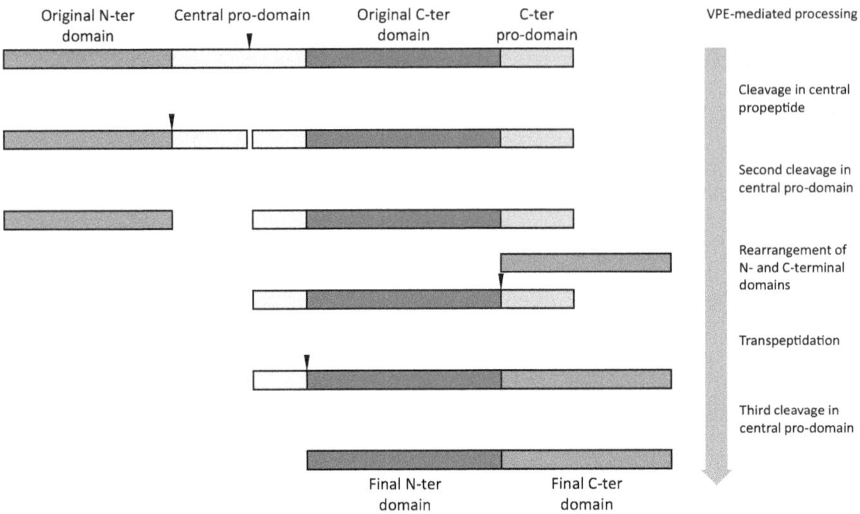

Figure 9 ***VPE-mediated ConA maturation.*** Schematic illustrating the hydrolysis and transpeptidation processing events required to produce mature ConA. Arrows indicate processing sites. *Figure adapted from Bowles et al. (1986) and Sheldon et al. (1996).* (See color plate)

5.2 Cyclization by a Modified Covalent Proteolysis Mechanism

The mechanisms and sequence requirements underpinning the transpeptidation ability of VPEs remain poorly understood. VPEs are cysteine proteases and as such contain a catalytic triad in which an acidic or polar residue aligns a histidine to deprotonate a cysteine, increasing its nucleophilicity. Catalysis proceeds by nucleophilic attack by the activated cysteine on the carbonyl of the peptide substrate to form a covalent acyl-enzyme intermediate (Figure 10(A)). The C-terminus of the cleaved substrate is released and a second nucleophilic attack occurs to resolve the intermediate and release the N-terminal peptide. In the more commonly observed proteolysis, this second nucleophile is a water molecule which is polarized by the histidine to hydrolyze the intermediate causing the final products to be the two peptide fragments (Figure 10(B)). However in cyclization, it is predicted that the N-terminus of the covalently bound peptide circles around and binds in the S′ pockets. The free amine of the peptide then acts as the second nucleophile to resolve the acyl-enzyme intermediate, yielding the mature cyclotide and linear C-terminal propeptide as the final products (Figure 10(C)).

The ratio of hydrolysis to transpeptidation is dependent on the relative strengths and local concentrations of water versus peptide amine to act as

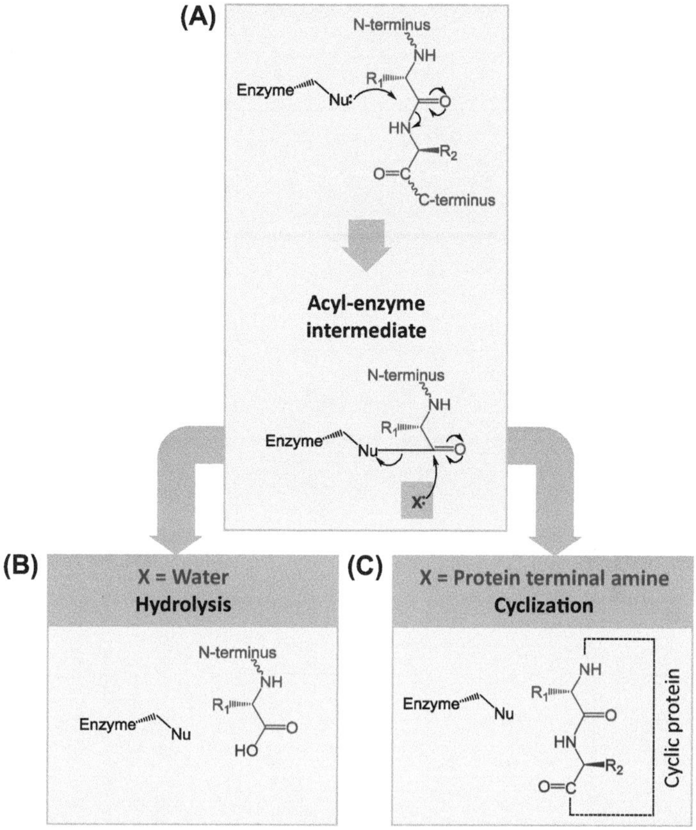

Figure 10 ***Chemical mechanism of cyclization by intramolecular transpeptidation.*** (A) The enzyme (black) contains a nucleophilic residue (Nu) in its active site which attacks the carbonyl of the substrate peptide (red) to form a covalent acyl-enzyme intermediate and eject the substrate's C-terminus. This intermediate is then attacked by a second nucleophile (X) to regenerate the free enzyme and release the final product. (B) In a typical protease, X is water and hydrolysis simply releases the remaining cleaved peptide. (C) Alternatively, the peptide's N-terminus enters the active site and the amine acts as the nucleophile X. This intramolecular transpeptidation yields a cyclic product. (See color plate)

the second nucleophile. Transpeptidation relies on the N-terminal amine outcompeting the vastly more abundant water to resolve the acyl-enzyme intermediate. It has been hypothesized that VPE achieves a high local concentration of the cyclotide N-terminus by binding a conserved tripeptide motif (represented by GLP in the prototypic cyclotide kB1) at the N-terminus of the peptide (Figure 11) (Gillon et al., 2008) and that this is aided by the structural constraints imposed by the disulphide bonded structure of cyclotides. In

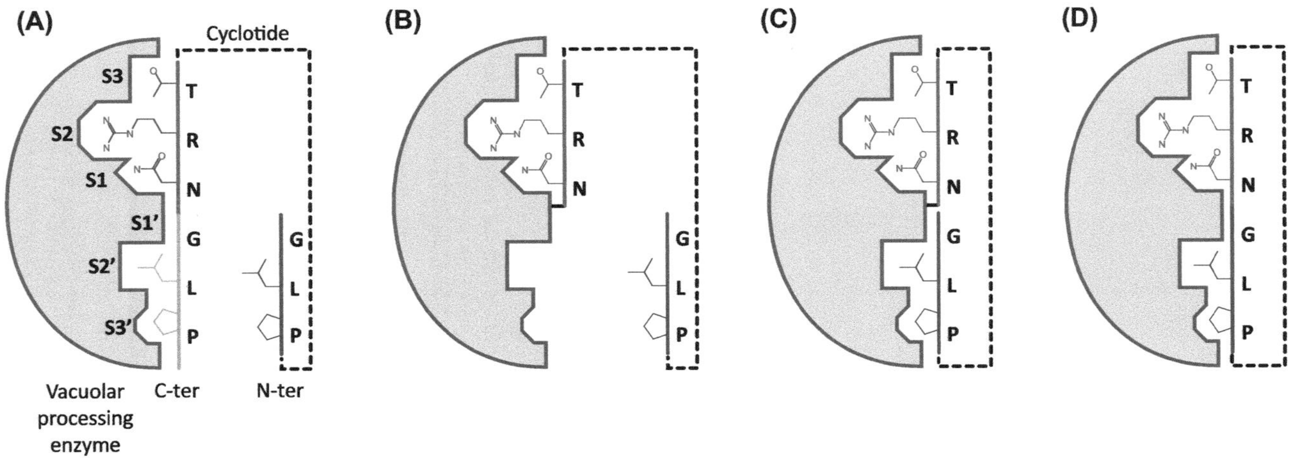

Figure 11 ***Proposed binding specificity for cyclization by VPE.*** (A) VPE specifically preferentially binds the substrate sequence TRNGLP in kB1. (B) Upon formation of the acyl-enzyme intermediate, the cleaved C-terminal peptide is released to expose the GLP binding pocket. (C) If the peptide N-terminus also contains the GLP motif, it binds to the exposed active site pocket. (D) The N-terminal amine outcompetes water as the nucleophile to resolve the acyl-enzyme intermediate into a cyclic product. (See color plate)

this scenario, the specificity of the enzyme S′ pockets for this tripeptide motif not only aids the binding of the initial substrate (Figure 11(A)), but also subsequent binding of the N-terminus (Figure 11(B) and (C)) for cyclization (Figure 11(D)). However, the finding that butelase 1 is able to accommodate a broad range of residues at the first position of this motif with little impact on yield of ligated product suggests that this N-terminal motif may not be as critical for cyclization as first thought (Nguyen et al., 2014).

Another factor that was previously thought to be important in pushing the equilibrium towards transpeptidation was the proximity of the N- and C-termini of the cyclotide precursor, maintained by the compact, disulphide-bonded structure. However butelase 1 does not require these structural constraints for efficient cyclization since a reduced and alkylated linear kB1 precursor was cyclized more readily than disulphide bonded linear kB1 (Nguyen et al., 2014). Furthermore, unrelated peptides derived from conotoxin, thanatin and histatin-3 that are not naturally backbone-cyclized were also cyclized following addition of compatible C- and N-termini.

There is evidence that as well as favoring transpeptidation, VPE also disfavors hydrolysis. Indeed, butelase 1 appears to have a low reactivity from water in its active site since, whilst still able to occur, hydrolysis of substrates lacking a free amine at the N-terminus is slow, indicating a preference for cyclization over hydrolysis (Nguyen et al., 2014). This relatively low hydrolysis rate indicates that water is either poorly polarized by the active site or hydrophobically excluded from it.

6. VACUOLAR PROCESSING ENZYMES (VPEs)

The preceding sections have detailed the evidence that VPEs can mature cyclotides to their native, backbone-cyclized product. Although VPEs are widely expressed in plants and animals, only four members of this class have been directly demonstrated to be able to ligate peptide bonds and, of these, only butelase 1 performs as an efficient cyclase (Bernath-Levin et al., 2015; Dall, Fegg, Briza, & Brandstetter, 2015; Nguyen et al., 2014; Sheldon et al., 1996). Whilst this number is likely to expand, given the circumstantial evidence of involvement of other VPEs in peptide cyclization (Mylne et al., 2012; Saska et al., 2007), the more commonly reported role is targeted proteolysis to mature precursor proteins and activate cellular pathways. It is unknown whether this disparity reflects subtle sequence differences that confer a preference for ligation or proteolysis or if dual capabilities are characteristic of all VPEs and dependent on substrate composition or conditions in vivo. In this context, this section will review the biosynthesis, structure and function of this important group of enzymes.

6.1 Isolation and Substrate Specificity

The first evidence of VPE activity in plants was obtained from cotyledons of *Vicia sativa* (common vetch) seedlings where cleavage C-terminal to Asn residues in the insulin A and B chains was observed (Shutov, Do, & Vaintraub, 1982). VPEs have since been isolated from many plants such as *Ricinus communis* (castor bean) (Hara-Nishimura et al., 1991), *Glycine max* (soybean) (Scott, Jung, Muntz, & Nielsen, 1992), and *A. thaliana* (Kinoshita, Nishimura, & Hara-Nishimura, 1995) as well as the protozoan *Schistosoma mansoni*, a parasitic blood fluke (Dalton & Brindley, 1990; Klinkert, Felleisen, Link, Ruppel, & Beck, 1989). Mammalian homologues (legumains), have also been identified in a number of organisms such as pigs (Chen et al., 1997), mice (Chen, Dando, et al., 1998) and humans (Dalton & Brindley, 1990; Klinkert et al., 1989). The proteinase responsible for this Asn-specific cleavage was later purified from maturing seeds of castor bean (Hara-Nishimura et al., 1991) and its vacuolar location demonstrated by immunocytochemical analysis (Hara-Nishimura, Takeuchi, & Nishimura, 1993). However, cell wall localization, together with a role in the maturation of cell wall proteins, has also been reported demonstrating that VPEs are not exclusively vacuolar (Linnestad et al., 1998). Similarly, mammalian legumains have been reported primarily in the lysosymes but also in the nucleus (Haugen et al., 2013; Patel et al., 2009) which has a neutral pH, suggesting complex regulation and involvement in diverse cellular processes.

Apart from the characteristic Asn and, to a lesser extent, Asp specificity, there is little consensus in the substrate specificity of VPEs from different sources. For example, soybean VPE had an absolute requirement for a P1 Asn (Jung et al., 1998), whilst other VPEs extend this specificity to Asp (Becker et al., 1995; Hiraiwa et al., 1999; Nguyen et al., 2014). Furthermore, substitution of the P2′ Leu with Ala in a native soybean VPE substrate appeared to be well tolerated (although it should be noted that no enzyme kinetics values are reported here) whereas the same substitution in the cyclotide precursor kB1 precluded its cyclization in transgenic plants (discussed in Section 4.3) (Conlan et al., 2012; Gillon et al., 2008). At the P3 position, kidney bean (*Phaseolus vulgaris*) VPE favored an amino acid with a bulky side-chain (Val, Phe, Tyr) (Rotari, Dando, & Barrett, 2001) and, similarly, sugarcane (*Saccharum officinarum*) VPE preferred Val at this site (Santos-Silva,

Soares-Costa, Gerald, Meneghin, & Henrique-Silva, 2012). However, again this is not applicable to all VPEs since mothbean VPE and a mammalian VPE from pig kidney did not exhibit the same specificity (Dando et al., 1999; Kembhavi, Buttle, Knight, & Barrett, 1993; Rotari et al., 2001). Additionally, butelase 1 could accommodate residues with different properties at this site, although processing efficiency of variants were not directly compared. A comparative study of the specificity of human and Schistosome legumain also revealed a lack of consensus in the optimal substrate sequences at the P2 and P3 positions (Mathieu et al., 2002). The precise substrate specificity VPEs from different sources is therefore variable, perhaps reflecting the multifunctional nature of VPEs and arguing for coevolution of substrate and enzyme, as has been reported for other substrate/enzyme combinations (Kolli, Lastere, & Schiffer, 2006). Interestingly, the finding that VPE-mediated peptide circularization occurs in distantly related plant lineages suggests that cyclization ability may have evolved independently multiple times, perhaps reflecting an enzyme that is particularly suited to being recruited to this function (Mylne et al., 2012).

6.2 Biosynthesis and Expression of Plant VPEs

Plant VPEs have been isolated from both maturing seeds and vegetative organs and share high sequence identity with each other and with mammalian homologues. They are synthesized as inactive precursors that have an ER signal peptide as well as short N-terminal and longer C-terminal prodomains that flank the mature endoproteinase domain (Figure 12). They also have a conserved tetra-peptide at the C-terminus that is essential for targeting to the vacuole (Kuroyanagi et al., 2002). The VPE precursors are trafficked through the secretory pathway to either the vacuole (Fischer et al., 2000; Hiraiwa et al., 1999; Schlereth, Standhardt, Mock, & Müntz, 2001) or the cell wall (Linnestad et al., 1998). Posttranslational processing and activation of the enzyme occurs immediately after the precursor has entered an acidic environment such as the vacuole or the cell wall (pH 5–6). This occurs by sequential self-processing C-terminal to selected Asn residues that yields the mature active enzyme (Hiraiwa, Nishimura, & Hara-Nishimura, 1997; Hiraiwa et al., 1999; Müntz & Shutov, 2002). Inactivity of the VPE precursor enables it to be packaged together with its substrates (including cyclotide precursors) into transport vesicles or ER bodies for transport to and activation in the vacuoles (Hara-Nishimura, Kinoshita, Hiraiwa, & Nishimura, 1998; Jung et al., 1998). Maintaining inactivity of the proteinase during subcellular trafficking is attributed to the C-terminal prodomain that acts as an

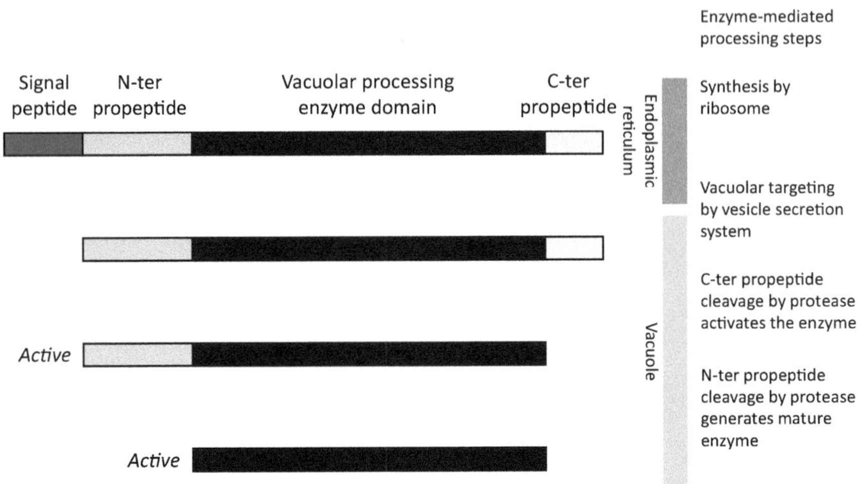

Figure 12 ***Maturation and activation of VPEs.*** A schematic representation of a VPE precursor. The signal peptide is removed posttranslationally in the ER. Once it reaches the acidic environment of the vacuole the C-terminal prodomain is removed and the enzyme becomes active. Subsequent cleavage of the N-terminal prodomain produces the fully mature active form.

auto-inhibitory domain by masking the active site and regulating the final processing step in the activation cascade (Dall & Brandstetter, 2013; Kuroyanagi et al., 2002). Interestingly, recent evidence has indicated that noncanonical self-processing of human legumain at a less acidic pH can impart a carboxypeptidase function, suggesting the complexity of regulation of this group of proteases is only beginning to be understood (Dall & Brandstetter, 2013).

Typically, different plant species contain multiple VPE homologues. These were initially classified as either seed-type, vegetative-type or uncharacterized according to sequence and expression patterns (Yamada, Shimada, Nishimura, & Hara-Nishimura, 2005). However, functional redundancies evident from selective VPE knock-outs and the finding that vegetative-type VPEs can be detected in developing seeds and vice-versa: seed-type VPEs have been detected in vegetative tissues (Gruis et al., 2004; Gruis et al., 2002; Julián, Gandullo, Santos-Silva, Diaz, & Martinez, 2013) suggests that this tissue-specific expression is not absolute. This functional overlap is unlikely to be complete since the loss of a given homologue was not always fully compensated by other VPEs implying that subtle sequence differences may dictate preferred substrates and, consequently, function. Whilst this may also be due to inadequate expression levels of the compensating homologue, it raises the possibility that the cyclizing ability recently attributed to VPEs is homologue specific rather than a capability intrinsic to all VPEs under the right conditions. Indeed, in extracts from the butelase 1 producing plant, *C. ternatea*, protein fractions that were active against the generic VPE substrate Z-AAN-MCA did not contain the cyclizing enzyme and, conversely, the butelase 1 containing fraction did not display activity against this generic substrate, providing support for this argument (Nguyen et al., 2014). To better understand the factors driving activity of VPEs as cyclases or proteases, the isolation and functional analysis of multiple VPE sequence variants from cyclotide producing plants will be required.

6.3 Structural Features of VPEs

VPEs are part of the CD clan of cysteine proteases and so share the same structure and mechanism as caspases and separases, despite their limited overall sequence identity (Hatsugai, Yamada, Goto-Yamada, & Hara-Nishimura, 2015; Rawlings, Barrett, & Bateman, 2012). The CD clan of cysteine proteases consists of a conserved core of secondary structural elements, with a set of variable loops at the bottom, and another at the top containing the binding and catalytic sites (Figure 13(A)). The active site contains an absolutely

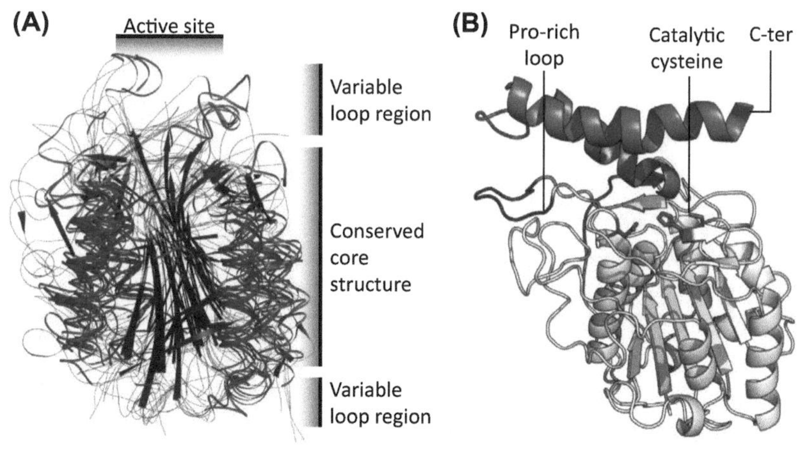

Figure 13 ***Structure of VPE and related enzymes.*** (A) Overlay of the known structures of CD clan proteases with α-helices in red, β-sheets in blue. (PDBs: 4AWA, 3H11, 4HQ0, 4JR0, 1CP3, 4F6P, 3V3K, 4HVA, 4JJ7, 4IEF, 3EEB, 3PA8). (B) Model of mature butelase 1 based on human pro-legumain (4FGU) with catalytic residues in red, VPE proline-rich loop in blue and C-terminal helices in gray. The exact orientation of the proline rich loop is unknown. (See color plate)

conserved catalytic dyad of Cys and His, with some members additionally using a third residue (Asn) or backbone carbonyl to form a catalytic triad (Nguyen et al., 2014; Rawlings et al., 2012; Watt et al., 1999). The VPEs, like most CD clan protease families, are synthesized as zymogens that undergo self-catalysis to release the mature active proteinase (Cohen, 1997; Hiraiwa et al., 1999; Kuroyanagi et al., 2002; Rawlings et al., 2012). Furthermore, VPEs can hydrolyze some caspase substrates, and their activity is blocked by some caspase inhibitors (Hatsugai et al., 2004; Kuroyanagi et al., 2005).

Within the CD clan, VPEs form a sister family (C13) to the caspase family (C14) due to a number of fundamental differences. Firstly, VPEs and caspases share limited sequence identity (Hatsugai, Kuroyanagi, Nishimura, & Hara-Nishimura, 2006). Secondly, VPEs are typically vacuolar and function optimally at acidic pH (Kinoshita et al., 1999) whereas caspases are cytosolic with an optimum pH of 7.5 (Yamin, Ayala, & Miller, 1996). Thirdly, the definition of a caspase is the absolute requirement for an Asp at the P1 position whereas VPEs prefer cleavage after an Asn residue although they can cleave next to Asp residues (Hatsugai et al., 2006; Woltering, van der Bent, & Hoeberichts, 2002).

The sequence and structural features of VPE that enable it to act as an efficient cyclase are currently unknown. Although the structure of a plant VPE is yet to be solved, the crystal structure of the related human legumain has been determined with and without its prodomains (Dall & Brandstetter, 2013), allowing a homology model to be built for the known cyclase butelase 1 (Figure 13(B)).

The main difference between human legumain and the model VPE structure is an extended polyproline loop (present in all plant VPEs) close to the P binding sites of the active site. However, because this is not present in the template structure, its exact structure and orientation is unknown. The C-terminal region of butelase 1 is predicted to be helical. This is reminiscent of the capping helices that impart cyclization ability on the unrelated cyclase PatG (discussed in Section 7.2) (Agarwal, Pierce, McIntosh, Schmidt, & Nair, 2012; Koehnke, Bent, Houssen, Zollman, & Morawitz, 2012) and it has been hypothesized that this region of butelase 1, which is cleaved off in active human legumain, may therefore be important for favoring transpeptidation over hydrolysis (Nguyen et al., 2014). A further potentially important feature is the different binding pocket surface charges of butelase 1 and human legumain. Whereas legumain lacks any apparent specificity at its S′ sites and has a charged surface, the region thought to comprise the S′

binding sites in butelase 1 is more hydrophobic in agreement with the apparent preference for substrate hydrophobicity in at least some of these positions. Ultimately, the importance of any or all of these features cannot be known without a combination of structural and functional analyses of wt and mutant enzymes.

6.4 VPE Functions

In addition to the recently confirmed role in cyclization (discussed in Section 5), plant VPEs also exhibit targeted proteolysis to control a variety of cellular processes. Early reports described the processing of proprotein precursors from developing seeds to their mature forms by VPEs (Hara-Nishimura et al., 1991; Hara-Nishimura & Nishimura, 1987; Scott et al., 1992; Shimada, Hiraiwa, Nishimura, & Hara-Nishimura, 1994) and the activation of other proteases to effect downstream outcomes such as protein degradation in senescing cells (Okamoto & Minamikawa, 1999; Rojo et al., 2003). Subsequently, a role for VPE in programmed cell death (PCD) was observed in VPE-deficient transgenic plants challenged with the tobacco mosaic virus (TMV) and the fungal toxin fumonisin B1 (Hatsugai et al., 2004; Kuroyanagi et al., 2005). In each case, only the wild type plants displayed evidence of PCD at the infection site. VPEs have since been implicated in the PCD response to other stimuli including elicitors (Gauthier et al., 2007; Zhang et al., 2010), ER stress (Qiang, Zechmann, Reitz, Kogel, & Schäfer, 2012) and heat stress (Li, Yue, & Xing, 2012) as well as developmental stimuli (Hatsugai et al., 2015; Nakaune et al., 2005; Teper-Bamnolker et al., 2012) Defining the functional profile of specific VPE homologues will provide insight into the mechanisms of activity and evolution of both the protease and cyclase activities attributable to these enzymes in vivo.

In animals, limited proteolytic processing by legumains has also been implicated in a variety of cellular processes. For example, legumain has been reported to play a role in host defense via regulation of antigen processing in the class II MHC pathway (Manoury et al., 1998; Matthews et al., 2010) and processing of Toll-like receptors, which are involved in pathogen detection and mobilization of the immune response (Ewald et al., 2011; Sepulveda et al., 2009). Recently, it has been reported that human legumain is capable of acting as a ligase via a mechanism independent of the catalytic cysteine (discussed in Section 7.3) that requires a C-terminal Asn (Dall et al., 2015). Interestingly, butelase 1 reportedly produced a small amount of cyclic product from substrates lacking any residues C-terminal to the P1 Asn suggesting this novel mechanism could also be applicable to butelase 1 (Nguyen et al., 2014). Given that this would require Asn-specific hydrolysis of cyclotide precursors *in planta*, followed by a separate cyclization step, it seems unlikely to be of biological relevance. The functional significance of this capability of human legumain remains unknown, however production of alternately spliced peptides for class II MHC presentation could be one possibility, as has been reported in the context of class I MHC presentation (Dalet, Vigneron, Stroobant, Hanada, & Van den Eynde, 2010; Hanada, Yewdell, & Yang, 2004; Vigneron et al., 2004; Warren et al., 2006).

Mammalian legumain has also been associated with some types of cancer since overexpression of legumain has been reported in colorectal and breast cancer tissue and has also been correlated with a more invasive phenotype (Liu, Sun, Huang, Janda, & Edgington, 2003). As reported in Section 6.2, in addition to the typical lysosomal localization, active mammalian legumain has been observed in the nucleus in two colorectal cell lines (Haugen et al., 2013). The difference in pH in these two compartments (acidic vacuole and neutral nucleus) is inconsistent with reports that active legumain was irreversibly denatured at neutral pH (Chen et al., 1997; Chen, Rawlings, & Stevens, 1998). However, this may be reconciled by the recent finding that a noncanonical processing event takes place at near neutral pH, imparting an asparaginyl-specific carboxypeptidase function (Dall & Brandstetter, 2013). Binding of legumain to integrin αVßIII has also been reported (Liu, Bajjuri, Liu, & Sinha, 2012) and this was shown to render legumain more stable at higher pH revealing multiple layers of regulation on this versatile enzyme (Dall & Brandstetter, 2013). It is possible that compartmentalization and the involvement of cofactors could also offer some level of control over the specific functions of plant VPEs as opposed to primary sequence differences. However the vacuolar localization of cyclotides and cyclizing ability of purified butelase 1 suggest that these factors do not play a role in controlling cyclization.

6.5 Summary

VPEs are present in a wide range of plants and animals and are involved in a range of cellular processes. They are self-processed from inactive zymogens and display characteristic Asx-specificity, although sequence requirements outside of this region are not well-conserved across species. They share structural homology with caspase-1, and exhibit a high degree of sequence conservation. VPEs display an array of functions that involve both protease and cyclase activity. Whilst not yet well understood, these distinct functions may be unique to specific VPE homologues. Further work will be required to fully elucidate the function and regulation of this interesting group of enzymes.

7. CIRCULARIZATION OF OTHER BACKBONE CYCLIZED PEPTIDES

Cyclotides are one of many unrelated families of cyclic peptides that are produced ribosomally and enzymatically cyclized. There are several common features amongst the biosynthesis of these different families, indicating that similar cyclization mechanisms have evolved repeatedly. Cyclic proteins so far identified are all produced with propeptide regions that are cleaved off during cyclization to energetically pay for the transpeptidation reaction. Transpeptidation is typically performed by a cysteine or serine protease cyclizing the peptide via an acyl-enzyme intermediate.

7.1 VPE Involvement in Cyclization of Other Cyclic Protein Families

In addition to the cyclotides, there is evidence for VPE involvement in the maturation of another class of backbone cyclized peptides: the Paw-S derived family. This group includes the sunflower trypsin inhibitor SFTI-I, a 14 residue cyclic peptide comprised of two β-strands cross-linked by a disulphide bond (Colgrave, Korsinczky, Clark, Foley, & Craik, 2010). Like the cyclotides, it is produced as a linear propeptide, however the SFTI-I encoding domain is buried within a functional preproalbumin, Paw-S1 (Mylne et al., 2011). This is particularly surprising given that sunflower trypsin inhibitors show remarkable sequence, structure and functional convergence with the inhibitory loop of the Bowman–Birk Inhibitors (BBIs) which are over 100 amino acids long (Luckett et al., 1999). It is likely that these cyclic peptides recently evolved de novo within the genus *Helianthus* since several seed preproalbumins encode unique extra proteins that are released during maturation of the albumin (Yamada, Shimada, & Nishimura, 1999). Like cyclotides, conserved features of the precursor sequence led to the identification of a VPE as the likely cyclase (Mulvenna, Foley, & Craik, 2005). This peptide also contains an Asn at the N-terminal P1 position and studies in VPE-null transgenic plants have demonstrated that proteolytic cleavage at both the C- and N-termini is mediated by a VPE (Mylne et al., 2011). Interestingly, an additional protease-mediated modification of SFTI occurs when it is inhibiting trypsin. In this case, the protease both cleaves the SFTI inhibitory loop and re-cyclizes it (Colgrave, Korsinczky, et al., 2010; Marx et al., 2003).

7.2 Cyclization by Other Enzymes

Segetalin A, a representative of another group of plant backbone cyclized peptides, the orbitides, is not cyclized by a VPE but by the serine protease PCY-1. This represents a broader example of convergent evolution since both serine and cysteine proteases act via a similar covalent mechanism. Cyclization by PCY-1 is also largely independent of length and composition of the target peptide outside of a C-terminal recognition sequence (Barber et al., 2013).

Similarly, a serine protease has also been implicated in the cyclization of the Cyanobactins; a family of short cyclic peptides from bacteria with particularly well-characterized biosynthesis (Agarwal et al., 2012). Understanding their biosynthesis has been aided by the organization of bacterial genomes, whereby genes that act in a pathway are commonly clustered into operons. For example, the cyclic peptide Patellamide A (from *Prochloron didemi*) is encoded by the PatE gene in an operon that also contains a processing protease (PatA), the cyclizing protease (PatG) and several posttranslational modification enzymes (e.g. PatD).

The cyclizing protease PatG is a subtilisin-like serine protease from the SB clan.[10] Structural analysis revealed the presence of an additional two helices above the active site compared with other subtilisin-like proteases (Agarwal et al., 2012; Koehnke et al., 2012). The proximity of these helices to the catalytic serine led to speculation that this addition imparts the cyclization ability of PatG by dictating the spatial position of the N- and C-termini. Indeed, when these helices are lacking, cyclization ability is lost (Koehnke et al., 2012). It is possible that the C-terminus of a VPE may control access to the active site in a similar manner. As with butelase 1 ligation of cyclotides, Pat G does not require that substrates be prearranged in a near cyclical conformation for cyclization to occur, although it is postulated that this may help offset the entropic costs of constraining an otherwise flexible substrate within the active site (Agarwal et al., 2012; McIntosh et al., 2010). PatG exhibits limited sequence specificity outside of the recognition motif and can cyclize a broad range of substrate sequences and lengths following addition of a short C-terminal sequence (McIntosh et al., 2010). Interestingly, there appears to be few N-terminal sequence requirements for this group of cyclic peptides suggesting that, provided access to the active site is not sterically hindered, the nature of the N-terminal sequence is unlikely to be crucial.

There is little mechanistic information available on how the enzyme achieves the active site desolvation presumably required for cyclization, however the apparent promiscuity of butelase 1, PCY1 and Pat G suggest that this is mediated primarily by the enzyme and the short C-terminal recognition sequence in the substrate. In the case of PatG, it is hypothesized that the C-terminal sequence motif remains bound to the active site post-cleavage from the main peptide, unable to be displaced by water, and is only disrupted by the incoming N-terminus of the peptide, leading to cyclization to resolve the acyl intermediate (Koehnke et al., 2012). This is consistent with the C-terminal sequence requirements reported for cyclotides.

Threonine proteases also perform proteolysis by a covalent mechanism and can perform transpeptidation (Conlan et al., 2011). For example, the proteasome (a multisubunit threonine protease) excises and ligates peptides together from the fibroblast growth factor 5 precursor protein (Dalet et al., 2010; Vigneron et al., 2004). Although there are currently no examples of cyclization by a threonine protease, it is conceivable that one may yet be discovered.

7.3 A Unique Ligation Mechanism

As a counterexample to the transpeptidation-based peptide ligation mechanisms described in Figure 10, there is evidence of an entirely different mechanism performed by human legumain. In addition to its canonical cysteine protease mechanism, legumain may be able to perform condensation ligation by a different mechanism using residues in the same active site (Figure 14). Adjacent to the His of the catalytic triad is an Asp residue (Asp_{147}) which forms a high-energy succinimide ring (Suc_{147}) by intramolecular condensation (PDB:4N6N). It has been proposed that this succinimide ring can act as an electrophile in a mechanism that involves conversion of the substrate C-terminal asparagine (Asn_{39i}) to a reactive succinimide (Suc_{39i}) which is then adjacent to the amino terminus of the second substrate (Dall et al., 2015).

A key feature of this succinimide mechanism is that it is not transpeptidation and so does not require the simultaneous cleavage of a propeptide. The energy to achieve ligation by condensation comes from the reactive succinimide ring. Additionally, the mechanism proceeds through a succinimide electrophile covalent intermediate which is entirely different from

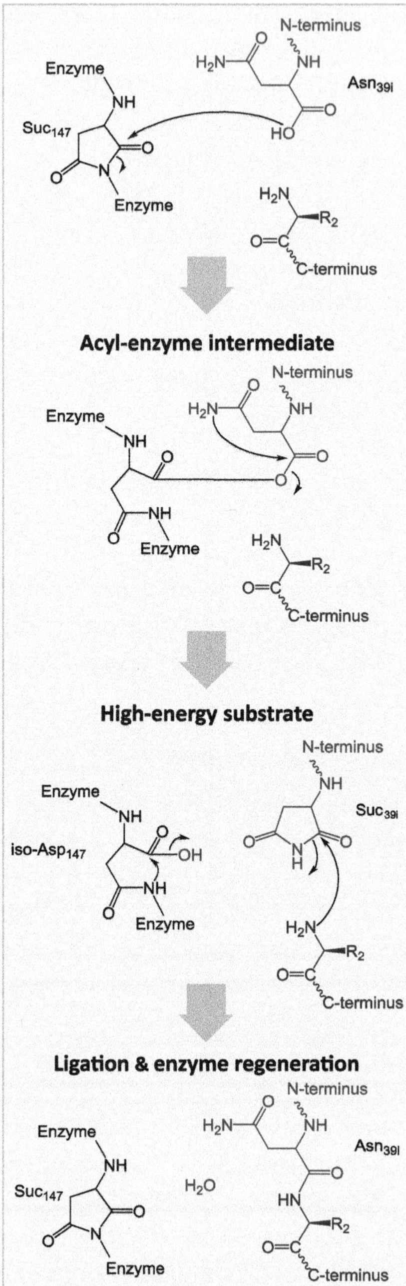

Figure 14 ***Proposed condensation ligation mechanism for Legumain.*** The enzyme (black) contains a high-energy succinimide (Suc_{147}) ring in its active site. This unnatural residue is the result of an intramolecular condensation of an aspartate residue. The

the canonical cysteine nucleophile covalent intermediate. It therefore represents an alternative specialized method of peptide ligation available to enzymes.

7.4 Enzymatic Cyclization in Biotechnology

Proteins that are naturally linear can be cyclized enzymatically using biotechnological tools to improve stability or increase activity. For example cyclized conotoxins (in development for pain relief) are less susceptible to proteolysis (Clark, Akcan, Kaas, Daly, & Craik, 2012), and cyclized histatin 1 promotes wound healing more efficiently (Oudhoff et al., 2009). Both nature and humans have repeatedly recruited proteases and run their native proteolysis reaction in reverse. Transpeptidation is achieved by coupling the cleavage of a C-terminal propeptide or activating group to intramolecular peptide bond formation to the N-terminus via an acyl-enzyme intermediate.

Sortase A is a bacterial transpeptidase (protease CL clan) which natively ligates proteins to the bacterial cell wall. It has been used in vitro to cyclize proteins whose N- and C-termini are spatially close (Bolscher et al., n.d.; Wu, Guo, & Guo, 2011). Analogous to the processing enzymes of naturally cyclic proteins, it performs transpeptidation via covalent catalysis using a cysteine and requires a small C-terminal propeptide of a single Gly residue which is cleaved during transpeptidation.

Similarly, subtiligase is an engineered mutant of the protease subtilisin (protease SB clan) that has been optimized to favor peptide ligation over hydrolysis. It can perform cyclization of activated peptides, however instead of a propeptide, the peptide is activated by a chemically added ester group which removes the need for a C-terminal propeptide (Jackson, Bumier, & Wellst, 1995). Once again, catalysis is covalent and the activating ester group is cleaved off during the transpeptidation reaction.

carboxylic acid of the first substrate peptide (red) performs a nucleophilic attack on the succinimide ring to form an acyl-enzyme intermediate. This intermediate is then attacked by the substrate asparagine (Asn_{39i}) to release the substrate. The enzyme now contains an iso-aspartic acid (iso-Asp_{147}) and the substrate contains a high-energy succinimide ring (Suc_{39i}). Finally, the substrate's high-energy succinimide is attacked by the amine of the second substrate (blue) to form a new peptide bond and the enzyme regenerates its succinimide by intramolecular condensation. Overall, this condensation reaction yields a ligated peptide product. (See color plate)

8. CONCLUSION

The biosynthesis of cyclotides is a multistage process occurring through several organelles and mediated by a number of enzymes. Expression, folding and disulphide formation occur in the ER followed by proteolytic processing in the vacuole. The final step, cleavage of the C-terminal propeptide, is coupled to protease-mediated intramolecular transpeptidation to form the mature cyclic product.

Characterization of the cyclization step has identified VPEs as the likely cyclase. The VPEs are members of the CD clan of cysteine proteases and are typically involved in maturation of proteins by proteolysis. Evidence for their role as cyclases comes both from genetic studies of mutants in vivo and from biochemical characterization of the enzymes in vitro. The cyclization mechanism is likely to be a modification of covalent proteolysis in which the N-terminus of the peptide binds into the VPE active site and performs the final nucleophilic attack instead of water. Characterization of cyclotide cyclization holds both evolutionary interest and potential as a biotechnological tool.

Cyclization by VPEs is an example of convergent evolution on several levels. Firstly, VPEs have also been evolutionarily recruited to cyclize other circular peptides such as SFTI. This may indicate that VPEs have features that make them particularly suitable proteases for conversion to cyclases such as their location, abundance, activity or specificity. Secondly, recruitment of VPEs parallels similar evolution of serine proteases to act as cyclases for orbitides and a threonine protease performing linear transpeptidation on fibroblast growth factor 5. It is likely that the two-step, mechanism of the covalent, nucleophilic proteases makes them particularly suitable for the evolutionary switch from hydrolysis to transpeptidation.

Production of cyclotides both in vivo and in vitro is of interest since they represent important candidates for production of transgenic plants as well as potential lead compounds for drug design. Development of transgenic plants using cyclotides to improve pathogen resistance requires introduction of the cyclotide precursor gene as well as genes for the processing enzymes if the target is not a native cyclotide producer. Likewise, the recombinant production of cyclotides for pharmaceutical applications requires an efficient method of cyclization, and the native cyclase is likely the most efficient option. Finally, thorough characterization of the cyclase may allow its use in cyclization and transpeptidation of nonnative substrates and so add it to the growing toolkit of protein biotechnology.

REFERENCES

Agarwal, V., Pierce, E., McIntosh, J., Schmidt, E. W., & Nair, S. K. (2012). Structures of cyanobactin maturation enzymes define a family of transamidating proteases. *Chemistry & Biology, 19*(11), 1411–1422. http://doi.org/10.1016/j.chembiol.2012.09.012.

Barber, C. J. S., Pujara, P. T., Reed, D. W., Chiwocha, S., Zhang, H., & Covello, P. S. (2013). The two-step biosynthesis of cyclic peptides from linear precursors in a member of the plant family Caryophyllaceae involves cyclization by a serine protease-like enzyme. *The Journal of Biological Chemistry, 288*(18), 12500–12510. http://doi.org/10.1074/jbc.M112.437947.

Basse, C. W. (July 2005). Dissecting defense-related and developmental transcriptional responses of maize during Ustilago maydis infection and subsequent tumor formation 1. *Plant Physiology, 138*, 1774–1784. http://dx.doi.org/10.1104/pp.105.061200.

Becker, C., Shutov, A. D., Nong, V. H., Senyuk, V. I., Jung, R., Horstmann, C., et al. (1995). Purification, cDNA cloning and characterization of proteinase B, an asparagine-specific endopeptidase from germinating vetch (*Vicia sativa* L.) seeds. *European Journal of Biochemistry, 228*(2), 456–462.

Bernath-Levin, K., Nelson, C., Elliott, A. G., Jayasena, A. S., Millar, A. H., Craik, D. J., et al. (2015). Peptide macrocyclization by a bifunctional endoprotease. *Chemistry & Biology, 22*(5), 571–582. http://dx.doi.org/10.1016/j.chembiol.2015.04.010.

Bolscher, J. G. M., Oudhoff, M. J., Nazmi, K., Antos, J. M., Guimaraes, C. P., Spooner, E., et al. (n.d.). Sortase A as a tool for high-yield histatin cyclization. *Federation of American Societies for Experimental Biology Journal*, http://doi.org/10.1096/fj.11-182212.

Bowles, D. J., Marcus, S. E., Pappin, D. J., Findlay, J. B., Eliopoulos, E., Maycox, P. R., et al. (1986). Posttranslational processing of concanavalin A precursors in jackbean cotyledons. *The Journal of Cell Biology, 102*(4), 1284–1297. http://doi.org/10.1083/jcb.102.4.1284.

Chen, J., Dando, P., Rawlings, N., Brown, M., Young, N., Stevens, R., et al. (1997). Cloning, isolation, and characterization of mammalian legumain, an asparaginyl endopeptidase. *The Journal of Biological Chemistry, 272*(12), 8090–8098. http://doi.org/10.1074/jbc.272.12.8090.

Chen, J. M., Dando, P. M., Stevens, R. A., Fortunato, M., & Barrett, A. J. (1998). Cloning and expression of mouse legumain, a lysosomal endopeptidase. *The Biochemical Journal, 335*(Pt 1), 111–117.

Chen, J. M., Rawlings, N. D., Stevens, R. A., & Barrett, A. J. (1998). Identification of the active site of legumain links it to caspases, clostripain and gingipains in a new clan of cysteine endopeptidases. *FEBS Letters, 441*(3), 361–365.

Clark, R. J., Akcan, M., Kaas, Q., Daly, N. L., & Craik, D. J. (2012). Cyclization of conotoxins to improve their biopharmaceutical properties. *Toxicon: Official Journal of the International Society on Toxinology, 59*(4), 446–455. http://doi.org/10.1016/j.toxicon.2010.12.003.

Cohen, G. M. (1997). Caspases: the executioners of apoptosis. *The Biochemical Journal, 326*(1), 1–16. http://dx.doi.org/10.1016/b978-0-12-386456-7.01411-8.

Colgrave, M. L., & Craik, D. J. (2004). Thermal, chemical, and enzymatic stability of the cyclotide kalata B1: the importance of the cyclic cystine knot. *Biochemistry, 43*(20), 5965–5975. http://doi.org/10.1021/bi049711q.

Colgrave, M. L., Huang, Y. H., Craik, D. J., & Kotze, A. C. (2010). Cyclotide interactions with the nematode external surface. *Antimicrobial Agents and Chemotherapy, 54*(5), 2160–2166. http://doi.org/10.1128/AAC.01306-09.

Colgrave, M. L., Korsinczky, M. J., Clark, R. J., Foley, F., & Craik, D. J. (2010). Sunflower trypsin inhibitor-1, proteolytic studies on a trypsin inhibitor peptide and its analogs. *Biopolymers, 94*(5), 665–672. http://doi.org/10.1002/bip.21415.

Colgrave, M. L., Kotze, A. C., Huang, Y., Grady, J. O., Simonsen, S. M., & Craik, D. J. (2008). Cyclotides: natural, circular plant peptides that possess significant activity against gastrointestinal nematode parasites of sheep †. *Biochemistry, 47*, 5581–5589.

Conlan, B. F., & Anderson, M. A. (December 2011). Circular micro-proteins and mechanisms of cyclization. *Current Pharmaceutical Design*. http://doi.org/10.2174/13816121179 8999410.

Conlan, B. F., Colgrave, M. L., Gillon, A. D., Guarino, R., Craik, D. J., & Anderson, M. A. (2012). Insights into processing and cyclization events associated with biosynthesis of the cyclic peptide kalata B1. *The Journal of Biological Chemistry, 287*(33), 28037–28046. http://doi.org/10.1074/jbc.M112.347823.

Conlan, B. F., Gillon, A. D., Barbeta, B. L., & Anderson, M. A. (2011). Subcellular targeting and biosynthesis of cyclotides in plant cells. *American Journal of Botany, 98*(12), 2018–2026. http://doi.org/10.3732/ajb.1100382.

Craik, D. J., Daly, N. L., Bond, T., & Waine, C. (1999). Plant cyclotides: a unique family of cyclic and knotted proteins that defines the cyclic cystine knot structural motif. *Journal of Molecular Biology, 294*(5), 1327–1336. http://doi.org/10.1006/jmbi.1999.3383.

Craik, D. J., & Malik, U. (2013). Cyclotide biosynthesis. *Current Opinion in Chemical Biology, 17*(4), 546–554. http://doi.org/10.1016/j.cbpa.2013.05.033.

Craik, D. J., Mylne, J. S., & Daly, N. L. (2010). Cyclotides: macrocyclic peptides with applications in drug design and agriculture. *Cellular and Molecular Life Sciences: CMLS, 67*(1), 9–16. http://doi.org/10.1007/s00018-009-0159-3.

Craik, D. J., Swedberg, J. E., Mylne, J. S., & Cemazar, M. (2012). Cyclotides as a basis for drug design. *Expert Opinion on Drug Discovery*, 7(3), 179–194. http://doi.org/10.1517/17460441.2012.661554.

Dalet, A., Vigneron, N., Stroobant, V., Hanada, K.-I., & Van den Eynde, B. J. (2010). Splicing of distant peptide fragments occurs in the proteasome by transpeptidation and produces the spliced antigenic peptide derived from fibroblast growth factor-5. *Journal of Immunology (Baltimore, Md.: 1950), 184*(6), 3016–3024. http://doi.org/10.4049/jimmunol.0901277.

Dall, E., & Brandstetter, H. (2013). Mechanistic and structural studies on legumain explain its zymogenicity, distinct activation pathways, and regulation. *Proceedings of the National Academy of Sciences of the United States of America, 110*(27), 10940–10945. http://doi.org/10.1073/pnas.1300686110/-/DCSupplemental.www.pnas.org/cgi/doi/10.1073/pnas.1300686110.

Dall, E., Fegg, J. C., Briza, P., & Brandstetter, H. (2015). Structure and mechanism of an aspartimide-dependent peptide ligase in human legumain. *Angewandte Chemie (International Ed. in English), 54*(10), 2917–2921. http://doi.org/10.1002/anie.201409135.

Dalton, J., & Brindley, P. (1990). Schistosome asparaginyl endopeptidase Sm32 in hemoglobin digestion. *Parasitology Today, 12*(3), 125.

Dando, P. M., Fortunato, M., Smith, L., Knight, C. G., McKendrick, J. E., & Barrett, A. J. (1999). Pig kidney legumain: an asparaginyl endopeptidase with restricted specificity. *The Biochemical Journal, 339*(Pt 3), 743–749. Retrieved from http://www.pubmedcentral.nih.gov/articlerender.fcgi?artid=1220212&tool=pmcentrez&rendertype=abstract.

Dörnenburg, H. (2010). Cyclotide synthesis and supply: from plant to bioprocess. *Biopolymers, 94*(5), 602–610. http://doi.org/10.1002/bip.21466.

Dutton, J. L., Renda, R. F., Waine, C., Clark, R. J., Daly, N. L., Jennings, C. V., et al. (2004). Conserved structural and sequence elements implicated in the processing of gene-encoded circular proteins. *The Journal of Biological Chemistry, 279*(45), 46858–46867. http://doi.org/10.1074/jbc.M407421200.

Ewald, S. E., Engel, A., Lee, J., Wang, M., Bogyo, M., & Barton, G. M. (2011). Nucleic acid recognition by Toll-like receptors is coupled to stepwise processing by cathepsins and

asparagine endopeptidase. *The Journal of Experimental Medicine, 208*(4), 643–651. http://doi.org/10.1084/jem.20100682.
Fehlbaum, P., Bulet, P., Chernysh, S., Briand, J., Roussel, J., Letellierii, L., et al. (1996). Structure-activity analysis of thanatin, a 21-residue inducible insect defense peptide with sequence homology to frog skin antimicrobial peptides. *Proceedings of the National Academy of Sciences of the United States of America, 93*, 1221–1225.
Fischer, J., Becker, C., Hillmer, S., Horstmann, C., Neubohn, B., Schlereth, A., et al. (2000). The families of papin- and legumain-like cysteine proteinases from embryonic axes and cotyledons of Vicia seeds: developmental patterns, intracellular localization and functions in globulin proteolysis. *Plant Molecular Biology, 43*(1), 83–101. http://doi.org/10.1023/A:1006456615373.
Gauthier, A., Lamotte, O., Reboutier, D., Bouteau, F., Pugin, A., & Wendehenne, D. (2007). Cryptogein-induced anion effluxes: electrophysiological properties and analysis of the mechanisms through which they contribute to the elicitor-triggered cell death. *Plant Signaling & Behavior, 2*(2), 86–95. http://doi.org/10.4161/psb.2.2.4015.
Gerlach, S. L., Burman, R., Bohlin, L., Mondal, D., & Go, U. (2010). Isolation, characterization, and bioactivity of cyclotides from the Micronesian plant *Psychotria leptothyrsa. Journal of Natural Products, 73*, 1207–1213.
Gillon, A. D., Saska, I., Jennings, C. V., Guarino, R. F., Craik, D. J., Anderson, M. A., et al. (2008). Biosynthesis of circular proteins in plants. *Plant Journal, 53*(3), 505–515.
Gruber, C. W., Cemazar, M., Anderson, M. A., & Craik, D. J. (2007). Insecticidal plant cyclotides and related cystine knot toxins. *Toxicon: Official Journal of the International Society on Toxinology, 49*(4), 561–575. http://doi.org/10.1016/j.toxicon.2006.11.018.
Gruber, C. W., Čemažar, M., Clark, R. J., Horibe, T., Renda, R. F., Anderson, M. A., et al. (2007). A novel plant protein-disulfide isomerase involved in the oxidative folding of cystine knot defense proteins. *The Journal of Biological Chemistry, 282*(28), 20435–20446. http://doi.org/10.1074/jbc.M700018200.
Gruber, C. W., Elliott, A. G., Ireland, D. C., Delprete, P. G., Dessein, S., Göransson, U., et al. (2008). Distribution and evolution of circular miniproteins in flowering plants. *The Plant Cell, 20*(9), 2471–2483. http://doi.org/10.1105/tpc.108.062331.
Gruis, D., Schulze, J., Jung, R., Gruis, D., Schulze, J., & Jung, R. (2004). Storage protein accumulation in the absence of the vacuolar processing enzyme family of cysteine proteases. *The Plant Cell, 16*(1), 270–290.
Gruis, D. F., Selinger, D. A., Curran, J. M., Jung, R., Gruis, D. F., Selinger, D. A., et al. (2002). Redundant proteolytic mechanisms process seed storage proteins in the absence of seed-type members of the vacuolar processing enzyme family of cysteine proteases. *The Plant Cell, 14*(11), 2863–2882.
Hanada, K., Yewdell, J. W., & Yang, J. C. (January 2004). Immune recognition of a human renal cancer antigen through post-translational protein splicing. *Nature, 427*, 1–5.
Hara-Nishimura, I., Inoue, K., & Nishimura, M. (1991). A unique vacuolar processing enzyme responsible for conversion of several proprotein precursors into the mature forms. *FEBS Letters, 294*(1–2), 89–93.
Hara-Nishimura, I., Kinoshita, T., Hiraiwa, N., & Nishimura, M. (1998). Vacuolar processing enzymes in protein-storage vacuoles and lytic vacuoles. *Journal of Plant Physiology, 152*(6), 668–674. http://doi.org/10.1016/S0176-1617(98)80028-X.
Hara-Nishimura, I., & Nishimura, M. (1987). Proglobulin processing enzyme in vacuoles isolated from developing pumpkin cotyledons. *Plant Physiology, 85*(2), 440–445.
Hara-Nishimura, I., Takeuchi, Y., & Nishimura, M. (1993). Molecular characterization of a vacuolar processing enzyme related to a putative cysteine proteinase of *Schistosoma mansoni. The Plant Cell, 5*(11), 1651–1659.

Hatsugai, N., Kuroyanagi, M., Nishimura, M., & Hara-Nishimura, I. (2006). A cellular suicide strategy of plants: vacuole-mediated cell death. *Apoptosis, 11*(6), 905–911. http://doi.org/10.1007/s10495-006-6601-1.

Hatsugai, N., Kuroyanagi, M., Yamada, K., Meshi, T., Tsuda, S., Kondo, M., et al. (2004). A plant vacuolar protease, VPE, mediates virus-induced hypersensitive cell death. *Science, 305*(5685), 855–858.

Hatsugai, N., Yamada, K., Goto-Yamada, S., & Hara-Nishimura, I. (April 2015). Vacuolar processing enzyme in plant programmed cell death. *Frontiers in Plant Science, 6*, 1–11. http://doi.org/10.3389/fpls.2015.00234.

Haugen, M. H., Johansen, H. T., Pettersen, S. J., Solberg, R., Brix, K., Flatmark, K., et al. (2013). Nuclear legumain activity in colorectal cancer. *PLoS One, 8*(1), e52980. http://doi.org/10.1371/journal.pone.0052980.

Henriques, S. T., Huang, Y.-H., Rosengren, K. J., Franquelim, H. G., Carvalho, F. A., Johnson, A., et al. (2011). Decoding the membrane activity of the cyclotide kalata B1: the importance of phosphatidylethanolamine phospholipids and lipid organization on hemolytic and anti-HIV activities. *The Journal of Biological Chemistry, 286*(27), 24231–24241. http://doi.org/10.1074/jbc.M111.253393.

Hiraiwa, N., Nishimura, M., & Hara-Nishimura, I. (1997). Expression and activation of the vacuolar processing enzyme in *Saccharomyces cerevisiae*. *Plant Journal, 12*(4), 819–829.

Hiraiwa, N., Nishimura, M., & Hara-Nishimura, I. (1999). Vacuolar processing enzyme is self-catalytically activated by sequential removal of the C-terminal and N-terminal propeptides. *FEBS Letters, 447*(2–3), 213–216. http://doi.org/10.1016/S0014-5793(99)00286-0.

Huang, Y.-H., Colgrave, M. L., Clark, R. J., Kotze, A. C., & Craik, D. J. (2010). Lysine-scanning mutagenesis reveals an amendable face of the cyclotide kalata B1 for the optimization of nematocidal activity. *The Journal of Biological Chemistry, 285*(14), 10797–10805. http://doi.org/10.1074/jbc.M109.089854.

Ireland, D. C., Colgrave, M. L., Nguyencong, P., Daly, N. L., & Craik, D. J. (2006). Discovery and characterization of a linear cyclotide from *Viola odorata*: implications for the processing of circular proteins. *Journal of Molecular Biology, 357*(5), 1522–1535. http://doi.org/10.1016/j.jmb.2006.01.051.

Jackson, D. Y., Bumier, J. P., & Wellst, J. A. (1995). Enzymatic cyclization of linear peptide esters using subtiligase. 819–820. http://doi.org/10.1021/ja00107a027.

Jennings, C., West, J., Waine, C., Craik, D., & Anderson, M. (2001). Biosynthesis and insecticidal properties of plant cyclotides: the cyclic knotted proteins from *Oldenlandia affinis*. *Proceedings of the National Academy of Sciences of the United States of America, 98*(19), 10614–10619.

Ji, Y., Majumder, S., Millard, M., Borra, R., Bi, T., Elnagar, A. Y., et al. (2013). In vivo activation of the p53 tumor suppressor pathway by an engineered cyclotide. *Journal of the American Chemical Society, 135*(31), 11623–11633. http://doi.org/10.1021/ja405108p.

Julián, I., Gandullo, J., Santos-Silva, L. K., Diaz, I., & Martinez, M. (2013). Phylogenetically distant barley legumains have a role in both seed and vegetative tissues. *Journal of Experimental Botany, 64*(10), 2929–2941. http://doi.org/10.1093/jxb/ert132.

Jung, R., Scott, M. P., Nam, Y. W., Beaman, T. W., Bassuner, R., Saalbach, I., et al. (1998). The role of proteolysis in the processing and assembly of 11S seed globulins. *The Plant Cell, 10*(3), 343–357.

Kembhavi, A. A., Buttle, D. J., Knight, G., & Barrett, A. J. (1993). The two cysteine endopeptidases of legume seeds: purification and characterization by use of specific fluorometric assays. *Archives of Biochemistry & Biophysics, 303*(2), 208–213.

Kinoshita, T., Nishimura, M., & Hara-Nishimura, I. (1995). Homologues of a vacuolar processing enzyme that are expressed in different organs in *Arabidopsis thaliana*. *Plant Molecular Biology, 29*(1), 81–89.

Kinoshita, T., Yamada, K., Hiraiwa, N., Kondo, M., Nishimura, M., & Hara-Nishimura, I. (1999). Vacuolar processing enzyme is up-regulated in the lytic vacuoles of vegetative tissues during senescence and under various stressed conditions. *Plant Journal, 19*(1), 43–53. http://doi.org/10.1046/j.1365-313X.1999.00497.x.

Klinkert, M. Q., Felleisen, R., Link, G., Ruppel, A., & Beck, E. (1989). Primary structures of Sm31/32 diagnostic proteins of *Schistosoma mansoni* and their identification as proteases. *Molecular and Biochemical Parasitology, 33*(2), 113–122. http://doi.org/10.1016/0166-6851(89)90025-X.

Koehnke, J., Bent, A., Houssen, W. E., Zollman, D., & Morawitz, F. (2012). The mechanism of patellamide macrocyclization revealed by the characterization of the PatG macrocyclase domain. *Nature Structural and Molecular Biology, 19*(8), 767–772. http://dx.doi.org/10.1038/nsmb.2340.

Kolli, M., Lastere, S., & Schiffer, C. A. (2006). Co-evolution of nelfinavir-resistant HIV-1 protease and the p1-p6 substrate. *Virology, 347*(2), 405–409. http://doi.org/10.1016/j.virol.2005.11.049.

Kuroyanagi, M., Nishimura, M., Hara-Nishimura, I., Kuroyanagi, M., Nishimura, M., & Hara-Nishimura, I. (2002). Activation of Arabidopsis vacuolar processing enzyme by self-catalytic removal of an auto-inhibitory domain of the C-terminal propeptide. *Plant & Cell Physiology, 43*(2), 143–151.

Kuroyanagi, M., Yamada, K., Hatsugai, N., Kondo, M., Nishimura, M., Hara-Nishimura, I., et al. (2005). Vacuolar processing enzyme is essential for mycotoxin-induced cell death in *Arabidopsis thaliana. The Journal of Biological Chemistry, 280*(38), 32914–32920.

Linnestad, C., Doan, D. N., Brown, R. C., Lemmon, B. E., Meyer, D. J., Jung, R., et al. (1998). Nucellain, a barley homolog of the dicot vacuolar-processing protease, is localized in nucellar cell walls. *Plant Physiology, 118*(4), 1169–1180. http://doi.org/10.1104/pp.118.4.1169.

Liu, C., Sun, C., Huang, H., Janda, K., & Edgington, T. (2003). Overexpression of legumain in tumors is significant for invasion/metastasis and a candidate enzymatic target for prodrug therapy. *Cancer Research, 63*(11), 2957–2964.

Liu, Y., Bajjuri, K., Liu, C., & Sinha, S. (2012). Targeting cell surface alpha(v)beta(3) integrins increases therapeutic efficacies of a legumain protease-activated auristatin prodrug. *Molecular Pharmaceutics, 9*(1), 168–175. http://doi.org/10.1016/j.biotechadv.2011.08.021.Secreted.

Li, Z., Yue, H., & Xing, D. (2012). MAP Kinase 6-mediated activation of vacuolar processing enzyme modulates heat shock-induced programmed cell death in Arabidopsis. *The New Phytologist, 195*(1), 85–96. http://doi.org/10.1111/j.1469-8137.2012.04131.x.

Luckett, S., Garcia, R. S., Barker, J. J., Konarev, A. V., Shewry, P. R., Clarke, A. R., et al. (1999). High-resolution structure of a potent, cyclic proteinase inhibitor from sunflower seeds. *Journal of Molecular Biology, 290*, 525–533. http://doi.org/10.1006/jmbi.1999.2891.

Manoury, B., Hewitt, E. W., Morrice, N., Dando, P. M., Barrett, A. J., & Watts, C. (1998). An asparaginyl endopeptidase processes a microbial antigen for class II MHC presentation. *Nature, 396*(6712), 695–699. http://doi.org/10.1038/25379.

Marx, U. C., Korsinczky, M. L. J., Schirra, H. J., Jones, A., Condie, B., Otvos, L., et al. (2003). Enzymatic cyclization of a potent bowman-birk protease inhibitor, sunflower trypsin inhibitor-1, and solution structure of an acyclic precursor peptide. *The Journal of Biological Chemistry, 278*(24), 21782–21789. http://doi.org/10.1074/jbc.M212996200.

Mathieu, M. A., Bogyo, M., Caffrey, C. R., Choe, Y., Lee, J., Chapman, H., et al. (2002). Substrate specificity of schistosome versus human legumain determined by P1-P3 peptide libraries. *Molecular and Biochemical Parasitology, 121*(1), 99–105. http://doi.org/10.1016/S0166-6851(02)00026-9.

Matthews, S. P., Werber, I., Deussing, J., Peters, C., Reinheckel, T., & Watts, C. (2010). Distinct protease requirements for antigen presentation in vitro and in vivo. *Journal of Immunology, 184*(5), 2423–2431. http://doi.org/10.4049/jimmunol.0901486.

McIntosh, J. A., Robertson, C. R., Agarwal, V., Nair, S. K., Bulaj, G. W., & Schmidt, E. W. (2010). Circular logic: nonribosomal peptide-like macrocyclization with a ribosomal peptide catalyst. *Journal of the American Chemical Society, 132*(44), 15499–15501. http://doi.org/10.1021/ja1067806.

Mulvenna, J. P., Foley, F. M., & Craik, D. J. (2005). Discovery, structural determination, and putative processing of the precursor protein that produces the cyclic trypsin inhibitor sunflower trypsin inhibitor 1. *The Journal of Biological Chemistry, 280*(37), 32245–32253. http://doi.org/10.1074/jbc.M506060200.

Mulvenna, J. P., Wang, C., & Craik, D. J. (2006). CyBase: a database of cyclic protein sequence and structure. *Nucleic Acids Research, 34*(Database issue), D192–D194. http://doi.org/10.1093/nar/gkj005.

Müntz, K., & Shutov, A. D. (2002). Legumains and their functions in plants. *Trends in Plant Science,* 7(8), 340–344. http://doi.org/10.1016/S1360-1385(02)02298-7.

Mylne, J. S., Chan, L. Y., Chanson, A. H., Daly, N. L., Schaefer, H., Bailey, T. L., et al. (2012). Cyclic peptides arising by evolutionary parallelism via asparaginyl-endopeptidase-mediated biosynthesis. *The Plant Cell, 24*(7), 2765–2778. http://doi.org/10.1105/tpc.112.099085.

Mylne, J. S., Colgrave, M. L., Daly, N. L., Chanson, A. H., Elliott, A. G., McCallum, E. J., et al. (2011). Albumins and their processing machinery are hijacked for cyclic peptides in sunflower. *Nature Chemical Biology,* 7(5), 257–259. http://doi.org/10.1038/nchembio.542.

Mylne, J. S., Wang, C. K., van der Weerden, N. L., & Craik, D. J. (2010). Cyclotides are a component of the innate defense of *Oldenlandia affinis*. *Biopolymers, 94*(5), 635–646. http://doi.org/10.1002/bip.21419.

Nakamura, K., & Matsuoka, K. (1993). Protein targeting to the vacuole in plant cells. *Plant Physiology, 101*(1), 1–5. http://doi.org/10.1104/pp.101.1.1.

Nakaune, S., Yamada, K., Kondo, M., Kato, T., Tabata, S., Nishimura, M., et al. (2005). A vacuolar processing enzyme, deltaVPE, is involved in seed coat formation at the early stage of seed development. *The Plant Cell, 17*(3), 876–887.

Nguyen, G., Wang, S., Qiu, Y., Hemu, X., Lian, Y., & Tam, J. P. (2014). Butelase 1 is an Asx-specific ligase enabling peptide macrocyclization and synthesis. *Nature Chemical Biology, 10*(9), 732–738. http://doi.org/10.1038/nchembio.1586.

Nguyen, G. K. T., Lian, Y., Pang, E. W. H., Nguyen, P. Q. T., Tran, T. D., & Tam, J. P. (2013). Discovery of linear cyclotides in monocot plant *Panicum laxum* of Poaceae family provides new insights into evolution and distribution of cyclotides in plants. *The Journal of Biological Chemistry, 288*(5), 3370–3380. http://doi.org/10.1074/jbc.M112.415356.

Nguyen, G. K. T., Lim, W. H., Nguyen, P. Q. T., & Tam, J. P. (2012). Novel cyclotides and uncyclotides with highly shortened precursors from *Chassalia chartacea* and effects of methionine oxidation on bioactivities. *The Journal of Biological Chemistry, 287*(21), 17598–17607. http://doi.org/10.1074/jbc.M111.338970.

Nguyen, G. K. T., Zhang, S., Wang, W., Wong, C. T. T., Nguyen, N. T. K., & Tam, J. P. (2011). Discovery of a linear cyclotide from the bracelet subfamily and its disulfide mapping by top-down mass spectrometry. *The Journal of Biological Chemistry, 286*(52), 44833–44844. http://doi.org/10.1074/jbc.M111.290296.

Okamoto, T., & Minamikawa, T. (1999). Molecular cloning and characterization of *Vigna mungo* processing enzyme 1 (VmPE-1), an asparaginyl endopeptidase possibly involved in post-translational processing of a vacuolar cysteine endopeptidase (SH-EP). *Plant Molecular Biology, 39*(1), 63–73. http://doi.org/10.1023/A:1006170518002.

Oudhoff, M. J., Kroeze, K. L., Nazmi, K., van den Keijbus, P. A. M., van 't Hof, W., Fernandez-Borja, M., et al. (2009). Structure-activity analysis of histatin, a potent wound healing peptide from human saliva: cyclization of histatin potentiates molar activity

1000-fold. *The FASEB Journal: Official Publication of the Federation of American Societies for Experimental Biology, 23*, 3928–3935. http://doi.org/10.1096/fj.09-137588.

Ovesen, R. G., Brandt, K. K., Göransson, U., Nielsen, J., Hansen, H. C. B., & Cedergreen, N. (2011). Biomedicine in the environment: cyclotides constitute potent natural toxins in plants and soil bacteria. *Environmental Toxicology and Chemistry/SETAC, 30*(5), 1190–1196. http://doi.org/10.1002/etc.496.

Patel, N., Krishnan, S., Offman, M. N., Krol, M., Moss, C. X., Leighton, C., et al. (2009). A dyad of lymphoblastic lysosomal cysteine proteases degrades the antileukemic drug L-asparaginase. *Journal of Clinical Investigation, 119*(7), 1964–1973. http://doi.org/10.1172/JCI37977.

Plan, M. R., Saska, I., Cagauan, A. G., & Craik, D. J. (2008). Backbone cyclised peptides from plants show molluscicidal activity against the rice pest *Pomacea canaliculata* (Golden Apple Snail). *Journal of Agricultural and Food Chemistry, 56*, 5237–5241.

Poth, A. G., Colgrave, M. L., Lyons, R. E., Daly, N. L., & Craik, D. J. (2011). Discovery of an unusual biosynthetic origin for circular proteins in legumes. *Proceedings of the National Academy of Sciences of the United States of America, 108*(25), 10127–10132. http://doi.org/10.1073/pnas.1103660108.

Poth, A. G., Colgrave, M. L., Philip, R., Kerenga, B., Daly, N. L., Anderson, M. A., et al. (2011). Discovery of cyclotides in the Fabaceae plant family provides new insights into the cyclization, evolution, and distribution of circular proteins. *ACS Chemical Biology, 6*(4), 345–355. http://doi.org/10.1021/cb100388j.

Poth, A. G., Mylne, J. S., Grassl, J., Lyons, R. E., Millar, A. H., Colgrave, M. L., et al. (2012). Cyclotides associate with leaf vasculature and are the products of a novel precursor in petunia (Solanaceae). *The Journal of Biological Chemistry, 287*(32), 27033–27046. http://doi.org/10.1074/jbc.M112.370841.

Pränting, M., Lööv, C., Burman, R., Göransson, U., & Andersson, D. I. (2010). The cyclotide cycloviolacin O2 from *Viola odorata* has potent bactericidal activity against Gram-negative bacteria. *The Journal of Antimicrobial Chemotherapy, 65*(9), 1964–1971. http://doi.org/10.1093/jac/dkq220.

Qiang, X., Zechmann, B., Reitz, M. U., Kogel, K.-H., & Schäfer, P. (February 2012). The mutualistic fungus *Piriformospora indica* colonizes Arabidopsis roots by inducing an endoplasmic reticulum stress-triggered caspase-dependent cell death. *The Plant Cell, 24*, 794–809. http://doi.org/10.1105/tpc.111.093260.

Rawlings, N. D., Barrett, A. J., & Bateman, A. (2012). MEROPS: the database of proteolytic enzymes, their substrates and inhibitors. *Nucleic Acids Research, 40*(Database issue), D343–D350. http://doi.org/10.1093/nar/gkr987.

Rojo, E., Zouhar, J., Carter, C., Kovaleva, V., Raikhel, N. V., Rojo, E., et al. (2003). A unique mechanism for protein processing and degradation in *Arabidopsis thaliana*. *Proceedings of the National Academy of Sciences of the United States of America, 100*(12), 7389–7394.

Rosengren, K. J., Daly, N. L., Plan, M. R., Waine, C., & Craik, D. J. (2003). Twists, knots, and rings in proteins. Structural definition of the cyclotide framework. *The Journal of Biological Chemistry, 278*(10), 8606–8616. http://doi.org/10.1074/jbc.M211147200.

Rotari, V. I., Dando, P. M., & Barrett, A. J. (2001). Legumain forms from plants and animals differ in their specificity. *Biological Chemistry, 382*(6), 953–959.

Santos-Silva, L. K., Soares-Costa, A., Gerald, L. T. S., Meneghin, S. P., & Henrique-Silva, F. (2012). Recombinant expression and biochemical characterization of sugarcane legumain. *Plant Physiology and Biochemistry: PPB/Société Française de Physiologie Végétale, 57*, 181–192. http://doi.org/10.1016/j.plaphy.2012.05.020.

Saska, I., Gillon, A. D., Hatsugai, N., Dietzgen, R. G., Hara-Nishimura, I., Anderson, M. A., et al. (2007). An asparaginyl endopeptidase mediates in vivo protein backbone cyclization. *The Journal of Biological Chemistry, 282*(40), 29721–29728. http://doi.org/10.1074/jbc.M705185200.

Schlereth, A., Standhardt, D., Mock, H. P., & Müntz, K. (2001). Stored cysteine proteinases start globulin mobilization in protein bodies of embryonic axes and cotyledons during vetch (*Vicia sativa* L.) seed germination. *Planta, 212*(5–6), 718–727. http://doi.org/10.1007/s004250000436.

Scott, M. P., Jung, R., Muntz, K., & Nielsen, N. C. (1992). A protease responsible for post-translational cleavage of a conserved Asn-Gly linkage in glycinin, the major seed storage protein of soybean. *Proceedings of the National Academy of Sciences of the United States of America, 89*(2), 658–662. http://doi.org/10.1073/pnas.89.2.658.

Sepulveda, F. E., Maschalidi, S., Colisson, R., Heslop, L., Ghirelli, C., Sakka, E., et al. (2009). Critical role for asparagine endopeptidase in endocytic Toll-like receptor signaling in dendritic cells. *Immunity, 31*(5), 737–748. http://doi.org/10.1016/j.immuni.2009.09.013.

Seydel, P., Gruber, C. W., Craik, D. J., & Dörnenburg, H. (2007). Formation of cyclotides and variations in cyclotide expression in *Oldenlandia affinis* suspension cultures. *Applied Microbiology and Biotechnology,* 77(2), 275–284. http://doi.org/10.1007/s00253-007-1159-6.

Sheldon, P. S., Keen, J. N., & Bowles, D. J. (1996). Post-translational peptide bond formation during conconavalin A processing in vitro. *The Biochemical Journal, 320*, 865–870.

Shimada, T., Hiraiwa, N., Nishimura, M., & Hara-Nishimura, I. (1994). Vacuolar processing enzyme of soybean that converts proproteins to the corresponding mature forms. *Plant & Cell Physiology, 35*(4), 713–718.

Shutov, A., Do, N., & Vaintraub, I. (1982). Purification and partial characterization of protease B from germinating vetch seeds. *Biokhimija, 47*, 814–821.

Simonsen, S. M., Sando, L., Rosengren, K. J., Wang, C. K., Colgrave, M. L., Daly, N. L., et al. (2008). Alanine scanning mutagenesis of the prototypic cyclotide reveals a cluster of residues essential for bioactivity. *The Journal of Biological Chemistry, 283*, 9805–9813. http://doi.org/10.1074/jbc.M709303200.

Teper-Bamnolker, P., Buskila, Y., Lopesco, Y., Ben-Dor, S., Saad, I., Holdengreber, V., et al. (2012). Release of apical dominance in potato tuber is accompanied by programmed cell death in the apical bud meristem. *Plant Physiology, 158*(4), 2053–2067. http://doi.org/10.1104/pp.112.194076.

Trabi, M., & Craik, D. (August 2004). Tissue-specific expression of head-to-tail cyclized miniproteins in Violaceae and structure determination of the root cyclotide *Viola hederacea* root cyclotide1. *The Plant Cell, 16*, 2204–2216. http://doi.org/10.1105/tpc.104.021790.1.

Trabi, M., Svangård, E., Herrmann, A., Göransson, U., Claeson, P., Craik, D. J., et al. (2004). Variations in cyclotide expression in viola species. *Journal of Natural Products, 67*(5), 806–810. http://doi.org/10.1021/np034068e.

Vigneron, N., Stroobant, V., Chapiro, J., Ooms, A., Degiovanni, G., Morel, S., et al. (2004). An antigenic peptide produced by peptide splicing in the proteasome. *Science (New York, N.Y.), 304*(5670), 587–590. http://doi.org/10.1126/science.1095522.

Wang, C. K. L., Colgrave, M. L., Gustafson, K. R., Ireland, D. C., Goransson, U., & Craik, D. J. (2008). Anti-HIV cyclotides from the Chinese medicinal herb *Viola yedoensis*. *Journal of Natural Products, 71*(1), 47–52. http://doi.org/10.1021/np070393g.

Warren, E. H., Vigneron, N. J., Gavin, M. A., Coulie, P. G., Stroobant, V., Dalet, A., et al. (2006). An antigen produced by splicing of noncontiguous peptides in the reverse order. *Science, 313*(5792), 1444–1447. http://doi.org/10.1126/science.1130660.

Watt, W., Koeplinger, K. A., Mildner, A. M., Heinrikson, R. L., Tomasselli, A. G., & Watenpaugh, K. D. (1999). The atomic-resolution structure of human caspase-8, a key activator of apoptosis. *Structure,* 7(9), 1135–1143. http://doi.org/10.1016/S0969-2126(99)80180-4.

Woltering, E. J., van der Bent, A., & Hoeberichts, F. A. (2002). Do plant caspases exist? *Plant Physiology, 130*(4), 1764–1769. http://doi.org/10.1104/pp.006338.

Wu, Z., Guo, X., & Guo, Z. (2011). Sortase A-catalyzed peptide cyclization for the synthesis of macrocyclic peptides and glycopeptides. *Chemical Communications, 47*(32), 9218–9220. http://doi.org/10.1039/c1cc13322e.

Yamada, K., Shimada, T., & Nishimura, M. (1999). Multiple functional proteins are produced by cleaving Asn-Gln bonds of a single precursor by vacuolar processing enzyme. *The Journal of Biological Chemistry*. http://doi.org/10.1074/jbc.274.4.2563.

Yamada, K., Shimada, T., Nishimura, M., & Hara-Nishimura, I. (2005). A VPE family supporting various vacuolar functions in plants. *Physiologia Plantarum, 123*(4), 369–375. http://doi.org/10.1111/j.1399-3054.2005.00464.x.

Yamin, T. T., Ayala, J. M., & Miller, D. K. (1996). Activation of the native 45-kDa precursor form of interleukin-1-converting enzyme. *The Journal of Biological Chemistry, 271*(22), 13273–13282. http://doi.org/10.1074/jbc.271.22.13273.

Zhang, H., Dong, S., Wang, M., Wang, W., Song, W., Dou, X., et al. (2010). The role of vacuolar processing enzyme (VPE) from *Nicotiana benthamiana* in the elicitor-triggered hypersensitive response and stomatal closure. *Journal of Experimental Botany, 61*(13), 3799–3812.

CHAPTER NINE

Chemical and Biological Production of Cyclotides

Yilong Li[*,a], **Tao Bi**[*,a], **Julio A. Camarero**[*,§,1]
[*]Department of Pharmacology and Pharmaceutical Sciences, University of Southern California, Los Angeles, CA, USA
[§]Department of Chemistry, University of Southern California, Los Angeles, CA, USA
[1]Corresponding author: E-mail: jcamarer@usc.edu

Contents

Abstract

Cyclotides are fascinating naturally occurring microproteins (≈30 residues long) present in several plant families, and display various biological properties such as protease inhibitory, antimicrobial, insecticidal, cytotoxic, anti-human immunodeficiency virus, and hormonelike activities. Cyclotides share a unique head-to-tail circular knotted topology of three disulfide bridges, with one disulfide penetrating through a macrocycle formed by the two other disulfides and interconnecting peptide backbones,

[a]Equal contribution.

Advances in Botanical Research, Volume 76
ISSN 0065-2296
http://dx.doi.org/10.1016/bs.abr.2015.08.006

forming what is called a cystine knot topology. This cyclic cystine knot framework gives the cyclotides exceptional rigidity, resistance to thermal and chemical denaturation, and enzymatic stability against degradation. Interestingly, cyclotides have been shown to be orally bioavailable, and other cyclotides have been shown to cross the cell membranes. Moreover, recent reports have also shown that engineered cyclotides can be efficiently used to target extracellular and intracellular protein–protein interactions, therefore making cyclotides ideal tools for drug development to selectively target protein–protein interactions. In this work, we will review all the available methods for production of these interesting proteins using chemical or biological methods.

1. INTRODUCTION

Cyclotides are fascinating microproteins (≈30 residues long) present in several families of plants (Poth, Colgrave, Lyons, Daly, & Craik, 2011). They share a unique head-to-tail circular knotted topology of three disulfide bridges, with one disulfide penetrating through a macrocycle formed by the two other disulfides and interconnecting peptide backbones, forming what is called a cyclic cystine knot (CCK) topology (Figure 1). This CCK framework gives cyclotides a compact, highly rigid structure (Daly et al., 2013; Puttamadappa, Jagadish, Shekhtman, & Camarero, 2010), which confers exceptional resistance to thermal/chemical denaturation, and enzymatic degradation (Colgrave & Craik, 2004; Garcia & Camarero, 2010). In fact, the use of cyclotide-containing plants in indigenous medicine first highlighted the fact that these peptides are resistant to boiling and are apparently orally bioavailable (Saether et al., 1995).

Despite the sequence diversity all cyclotides share the same CCK motif (Figure 1). Hence, cyclotides can be considered as natural combinatorial peptide libraries structurally constrained by the cystine knot scaffold and head-to-tail cyclization but in which hypermutation of essentially all residues is permitted with the exception of the strictly conserved cysteines that comprise the knot (Austin, Kimura, Woo, & Camarero, 2010; Huang, Colgrave, Clark, Kotze, & Craik, 2010; Simonsen et al., 2008).

More than 280 cyclotides have been isolated from different plant families, including Rubiaceae, Vioalceae, Cucurbitaceae, Fabaceae, and Solanaceae (Chiche et al., 2004; Gruber et al., 2008; Nguyen et al., 2011; Poth et al., 2011, 2010, 2012). It has been estimated, however, that around 50,000 cyclotides might exist (Gruber et al., 2008; Zhang et al., 2014). All the cyclotides reported so far from the Violaceae and Rubiaceae families are biosynthesized via processing from dedicated genes that in some cases encode multiple copies of the same cyclotide, and in others, mixtures of

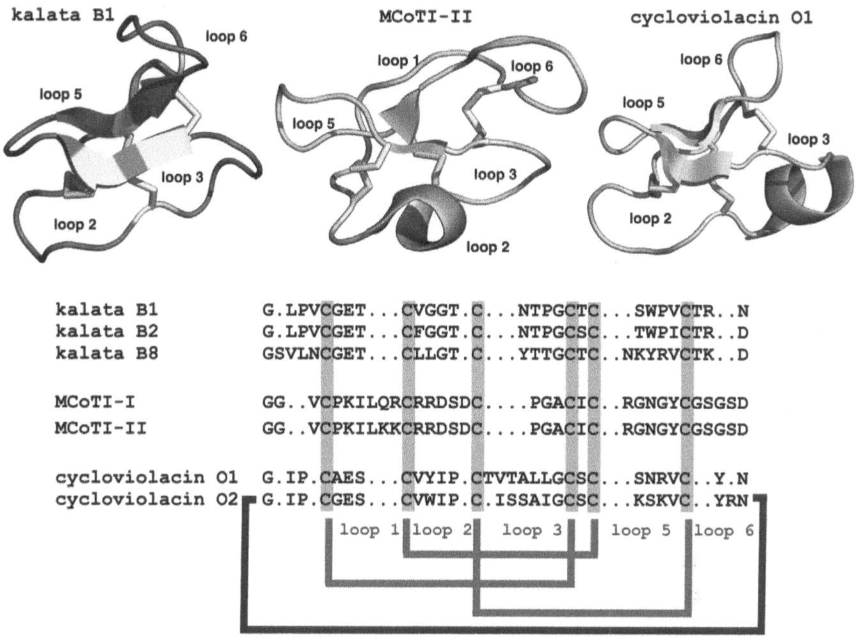

Figure 1 Primary and tertiary structures of cyclotides from the Möbius (kalata B1, pdb code: 1NB1), bracelet (cycloviolacin O1, pdb code: 1NBJ), and trypsin inhibitor (MCoTI-II, pdb code: 1IB9) subfamilies. The sequence of kalata B8, a novel hybrid cyclotide isolated from the plant *Otocinclus affinis* is also shown. Conserved cysteine residues are marked in yellow (light gray in print versions) and disulfide connectivities in red (gray in print versions). The circular backbone topology is shown with a blue (dark gray in print versions) line.

different cyclotide sequences (Dutton et al., 2004). Cyclotides from the Fabaceae family, however, are biosynthesized from evolved albumin-1 genes (Nguyen et al., 2011; Poth et al., 2011). These precursor genes undergo post-translational processing to generate a circular peptide by a mechanism that likely involves a transpeptidation reaction mediated by asparaginyl endoproteinase (AEP)-like proteases (Gillon et al., 2008; Mylne et al., 2012, 2011; Saska et al., 2007). For example, an AEP-related asparagine/aspartate (Asx) peptide ligase isolated from *Clitoria ternatea* was recently shown to be able to cyclize polypeptides, including cyclotides, in vitro (Nguyen et al., 2014).

Cyclotides can be also produced chemically using solid-phase peptide synthesis (SPPS), chemoenzymatically, or expressed in microorganisms using deoxyribonucleic acid recombinant techniques. The present review will describe the advantages and disadvantages of these approaches as well as their most recent applications (Tables 1 and 2).

Table 1 Summary of the Chemical Approaches Used to Produce Cyclotides

Cyclization	Chemistry	Linker	Pros and Cons	Application	References
NCL	Boc	Mercato-propionamide	• Thioester linker not completely stable • Fast	• Synthesis of many different cyclotides	Daly et al. (1999), Tam et al. (1999)
NCL	Fmoc	Sulfonamide	• Totally stable linker • Longer cleavage times • Not compatible with Met residues	• Synthesis of many different cyclotides • Synthesis of libraries using parallel synthesis with "tea bags"	Ingenito et al. (1999), Shin et al. (1999)
NCL	Fmoc	Dbz (3,4-diaminobenzoic acid)	• Linker can provide branched products during synthesis	• Synthesis of several kalata cyclotides	Blanco-Canosa and Dawson (2008), Gunasekera et al. (2013), Koehbach et al. (2013)
NCL	Fmoc	Trityl	• Linker completely stable • Limited solubility and aggregation issues with partially protected peptides • Thioester formation can lead to epimerization of C-terminal residue	• Synthesis of the bracelet cyclotide cycloviolacin O_2	Aboye et al. (2008)
Head-to-tail chemical ligation	Fmoc	2-Chlorotrityl	• Linker completely stable • Limited solubility and aggregation issues with partially protected peptides • Thioester formation can lead to epimerization of C-terminal residue	• Synthesis of different cyclotides from the Möbius, bracelet and trypsin inhibitory families	Cheneval et al. (2014)

SrtA	Fmoc	2-Chlorotrityl	• High efficiency with evolved SrtA (m^4SrtA) • Sequence requirements for SrtA recognition • SrtA recognition sequence of seven residues remains in the cyclotide after cyclization	• Synthesis of cyclotide kalata B1 analogs	Jia et al. (2014)
Trypsin	Fmoc	Wang	• High efficiency • Cyclization requires the linear precursor to be natively folded • Only works with trypsin inhibitory cyclotides • Introduction of mutations may compromise cyclization efficiency	• Synthesis of MCoTI cyclotides	Thongyoo et al. (2008)
Butelase 1	Fmoc	Wang	• High efficiency • Cyclization does not require conformational assistance by disulfide bonds • Butelase 1 can be only obtained from natural sources so far	• Synthesis of kalata B1	Nguyen et al. (2014)

Table 2 Examples of Cyclotides Produced Using Ribosomal Expression Methods

Cyclotide	Method	In vitro/ In vivo	Pros and Cons	Application	References
Kalata B1	Thiol-induced N→S acyl shift	In vitro	• Requires GC at ligation site • Harsh cyclization conditions	• Natural product	Cowper et al. (2013)
Kalata B1	EPL	Both	• Requires N-terminal Cys residues only • Yields depend on intein used	• Natural product	Kimura et al. (2006)
MCoTI-I	EPL	Both	• Requires N-terminal Cys residues only • Yields depend on intein used	• Library scaffold • Nonnatural amino acid	Austin et al. (2009), Jagadish et al. (2013)
MCoTI-II	EPL	Both	• Requires N-terminal Cys residues only • Yields depend on intein used	• Library scaffold	Camarero et al. (2007)
MCo-PMI	EPL	In vitro	• Requires N-terminal Cys residues only • Yields depend on intein used	• Anticancer	Ji et al. (2013)
MCoTI-I	PTS	In vivo	• Highly efficient (10 times better than EPL) • Cyclization can be carried out in Ser/Cys residues • Spontaneous self-processing cyclization, no need for enzymes	• Library scaffold • Nonnatural amino acid	Jagadish et al. (2013)
rMCoTI-II	SrtA	In vitro	• Requires multiple proteolytic steps • SrtA recognition sequence of 7 residues remains in the cyclotide after cyclization	• Large-scale production	Stanger et al. (2014)

2. CHEMICAL SYNTHESIS OF CYCLOTIDES

Cyclotides are small peptides, approximately 30 amino acids long, and therefore can be readily synthesized by chemical methods using SPPS (Marglin & Merrifield, 1970). Chemical synthesis using a solid-phase approach has been utilized to generate native cyclotide structures as well as grafted analogs (Aboye, Clark, Craik, & Göransson, 2008; Aboye et al., 2012; Daly, Love, Alewood, & Craik, 1999; Ji et al., 2013; Tam & Lu, 1997, 1998; Thongyoo, Tate, & Leatherbarrow, 2006). Table 1 summarizes the different synthetic approaches used thus far for the chemical synthesis of cyclotides.

2.1 Cyclization by Intramolecular Native Chemical Ligation

The most commonly used method for the backbone cyclization of the linear cyclotide precursor employs an intramolecular native chemical ligation (NCL) (Dawson, Muir, Clark-Lewis, & Kent, 1994), in which the peptide sequence contains an N-terminal cysteine and an α-thioester group at the C-terminus (Figure 2) (Camarero & Mitchell, 2005; Camarero & Muir, 1997, 1999). This ligation does not compromise the chirality of the C-terminal residue at the ligation site and more importantly can be accomplished in aqueous buffers at physiological pH. Although in principle any of the six Cys residues present in the cyclotide can be used, some studies on the naturally occurring cyclotide kalata B1 and MCoTI-I/II have indicated that the Cys residues located in loop 3 and 6 usually provide better yields (Ji et al., 2013; Kimura, Tran, & Camarero, 2006). Both tert-butyloxycarbonyl (Boc)-based and 9-fluorenylmethoxycarbonyl (Fmoc)-based chemistries have been used to incorporate C-terminal thioesters during peptide chain assembly (Beligere & Dawson, 1999; Camarero, Adeva, & Muir, 2000; Camarero, Cotton, Adeva, & Muir, 1998) or using safety catch-based linkers (Fmoc) (Blanco-Canosa & Dawson, 2008; Camarero, Hackel, de Yoreo, & Mitchell, 2004; Camarero & Mitchell, 2005; Ingenito, Bianchi, Fattori, & Pessi, 1999; Shin et al., 1999). Once the peptide is cleaved from the resin, the linear cyclotide precursors can be cyclized and folded sequentially. More recently, it has been shown that the cyclization and folding can be carried out as a single-pot reaction by using glutathione (GSH) as a thiol additive (Contreras, Elnagar, Hamm-Alvarez, & Camarero, 2011).

2.1.1 *Synthesis of α-Thioester Peptides by tert-Butyloxycarbonyl -Based Chemistry*

The most commonly used approach for the production of α-thioester linear cyclotide precursors makes use of the 3-mercaptopropionamide linker

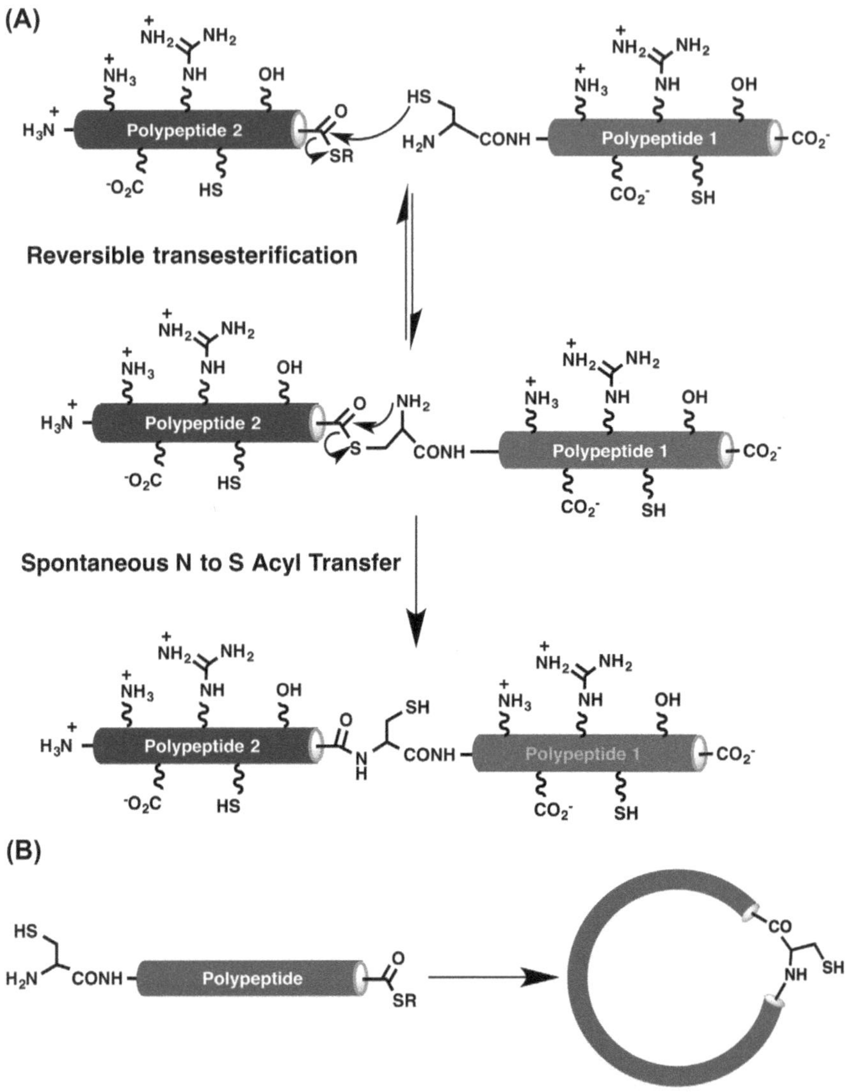

Figure 2 Backbone cyclization of polypeptides using native chemical ligation. (A) Principle of NCL. (B) Intramolecular NCL leads to the formation of a backbone-cyclized polypeptide.

(Camarero & Muir, 1997). This linker is stable to the conditions employed during Boc-based peptide synthesis and provides the desired α-thioester peptide in good yields. This scheme was first used by Craik and Tam for the chemical synthesis of cyclotides kalata B1, circulins A/B, and cyclopsychotride to study the folding pathways and antimicrobial properties of cyclotides, respectively (Daly et al., 1999; Tam, Lu, Yang, & Chiu, 1999). Synthetic cyclotides were used to study their antimicrobial activities (Tam et al., 1999) and to analyze the folding pathways of kalata B1 (Daly et al., 1999). Despite the limitations of Boc-based peptide synthesis requiring repeated use of corrosive trifluoroacetic acid (TFA) and highly toxic anhydrous hydrogen fluoride in the final cleavage step, which make it difficult to parallelize for automated high-throughput synthesis, this approach is still widely used for the synthesis of cyclotides. For example, this approach was recently used to engineer an orally active cyclotide based on the sequence cyclotide kalata B1 for treating inflammatory pain in animal models (Tam et al., 1999). Other bioactive cyclotides engineered to target extracellular receptors have also been recently synthesized using this approach (Chan et al., 2011; Eliasen et al., 2012; Gunasekera et al., 2008).

2.1.2 Synthesis of α-Thioester Peptides by 9-Fluorenylmethoxycarbonyl-Based Chemistry

The limitations associated with the use of Boc-based chemistry for the production of α-thioester linear cyclotide precursors spurred the development of different approaches for the synthesis of α-thioester peptides using the milder Fmoc-based chemistry (Camarero & Mitchell, 2005). The use of thioester-based linkers, however, is seriously limited due to the poor stability of the thioester functionality to strong nucleophiles such as piperidine, which are used repeatedly for the deprotection the Nα-Fmoc group during the peptide assembly.

2.1.2.1 Trityl Resin

The first attempt to synthesize peptide α-thioesters using Fmoc-based SPPS was reported by Futaki (Futaki, Sogawa, Maruyama, Asahara, & Niwa, 1997). In this approach, the corresponding peptide α-thioesters were prepared in solution using a partially protected precursor. The protected peptide precursor was assembled on a 4-chlorotrityl (4-Cl-Trt) resin using standard Fmoc-based chemistry. After cleavage from the acid-labile resin, the C-terminal carboxylic group of the partially protected peptide was thioesterified with the appropriate thiol and coupling reagent. This approach was recently used for the synthesis of the bracelet cyclotide cycloviolacin O_2 (Aboye et al., 2008). In this work the α-thioester peptide obtained by Fmoc SPPS was cyclized by NCL and oxidatively folded using redox buffer containing GSH/cysteamine and 35% dimethylsulfoxide at pH 8.5. The folding pathway was characterized and was found to include predominately fully oxidized intermediates that slowly converted to the native peptide structure. It is important to note, however, that a major drawback of this method, besides the poor solubility of protected peptides, is epimerization of C-terminal residues other than Gly during the thioesterification reaction. A recent study (von Eggelkraut-Gottanka, Klose, Beck-Sickinger, & Beyermann, 2003) has shown that epimerization strongly depends on the conditions as well as the coupling reagent used. The best results were obtained using phosphonium salts, which only gave $\approx 2\%$ epimerization at the C-terminal amino acid.

2.1.2.2 Sulfonamide Linker

Ingenito et al. (1999) and Shin et al. (1999) independently reported in 1999 the Fmoc-based SPPS of C-terminal thioesters using Kenner's acylsulfonamide "safety-catch" linker (Kenner, McDermott, & Sheppard, 1971) as modified by Backes and Ellman (1999) and Backes, Virgilio, and Ellman (1996). This acylsulfonamide linker is completely stable to basic or strongly nucleophilic conditions and can be activated by treatment with either trimethylsilyldiazomethane (TMS-CHN_2) or iodoacetonitrile to provide an Nα-alkyl acylsulfonamide, which is susceptible to nucleophilic attack. Activation with TMS-CHN_2 gives usually better thiolysis yields (Ingenito et al., 1999). This approach has been used for the generation of α-thioester linear cyclotide precursors. For example, Leatherbarrow used this approach to synthesize cyclotide MCoTI-II. The cyclization site in this work was chosen at the end of loop 3 (Figure 1). Cyclotides MCoTI-I/II are powerful trypsin inhibitors that have recently been isolated from the dormant seeds of *Momordica cochinchinensis*, a plant member of the *Cucurbitaceae* family (Hernandez et al., 2000). Although MCoTI cyclotides do not share significant sequence homology with other cyclotides beyond the presence of the three cystine bridges, structural analysis by nuclear magnetic resonance (NMR) has shown that they adopt a similar backbone cyclic cystine knot topology (Felizmenio-Quimio, Daly, & Craik, 2001; Heitz et al., 2001). MCoTI cyclotides, however, show high sequence homology with related cystine knot squash trypsin inhibitors (Hernandez et al., 2000) and therefore represent interesting molecular scaffolds for drug design (Reiss et al., 2006; Werle, Kafedjiiski, Kolmar, & Bernkop-Schnurch, 2007). The same group also used this approach to generate MCoTI-based cyclotides with potent β-tryptase and human leukocyte elastase (Gray et al., 2014; Thongyoo, Bonomelli, Leatherbarrow, & Tate, 2009).

Our group has also recently used a similar approach for the chemical synthesis of several engineered MCoTI cyclotides able to target both extracellular (CXCR4 antagonist) (Aboye et al., 2012) and intracellular (p53 agonist) (Ji et al., 2013) protein—protein interactions with low nanomolar range activities. In both cases the N-terminal region of loop 6 was used for the cyclization. The linear thioester precursors were cyclized and folded directly in a "one-pot" reaction using reduced GSH as redox agent and thiol cofactor in phosphate buffer at pH 7.2 resulting in very good yields (Aboye et al., 2012; Ji et al., 2013). We have also recently used this synthetic scheme for the rapid parallel synthesis of cyclotides using a "tea-bag" approach (Aboye, Kuang, Neamati, & Camarero, 2015). This work also includes an efficient purification procedure to rapidly remove nonfolded or partially folded cyclotides from the cyclization—folding crude. This procedure can be easily used in parallel for the purification of individual compounds, and more importantly for the purification of cyclotide mixtures also. This makes this approach potentially compatible with the synthesis of positional scanning libraries in order to perform efficient screening of large libraries. These protocols were successfully used for the production of a small library of bioactive cyclotides based on the cyclotide MCo-CVX-5c, which is a potent CXCR4 antagonist recently developed in our group (Aboye et al., 2012).

2.1.2.3 3,4-Diaminobenzoic acid (Dbz) Linker

This "safety catch" linker was recently developed by Dawson for the production of α-thioester peptides by Fmoc-based SPPS (Blanco-Canosa & Dawson, 2008). This linker is stable to the conditions used in Fmoc-based SPPS and can be activated at the end of the synthesis to yield the corresponding α-thioester after thiolytic cleavage followed by TFA deprotection of the side chain protecting groups. This approach has been recently used for the chemical synthesis of cyclotides kalata B1 and B7 (Gunasekera, Aboye, Madian, El-Seedi, & Goransson, 2013; Koehbach et al., 2013).

2.2 Cyclization by Intramolecular Head-To-Tail Chemical Ligation

Craik has recently described the chemical synthesis of cyclotides by Fmoc-based SPPS by direct chemical coupling of the Nα-amino and C-terminal carboxylic groups using uronium salts as coupling reagent. Due to the lack of chemoselectivity, this reaction requires to be performed with all the side chain protecting groups. In this approach, the linear precursor is assembled on the acid-labile 2-chlorotrityl chloride resin. Treatment of the peptide resin with mild acid conditions (1% TFA) allows cleaving the peptide with all the side chain protecting groups intact. This partially protected peptide is then cyclized in solution using uronium salts as coupling reagent and dimethylformamide as solvent. Once the backbone cyclization has been performed the side chain protecting groups are removed and the unprotected reduced cyclotides are folded oxidatively in aqueous buffers. This approach has been used for the synthesis of different cyclotides, including kalata B1 and its enantiomer D-kalata B1, MCoTI-II, and parigidin-br1.

One of the main advantages of this approach is its simplicity as it employs standard SPPS reagents. It also allows choosing different cyclization sites without the need to have an extra N-terminal Cys residue, required for NCL-mediated cyclization. On the other hand, protected peptides are difficult to handle and characterize and they have tendency to aggregate and precipitate. The cyclization using uronium salts can also provide unwanted reactions between the Nα-amino group and the uronium salt leading to the capping of the amino functionality by forming a tetramethylguanidine adduct. The use of phosphonium-based coupling reagents (e.g., benzotriazol-1-yl-oxytripyrrolidinophosphonium hexafluorophosphate) should minimize this unwanted side reaction. In addition, special precaution should be taken when choosing the cyclization site as racemization (epimerization) of the C-terminal residue during the cyclization can happen. This should be easily prevented by using cyclization sites such as Gly-Xaa or Pro-Xaa (where Xaa denotes any amino acid).

2.3 Cyclization by Chemoenzymatic-Mediated Ligation

Proteases and transpeptidases have been recently used for the production of backbone-cyclized polypeptides, including cyclotides among other circular polypeptides, using linear synthetic peptides as substrates (Aboye & Camarero, 2012).

2.3.1 Trypsin-Mediated Backbone Cyclization

The serine protease trypsin has been shown to be able to cyclize the naturally occurring cyclic peptide sunflower trypsin inhibitor 1 (SFTI-1) (Marx et al., 2003). SFTI-1 is a potent trypsin inhibitor with a K_i value in the nanomolar range. The enzyme trypsin was able to catalyze the formation of a peptide bond between the α-amino and α-carboxylic groups of the P1′ and P1 residues, respectively. A similar approach has also been reported for the enzymatic cyclization of MCoTI cyclotides (Thongyoo, Roque-Rosell, Leatherbarrow, & Tate, 2008). In this work a folded linear cyclotide precursor bearing the P1 and P1′ residues at the C- and N-termini, respectively, was used as a viable substrate for trypsin-mediated cyclization, thus enabling synthesis of the cyclic backbone without the need for a C-terminal α-thioester. The linear peptide was produced by standard Fmoc-based SPPS and was folded in 1 mM GSH and 100 mM NH_4HCO_3 buffer at pH 7.8 containing 50% MeCN. The cyclization step was performed using trypsin immobilized on sepharose beads in 100 mM sodium phosphate buffer at pH 7.4 for 15 min at 37 °C. After washing the trypsin beads with the same buffer, a simple treatment of the beads with 0.1% aqueous TFA (pH ≈ 3) cyclotide MCoTI-II with 92% yield was produced (Thongyoo et al., 2008). More recently, Craik has also used this approach for the synthesis of different MCoTI-based cyclotide mutants that were used to study their mechanism of cellular uptake (Cascales et al., 2011).

The use of trypsin-mediated cyclization provides a very efficient route for the production of cyclotides with trypsin inhibitory properties. It is worth noting, however, that although the cyclization yield for naturally occurring cyclotide MCoTI-II was extremely efficient (≈92%), the introduction of mutations that affect the binding to the proteolytic enzyme may lower the cyclization yield. For example, it has been shown that introduction of mutations into the trypsin-binding loop of SFTI-1 can affect the equilibrium between the cyclized and linear form of SFTI-1 when incubated with trypsin (Austin et al., 2010), which should affect the cyclization yield of this peptide.

2.3.2 Butelase-Mediated Backbone Cyclization

For the past few years mounting evidence has suggested that the biosynthesis of cyclotides involves an asparaginyl endopeptidase-mediated transpeptidation step during the cyclization process of these circular peptides in plants (Mylne et al., 2011; Saska et al., 2007). Tam has recently provided an important step toward fully understanding how cyclotides are biosynthesized in plants. His group reported the isolation and characterization of a new Asx-specific peptide ligase from the cyclotide-producing plant *C. ternatea* of the Fabaceae family (Nguyen et al., 2014). This new ligase, called butelase 1, was able to efficiently cyclize different polypeptides, including cyclotide kalata B1, with yields greater than 95%. Butelase 1 shares 71% sequence identity and the same catalytic triad with asparaginyl endopeptidase legumain but was unable to hydrolyze the protease substrate of legumain. Instead, butelase 1 was an efficient peptide ligase with K_{cat} values of up to 17 s^{-1} and catalytic efficiencies as high as 542,000 M^{-1} s^{-1}, making it one of fastest peptide ligase known so far. Butelase 1 also displayed a broad specificity for the N-terminal amino acids of the peptide substrate requiring only the C-terminal NHV or DHV tripeptide motif for efficient backbone cyclization. The ligase, however, was more efficient with substrates bearing the NHV tripeptide motif (Nguyen et al., 2014). Butelase 1-mediated cyclization of kalata B1 was carried out using the natural Asn residue in loop 6, which has been suggested as the natural cyclization site in plants. It is worth noting that in contrast to trypsin-mediated cyclizations (see above), butelase 1 does not require conformational assistance by disulfide bonds for efficient cyclization. For example, an S-alkylated version of linear kalata B1 peptide bearing the tripeptide NHV was also efficiently cyclized by the ligase. Intriguingly, kinetic analysis revealed that the S-alkylated version of linear kalata B1 was cyclized around 50 times faster than the corresponding natively folded linear polypeptide. The use of butelase 1 to cyclize cyclotides has only been tested for the cyclization of kalata B1, and surprisingly not with any of the naturally occurring cyclotides found in *C. ternatea*. However, the high sequence homology of kalata B1 with the different cyclotides isolated from *C. ternatea* (Poth et al., 2011, 2010), especially at the ligation site in loop 6, suggests that cyclization should also be efficient with *C. ternatea* cyclotides. In fact, butelase 1 has shown to efficiently cyclize different peptides including SFTI-1, conotoxin MrIA, the insect antimicrobial peptide thanatin, and the human salivary antimicrobial peptide histatin (Nguyen et al., 2014).

Butelase 1-mediated cyclizations are very efficient and have minimal sequence requirements at the ligation site, and together with all the features mentioned above make this ligase a very promising tool for the production of cyclic polypeptides. All the studies performed with butelase 1, however, have been carried out using enzyme isolated from natural sources. Attempts to produce biologically active butelase using recombinant heterologous expression have not been successful so far. This may limit its future potential in biotechnological applications.

2.3.3 Sortase-Mediated Backbone Cyclization

It is well known that prokaryotes use protease-catalyzed protein splicing to attach proteins to peptidoglycan in their cell wall envelope (Marraffini, DeDent, & Schneewind, 2006). For example, sortases are transpeptidase enzymes found in most Gram-positive bacteria that are specialized in this task. Among the different sortases characterized so far, the *Staphylococcus aureus* sortase A (SrtA) is one of the most widely used for biotechnological applications (Katoh, Goto, Reza, & Suga, 2011; Popp & Ploegh, 2011). SrtA specifically recognizes the hexapeptide motif LPTXTG (where X can be any residue), cleaving the peptide bond between the Thr and Gly residues. SrtA active site contains a Cys residue that generates a transient thioacyl enzyme intermediate during the cleavage reaction. This branched thioester intermediate then reacts with the α-amino group of poly-Gly (a tri-Gly sequence in the case of SrtA) substrates. In the absence of any poly-Gly substrate the thioester group is simply hydrolyzed.

SrtA has been recently used for the generation of circular green fluorescent protein using an intramolecular version of sortase-mediated ligation (SML) (Figure 3) (Antos et al., 2009; Parthasarathy, Subramanian, & Boder, 2007). Other proteins/polypeptides cyclized using SML include dihydrofolate

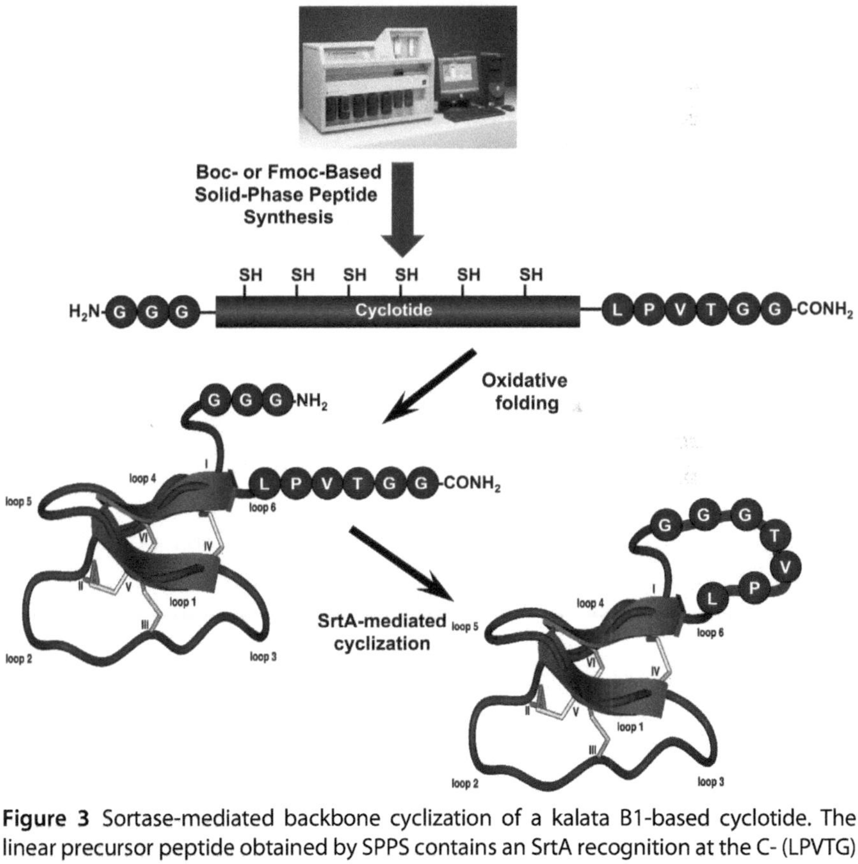

Figure 3 Sortase-mediated backbone cyclization of a kalata B1-based cyclotide. The linear precursor peptide obtained by SPPS contains an SrtA recognition at the C- (LPVTG) and an N-termini (GGG). The linear precursor is oxidatively folded and backbone-cyclized using SrtA. Note that after the ligation the sequence LPVTTGGG remains on the cyclotide.

reductase and pleckstrin homology domains (Tsukiji & Nagamune, 2009), as well as circular versions of cytokines and histatin 1 (Bolscher et al., 2011). Craik has also recently used this approach for the backbone cyclization of different classes of disulfide-rich peptides including an analog of the cyclotide kalata B1 (Jia et al., 2014). It is worth noting, however, that due to the sequence requirements for SrtA, the heptapeptide motif (LPVTGGG) is left at the ligation site (Figure 3). This should be taken into consideration when using this approach for the cyclization of small cyclic polypeptides. In this work the corresponding kalata B1 linear precursor was chemically synthesized using standard Fmoc-based SPPS using loop 6 for the cyclization site. This loop has been shown to be relatively more flexible than the rest of the cyclotide loops (Daly et al., 2013; Puttamadappa et al., 2010; Puttamadappa, Jagadish, Shekhtman, & Camarero, 2011) and it has been shown to tolerate loop extension without negatively affecting the native cyclotide fold (Gunasekera et al., 2008; Wong et al., 2012). To facilitate the SrtA-mediated cyclization, the tripeptides GGG and TGG were added to the N- and C-termini of the linear precursor, respectively. The linear precursor was oxidatively folded and cyclized using SrtA to provide a natively folded kalata B1 analog.

3. BIOLOGICAL SYNTHESIS OF CYCLOTIDES

Recent advances in the fields of protein engineering and molecular biology make possible the biological synthesis of backbone-cyclized polypeptides using standard heterologous expression systems (Aboye & Camarero, 2012; Sancheti & Camarero, 2009). The different approaches described thus far for the biological synthesis of folded cyclotides include the use of thioester-mediated cyclization generated by modified inteins (also called expressed protein ligation (EPL)) (Camarero, Kimura, Woo, Shekhtman, & Cantor, 2007; Kimura & Camarero, 2005) or thiol-induced selective N→S acyl transfer (Cowper, Craik, & Macmillan, 2013) and intein-mediated protein trans-splicing (PTS) (Borra, Dong, Elnagar, Woldemariam, & Camarero, 2012) (see reference (Perler, 2002) for an excellent searchable database of intein literature including circular peptides and protein) (Table 2).

3.1 Intein-Mediated Backbone Cyclization

Developments in the field of protein engineering and the discovery of intein-mediated protein splicing have made possible the generation of recombinant proteins containing an N-terminal Cys residue and/or a C-terminal α-thioester functionality. These important developments have made possible to perform NCL reactions between recombinant and/or synthetic polypeptides, hence allowing access to a multitude of chemically engineered proteins including backbone-cyclized polypeptides (see reference (Berrade & Camarero, 2009) for a recent review on the use of inteins for protein engineering).

An intramolecular version of intein-mediated ligation (also called EPL) has been used for the generation of many different types of backbone-cyclized polypeptides and proteins (see reference (Aboye & Camarero, 2012) for a recent review on the use of EPL for protein cyclization). Our group has recently used this approach for the recombinant expression of the cyclotide kalata B1 in *Escherichia coli* (Figure 4) (Kimura et al., 2006). In this work, several cyclotide linear precursors were fused to a modified *Saccharomyces cerevisiae* (*Sce*) vacuolar membrane ATPase intein at the C-terminus and to a methionine residue at the N-terminus. The N-terminal Met residue is removed in situ by the endogenous bacterial methionine aminopeptidase providing the required N-terminal Cys residue for the cyclization step. The intein fusion construct was then expressed in *E. coli* and purified. The recombinant precursor was cyclized and oxidatively folded in a "one-pot" reaction in aqueous buffer at pH 7.2 containing reduced GSH (Figure 5). This approach was also used for the production of different kalata B1 mutants (Figure 5). Some of the mutants were designed to test the effect of replacement of different β-branched residues adjacent to Cys residues in loops 1, 2, and 6 by the highly flexible Gly residue. The replacement of the hydrophobic residue Trp in loop 5 by the more hydrophilic Arg and Ser residues and the less hydrophobic Ala residue was also explored. Finally, the Pro residue in loop 3, which has been found to be important for defining the β-turn in this loop (Rosengren, Daly, Plan, Waine, & Craik, 2003), was also mutated to Ala. As shown in Figure 5, the "one-pot" cyclization/folding reaction was very efficient in all the cases with yields ranging from 20% to 60%, hence highlighting the robustness of this methodology.

Our group has also pioneered the use of intein-mediated backbone cyclization for the biosynthesis of fully folded cyclotides inside bacterial cells using standard heterologous expression systems (Austin, Wang, Puttamadappa, Shekhtman, & Camarero, 2009; Camarero et al., 2007). We have recently used this approach for the generation of a small library based on the cyclotide MCoTI-I where all the residues except Cys and those located in loop 6 were replaced by different residues. In this work, we used the more efficient

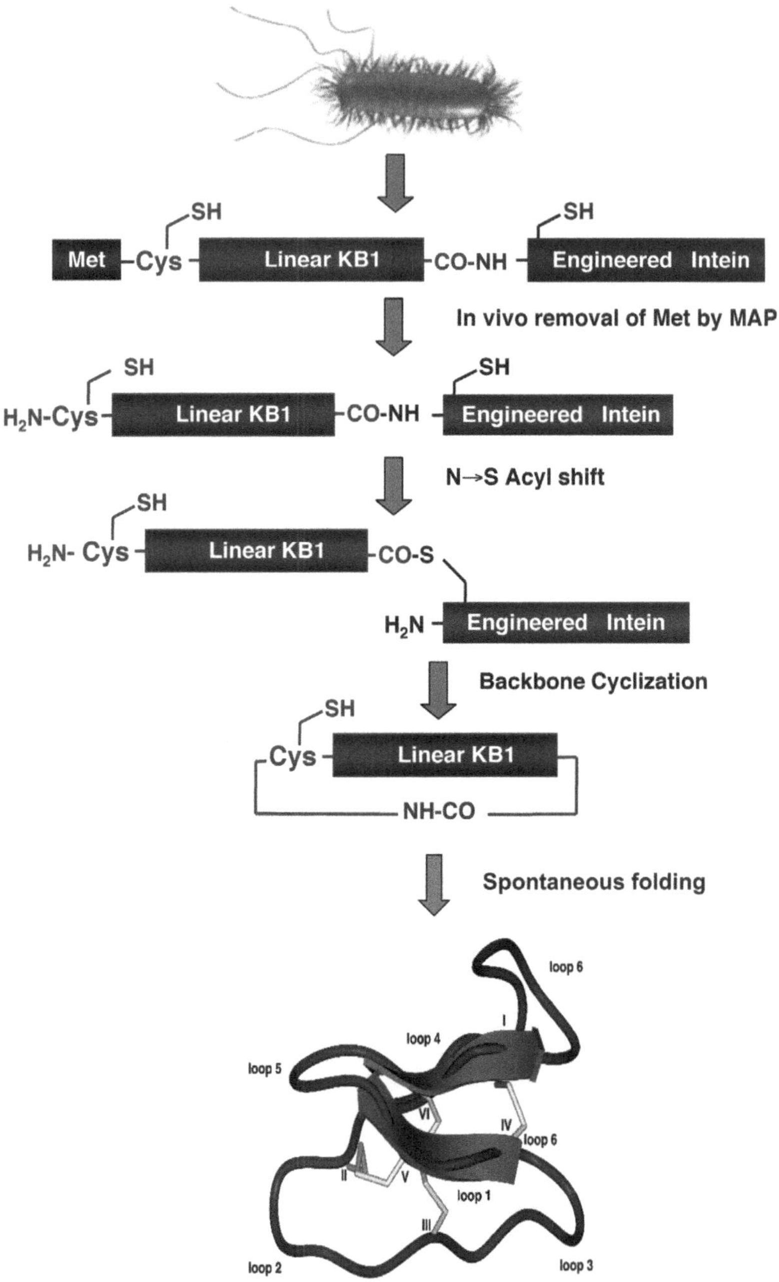

Figure 4 Intein-mediated backbone cyclization of a cyclotide linear precursor. The intein–cyclotide construct is recombinantly expressed in *E. coli* and can be cyclized and folded either in vitro (Kimura et al., 2006) or in vivo (Austin et al., 2009; Camarero et al., 2007). MAP, methionine aminopeptidase. (See color plate)

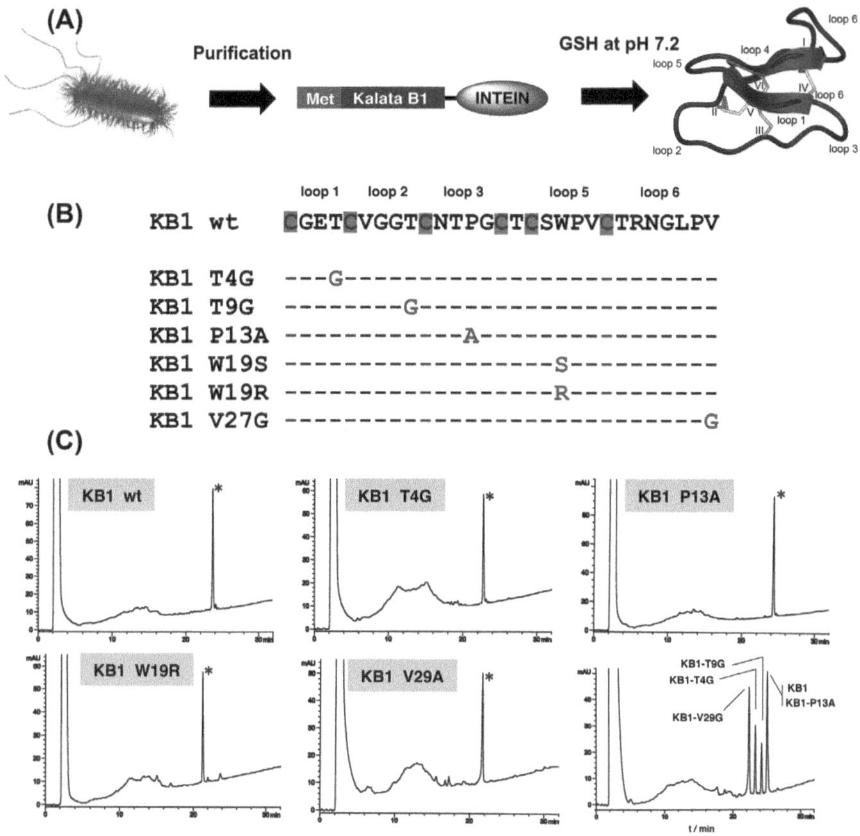

Figure 5 In vitro production of different kalata B1 cyclotides using an intein-mediated backbone cyclization (Kimura et al., 2006). (A) The intein–kalata constructs were expressed in *E. coli*, purified, and cyclized/folded in aqueous buffer at pH 7.2 containing reduced GSH for 24 h. (B) Sequences of all the katala B mutants cyclized using this approach. (C) Analytical high-performance liquid chromatography traces of the cyclization/folding crude after 18 h. The peak corresponding to the folded cyclotide is marked with an asterisk. (See color plate)

Mycobacterium xenopi (*Mxe*) GyrA intein instead. The resulting library was screened for biological activity and the ability to adopt a cyclotide fold indicating that most of the residues in loops 1, 2, 3, and 5 could be replaced without significantly affecting the ability to adopt a native cyclotide topology (Austin et al., 2009). Again, these results highlight the robustness of the cyclotide molecular framework, demonstrating the feasibility of in-cell generation of large genetically encoded libraries of cyclotides that could be screened inside the cell using high-throughput selection techniques.

Intein-mediated backbone cyclization of polypeptides has also been used for the biosynthesis of other naturally occurring disulfide-rich backbone-cyclized polypeptides such as the Bowman–Birk inhibitor SFTI-1 (Austin et al., 2010), and the θ-defensins Rhesus theta defensin-1 (Gould et al., 2012) and Human theta defensin-2 (Conibear et al., 2014), hence showing the general applicability of this method for the production of naturally occurring or engineered circular polypeptides.

3.2 Thiol-Induced N→S Acyl Backbone Cyclization

Recent work by Macmillan has shown that a polypeptide having a single C-terminal Cys residue is capable under special conditions of initiating an N→S acyl shift to provide an α-thioester intermediate that can be used for NCL (Kang, Richardson, & Macmillan, 2009). This reaction has been shown to be selective depending on the nature of the preceding Cys residue, being favored in cases the C-terminal Cys is preceded by a Gly, Cys, or His residue. According to the same authors, the rate of the molecular rearrangement is also significantly slower when the Cys residue is not positioned at the C-terminus, which could allow regioselective control for the formation of α-thioesters in polypeptides containing multiple Cys residues (Richardson, Chan, Blanc, Saadi, & Macmillan, 2010). To test it, Macmillan and Craik used this approach to generate the cyclotide kalata B1 from a linear precursor produced by heterologous expression in *E. coli* (Cowper et al., 2013). In this work the cyclization site was selected at the C-terminal of loop 3 given the presence of required GC dipeptide motif in this loop (Figure 1). The linear cyclotide precursor was fused to the C-terminus of the thioredoxin (Trx) tag to improve its expression yield and solubility. To allow the generation of the required N-terminal Cys residue for the cyclization step, the Trx tag was linked through a tobacco etch virus (TEV) protease recognition sequence. After purification of the precursor Trx-kalata B1 construct, the Trx tag was removed with TEV protease. Subsequent in vitro cyclization of the purified linear kalata B1 precursor was accomplished by incubating at 45 °C in the presence of 10% (w/v) sodium 2-mercaptoethane sulfonate for 48 h. The reduced cyclotide was then oxidatively folded. The final product was fully characterized by NMR to show that it was adopting a native cyclotide fold. Although this approach seems promising, it still remains to be seen how applicable it is for the synthesis of other cyclotides that do not contain a suitable Xaa-Cys motif, where Xaa is a Gly, Cys, or His residue.

3.3 Sortase-Induced Backbone Cyclization

As mentioned earlier, SrtA has been used for the backbone cyclization of synthetic linear cyclotide precursors (see Section 2.3.3). In a similar fashion, Hannoush recently expressed a glutathione S-transferase (GST) fusion protein containing a linearized version of the cyclotide MCoTI-II (Stanger et al., 2014). To allow the SrtA-mediated backbone cyclization, the SrtA recognition motifs GGG and LPETGG were added to the N- and C-termini of the linear cyclotide sequence, respectively. A TEV protease cleavage site was also inserted between GST tag and the GGG motif in order to remove the His-GST tag after Ni^{2+}-nitriloacetic acid purification and to enable subsequent cyclization by SrtA (Figure 6). After expression of the GST construct in *E. coli*, the GST tag was removed by treatment with TEV protease under oxidative conditions. This step provided the required N-terminal GGG motif for SrtA-mediated cyclization and folded the linear precursor into a native Cys knot fold. It is worth noting, that in this work the cyclization was carried out with an optimized SrtA enzyme with improved catalytic activity (m^4SrtA) (Chen, Dorr, & Liu, 2011). This evolved enzyme has a 140-fold increase in k_{cat}/K_m compared to the wild-type form. Given that the cyclization using SrtA leaves the heptapeptide motif LPETGGG at the cyclization, the cyclization was performed at loop 6. As in other

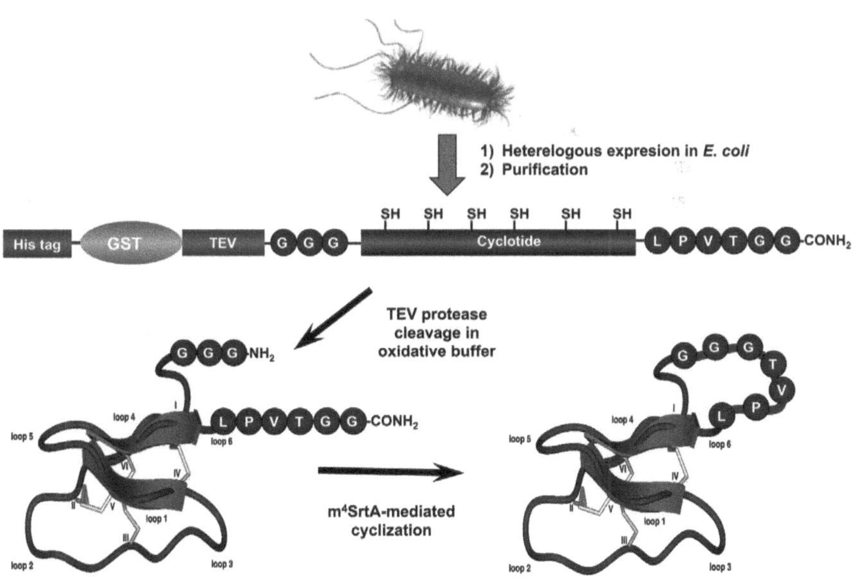

Figure 6 Strategy used for the biosynthesis of a cyclotide using SrtA-induced backbone cyclization (Stanger et al., 2014). (See color plate)

cyclotides, loop 6 in MCoTI cyclotides has been shown to be more flexible than the rest of cyclotide framework (Puttamadappa et al., 2010, 2011) and therefore able to accept long peptide grafts without negatively affecting the cyclotide fold (Aboye et al., 2012; Chan et al., 2011; Ji et al., 2013).

3.4 Backbone Cyclization Using Intein-Mediated Protein Trans-Splicing

Our group has recently used PTS for the production of naturally occurring and engineered cyclotides (Jagadish et al., 2013). PTS is a posttranslational modification similar to protein splicing with the difference that the intein self-processing domain is split into two fragments (N-intein and C-intein). The split intein fragments are not active individually; however, they can bind to each other with high specificity under appropriate conditions to form an active protein splicing or intein domain in trans. By rearranging the order of the intein fragments, i.e., by fusing the N- and C-intein fragments to the C- and N-termini of the polypeptide to cyclize, the trans-splicing reaction results in the formation of a backbone-cyclized polypeptide (Figure 7).

This approach was first reported by the Benkovic group using the naturally occurring *Synechocystis* sp. (*Ssp*) PCC6803 DnaE split intein, and it has been used for the preparation of many small backbone-cyclized polypeptides as well as large genetically encoded libraries of small cyclic peptides in bacterial cells (Abel-Santos, Scott, & Benkovic, 2003; Scott, Abel-Santos, Jones, & Benkovic, 2001; Scott, Abel-Santos, Wall, Wahnon, & Benkovic, 1999). PTS has also been used for the production of recombinant backbone-cyclized proteins. In a recent study, the artificially split *Ssp* PCC6803 DnaB mini-intein was used for the cyclization of the TEM-1 β-lactamase in the bacterial periplasm (Deschuyteneer et al., 2010).

It should be noted that use of the *Ssp* DnaE or DnaB split inteins requires the presence of specific amino acids at both extein–intein junctions for efficient protein splicing to happen (Tavassoli & Benkovic, 2007). Our group has overcome this limitation by using the *Nostoc punctiforme* PCC73102 (*Npu*) DnaE split intein (Iwai, Zuger, Jin, & Tam, 2006). The *Npu* DnaE split intein is one of the most efficient split inteins reported thus far. This intein has one of the highest reported rates of PTS ($\tau_{1/2} \sim 60$ s), a high splicing efficiency ($\approx 80\%$), and high sequence tolerance at the extein–intein junctions (Borra et al., 2012; Jagadish et al., 2013; Zettler, Schutz, & Mootz, 2009). We have used this split intein for the in-cell production of several MCoTI-based cyclotides (Jagadish et al., 2013). In-cell expression was highly efficient providing intracellular concentrations in the range of

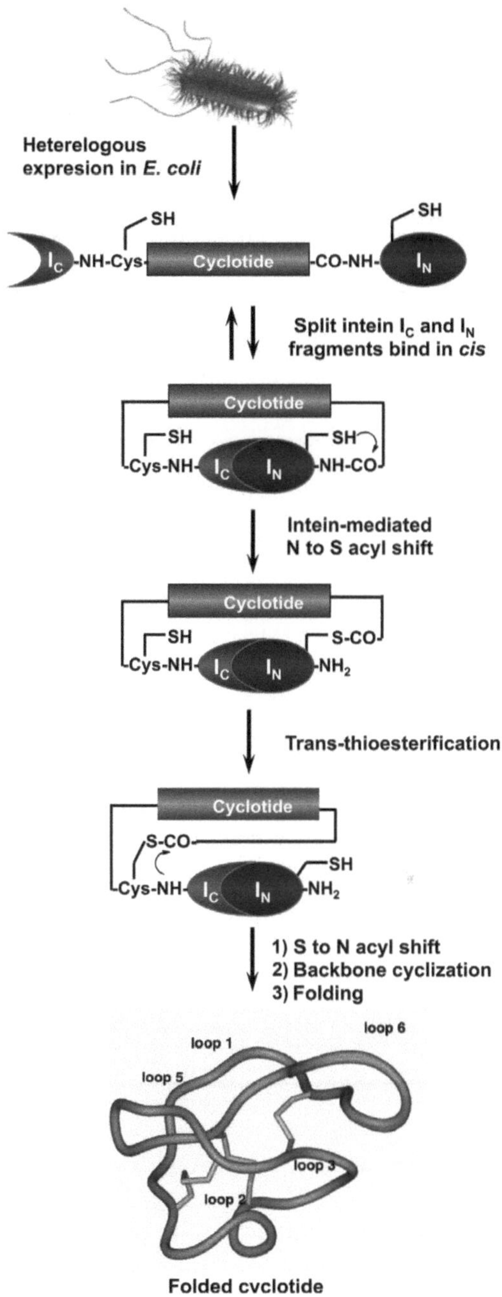

Figure 7 In-cell expression of native folded cyclotides using intein-mediated PTS (Jagadish et al., 2013). (See color plate)

20–40 μM. The high efficiency of *Npu* DnaE PTS-mediated cyclization in combination with the use of nonsense suppressing orthogonal transfer ribonucleic acid (tRNA)/synthetase technology has also made possible the production of cyclotides containing unnatural amino acids inside live bacterial cells (Jagadish et al., 2013). We used this is approach to incorporate *p*-azido-phenylalanine into loop 2 of cyclotide MCoTI-I, which reacted in cell through copper-free click chemistry with dibenzocyclooctyne-containing fluorescent probes to provide in-cell fluorescently labeled cyclotides (Jagadish et al., 2013). This exciting result opens the possibility for in-cell screening of genetically encoded libraries of cyclotides for the rapid selection of novel cyclotide sequences able to bind a specific bait protein using high-throughput cell-based optical screening approaches.

More recently, we have shown that this approach can be also used for the expression of folded cyclotides in eukaryotic microorganisms such as yeast *S. cerevisiae* (Jagadish et al., 2015). In this work, in-cell PTS was used to generate a novel bioactive cyclotide that was able to inhibit α-synuclein-induced cytotoxicity in live yeast cells, therefore allowing us for the first time to perform a phenotypic screen to select cyclotide sequences with biological activity from inactive cyclotides in a yeast synucleopathy model. These results make possible to perform in-cell phenotypic screens against other cellular pathologies using cyclotide-based libraries. Using eukaryotic microorganisms for screening purposes should not only provide a more biologically relevant cellular background for phenotypic screening but also facilitate the production of cyclotide-based libraries containing posttranslational modifications such as phosphorylation and/or glycosylation, which are not available in bacterial expression systems.

4. SUMMARY AND CONCLUDING REMARKS

In this chapter we have reviewed different chemical and biological methods for the production of cyclotides. Most of the methods reviewed make use of an intramolecular version of NCL. NCL requires only the presence of an N-terminal Cys residue and a α-thioester function. Cyclotides present six Cys residues that in theory could be used as cyclization sites. According to the published data, the Cys residues at the C- and N-terminal regions of loops 3 and 6 have been successfully used in the production of several cyclotides. The NCL reaction is also fully compatible with physiological conditions, i.e., aqueous buffers at pH $\approx$ 7, and does not compromise the chirality of the C-terminal residue during the cyclization step. In addition, if GSH is used as thiol cofactor, the cyclization and folding can be accomplished in a single "one-pot" reaction (Austin et al., 2010; Kimura et al., 2006). The linear precursors required for NCL-mediated cyclization can be also obtained by either pure synthetic (using both Boc- and Fmoc-based SPPS) or recombinant (using modified inteins or a thiol-induced N→S acyl shift reaction) methods. Synthetic methods allow the synthesis of cyclotides containing nonnatural amino acids (Aboye et al., 2012) or even the production of D-cyclotides (Sando et al., 2011). On the other hand, the use of pure recombinant expression methods allows scalability for cyclotide production and more importantly in-cell production of cyclotides (Austin et al., 2010; Camarero et al., 2007; Jagadish et al., 2013). In-cell production makes possible the generation of large genetically encoded libraries of cyclic peptides inside live cells that can be rapidly screened for the selection of novel sequences able to modulate or inhibit the biological activities of particular biomolecular targets.

Having access to the recombinant production of cyclotides using bacterial heterologous expression also facilitates the introduction of NMR-active isotopes such as ^{15}N and/or ^{13}C in a very inexpensive fashion (Puttamadappa et al., 2010, 2011). This should facilitate the use of heteronuclear NMR spectroscopy to study structure—activity relationships of biologically active circular polypeptides and their molecular targets. This was recently demonstrated in the elucidation of the solution structure of the first cyclotide engineered to bind the p53-binding domain of the oncogene Hdm2 (Figure 8) (Ji et al., 2013).

The use of PTS-mediated backbone cyclization in combination with nonsense suppressing orthogonal tRNA/synthetase technology has also made possible the introduction of nonnatural amino acids into in-cell expressed backbone-cyclized polypeptides (Jagadish et al., 2013). Expansion of the genetic code allows for the incorporation of novel chemical functionalities that can be used in molecular evolution and cell-based assays for the selection of novel cyclic peptides with improved biological or novel biophysical activities (Jagadish et al., 2013; Young et al., 2011). More recently PTS has also been used for in-cell expression of folded bioactive cyclotides in eukaryotic microorganisms such as yeast *S. cerevisiae* (Jagadish et al., 2015). *Saccharomyces cerevisiae* is a versatile and robust model system for eukaryotic cellular biology and several human pathologies derived from protein misfolding have been successfully modeled in this microorganism. Using a yeast synucleopathy model (Kritzer et al., 2009) in combination with PTS we

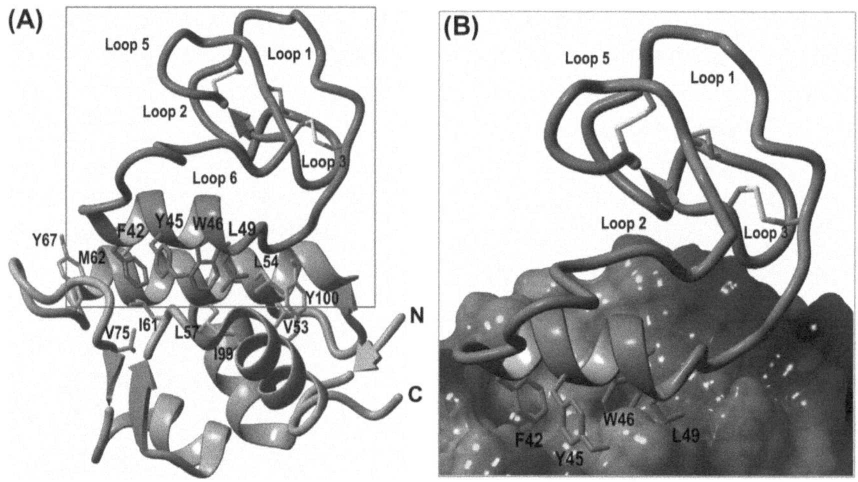

Figure 8 Solution structure of the complex formed by engineered cyclotide MCo-PMI and the p53-binding domain of Hdm2 (pdb code: 2M86) (Ji et al., 2013). (A) Ribbon representation of MCo-PMI (purple) and Hdm2 (cyan blue) complex. The side chains of Phe42, Trp46, and Leu49 in MCo-PMI and the Hdm2 residues shaping the hydrophobic binding pocket are shown as sticks in purple and cyan blue, respectively. (B) Close-up view of the binding interface within the Hdm2-MCo-PMI complex. The electrostatic potential at the molecular surface of Hdm2 is shown as positive in blue, negative in red, and noncharged in white. (See color plate)

were recently able to express a cyclotide inhibitor of α-synuclein cytotoxicity. This made possible for the first time to perform a phenotypic screen to select bioactive cyclotides in a eukaryotic system. The use of eukaryotic microorganisms, such as yeast, for screening purposes should provide a more biologically relevant cellular background for phenotypic screening and also facilitate the production of cyclotide-based libraries containing posttranslational modifications such as phosphorylation and/or glycosylation, which are not available in bacterial expression systems.

Cyclotides are small globular microproteins with a unique head-to-tail cyclized backbone, which is stabilized by three disulfide bonds. The number and positions of cysteine residues are conserved throughout the family, forming a CCK motif (Craik, Daly, Bond, & Waine, 1999) that acts as a highly stable and versatile framework on which hypervariable loops are arranged (Figure 1). This CCK scaffold affords exceptional resistance to thermal and chemical denaturation as well as enzymatic degradation. This is particularly important for oral bioavailability. In fact, the use of cyclotide-containing plants in indigenous medicine first highlighted the fact that the peptides are resistant to boiling and are orally bioavailable (Saether et al., 1995). An orally bioavailable bradykinin B1 cyclotide receptor antagonist has also been recently reported (Wong et al., 2012). Cyclotides have also shown excellent tolerance to sequence variations. This combined with the possibility to produce them using heterologous expressions systems makes them ideal substrates for molecular evolution strategies to enable generation and selection of compounds with optimal binding and inhibitory characteristics. Finally, cyclotides can also be efficiently produced using standard SPPS techniques thereby allowing the introduction of nonnatural amino acids and chemical modifications to improve their pharmacological properties.

In summary, all these features open new opportunities for the development of the cyclotide molecular scaffold for novel and exciting biotechnological and biomedical applications.

ACKNOWLEDGMENTS

This work was partially supported by National Institutes of Health Research Grant R01-GM090323 (JAC) and Bristol-Myers Squibb.

REFERENCES

Abel-Santos, E., Scott, C. P., & Benkovic, S. J. (2003). Use of inteins for the in vivo production of stable cyclic peptide libraries in *E. coli*. *Methods in Molecular Biology, 205*, 281–294.

Aboye, T., Kuang, Y., Neamati, N., & Camarero, J. A. (2015). Rapid parallel synthesis of bioactive folded cyclotides by using a tea-bag approach. *ChemBioChem, 16*(5), 827–833.

Aboye, T. L., & Camarero, J. A. (2012). Biological synthesis of circular polypeptides. *Journal of Biological Chemistry, 287*(32), 27026–27032.

Aboye, T. L., Clark, R. J., Craik, D. J., & Göransson, U. (2008). Ultra-stable peptide scaffolds for protein engineering-synthesis and folding of the circular cystine knotted cyclotide cycloviolacin O_2. *ChemBioChem, 9*(1), 103–113.

Aboye, T. L., Ha, H., Majumder, S., Christ, F., Debyser, Z., Shekhtman, A., et al. (2012). Design of a novel cyclotide-based CXCR4 antagonist with anti-human immunodeficiency virus (HIV)-1 activity. *Journal of Medicinal Chemistry, 55*(23), 10729–10734.

Antos, J. M., Popp, M. W., Ernst, R., Chew, G. L., Spooner, E., & Ploegh, H. L. (2009). A straight path to circular proteins. *Journal of Biological Chemistry, 284*(23), 16028–16036.

Austin, J., Kimura, R. H., Woo, Y. H., & Camarero, J. A. (2010). In vivo biosynthesis of an Ala-scan library based on the cyclic peptide SFTI-1. *Amino Acids, 38*(5), 1313–1322.

Austin, J., Wang, W., Puttamadappa, S., Shekhtman, A., & Camarero, J. A. (2009). Biosynthesis and biological screening of a genetically encoded library based on the cyclotide MCoTI-I. *ChemBioChem, 10*(16), 2663–2670.

Backes, B. J., & Ellman, J. A. (1999). An alkanesulfonamide "safety-catch" linker for solid-phase synthesis. *Journal of Organic Chemistry, 64*(64), 2322–2330.

Backes, B. J., Virgilio, A. V., & Ellman, J. A. (1996). Activation method to prepare a highly reactive acylsulfonamide "safety-catch" linker for solid-phase synthesis. *Journal of the American Chemical Society, 118*, 3055–3056.

Beligere, G. S., & Dawson, P. E. (1999). Conformationally assisted ligation using C-terminal thioester peptides. *Journal of the American Chemical Society, 121*, 6332–6333.

Berrade, L., & Camarero, J. A. (2009). Expressed protein ligation: a resourceful tool to study protein structure and function. *Cellular and Molecular Life Sciences, 66*(24), 3909–3922.

Blanco-Canosa, J. B., & Dawson, P. E. (2008). An efficient Fmoc-SPPS approach for the generation of thioester peptide precursors for use in native chemical ligation. *Angewandte Chemie, International Edition in English, 47*(36), 6851–6855.

Bolscher, J. G., Oudhoff, M. J., Nazmi, K., Antos, J. M., Guimaraes, C. P., Spooner, E., et al. (2011). Sortase A as a tool for high-yield histatin cyclization. *FASEB Journal, 25*(8), 2650–2658.

Borra, R., Dong, D., Elnagar, A. Y., Woldemariam, G. A., & Camarero, J. A. (2012). In-cell fluorescence activation and labeling of proteins mediated by FRET-quenched split inteins. *Journal of the American Chemical Society, 134*(14), 6344–6353.

Camarero, J. A., Adeva, A., & Muir, T. W. (2000). 3-Thiopropionic acid as a highly versatile multidetachable thioester resin linker. *Letters in Peptide Science,* 7(1), 17–21.

Camarero, J. A., Cotton, G. J., Adeva, A., & Muir, T. W. (1998). Chemical ligation of unprotected peptides directly from a solid support. *Journal of Peptide Research, 51*(4), 303–316.

Camarero, J. A., Hackel, B. J., de Yoreo, J. J., & Mitchell, A. R. (2004). Fmoc-based synthesis of peptide alpha-thioesters using an aryl hydrazine support. *Journal of Organic Chemistry, 69*(12), 4145–4151.

Camarero, J. A., Kimura, R. H., Woo, Y. H., Shekhtman, A., & Cantor, J. (2007). Biosynthesis of a fully functional cyclotide inside living bacterial cells. *ChemBioChem, 8*(12), 1363–1366.

Camarero, J. A., & Mitchell, A. R. (2005). Synthesis of proteins by native chemical ligation using Fmoc-based chemistry. *Protein & Peptide Letters, 12*(8), 723–728.

Camarero, J. A., & Muir, T. W. (1997). Chemoselective backbone cyclization of unprotected peptides. *Chemical Communications, 1997*, 202–219.

Camarero, J. A., & Muir, T. W. (1999). Biosynthesis of a head-to-tail cyclized protein with improved biological activity. *Journal of the American Chemical Society, 121*, 5597–5598.

Cascales, L., Henriques, S. T., Kerr, M. C., Huang, Y. H., Sweet, M. J., Daly, N. L., et al. (2011). Identification and characterization of a new family of cell-penetrating peptides: cyclic cell-penetrating peptides. *Journal of Biological Chemistry, 286*(42), 36932–36943.

Chan, L. Y., Gunasekera, S., Henriques, S. T., Worth, N. F., Le, S. J., Clark, R. J., et al. (2011). Engineering pro-angiogenic peptides using stable, disulfide-rich cyclic scaffolds. *Blood, 118*(25), 6709–6717.

Chen, I., Dorr, B. M., & Liu, D. R. (2011). A general strategy for the evolution of bond-forming enzymes using yeast display. *Proceedings of the National Academy of Sciences of the United States of America, 108*(28), 11399–11404.

Cheneval, O., Schroeder, C. I., Durek, T., Walsh, P., Huang, Y. H., Liras, S., et al. (2014). Fmoc-based synthesis of disulfide-rich cyclic peptides. *Journal of Organic Chemistry, 79*(12), 5538–5544.

Chiche, L., Heitz, A., Gelly, J. C., Gracy, J., Chau, P. T., Ha, P. T., et al. (2004). Squash inhibitors: from structural motifs to macrocyclic knottins. *Current Protein and Peptide Science, 5*(5), 341–349.

Colgrave, M. L., & Craik, D. J. (2004). Thermal, chemical, and enzymatic stability of the cyclotide kalata B1: the importance of the cyclic cystine knot. *Biochemistry, 43*(20), 5965–5975.

Conibear, A. C., Wang, C. K., Bi, T., Rosengren, K. J., Camarero, J. A., & Craik, D. J. (2014). Insights into the molecular flexibility of theta-defensins by NMR relaxation analysis. *Journal of Physical Chemistry B, 118*(49), 14257–14266.

Contreras, J., Elnagar, A. Y., Hamm-Alvarez, S. F., & Camarero, J. A. (2011). Cellular uptake of cyclotide MCoTI-I follows multiple endocytic pathways. *Journal of Controlled Release, 155*(2), 134–143.

Cowper, B., Craik, D. J., & Macmillan, D. (2013). Making ends meet: chemically mediated circularization of recombinant proteins. *ChemBioChem, 14*(7), 809–812.

Craik, D. J., Daly, N. L., Bond, T., & Waine, C. (1999). Plant cyclotides: a unique family of cyclic and knotted proteins that defines the cyclic cystine knot structural motif. *Journal of Molecular Biology, 294*(5), 1327–1336.

Daly, N. L., Love, S., Alewood, P. F., & Craik, D. J. (1999). Chemical synthesis and folding pathways of large cyclic polypeptides: studies of the cystine knot polypeptide kalata B1. *Biochemistry, 38*(32), 10606–10614.

Daly, N. L., Thorstholm, L., Greenwood, K. P., King, G. J., Rosengren, K. J., Heras, B., et al. (2013). Structural insights into the role of the cyclic backbone in a squash trypsin inhibitor. *Journal of Biological Chemistry, 288*(50), 36141–36148.

Dawson, P. E., Muir, T. W., Clark-Lewis, I., & Kent, S. B. H. (1994). Synthesis of proteins by native chemical ligation. *Science, 266*, 776–779.

Deschuyteneer, G., Garcia, S., Michiels, B., Baudoux, B., Degand, H., Morsomme, P., et al. (2010). Intein-mediated cyclization of randomized peptides in the periplasm of *Escherichia coli* and their extracellular secretion. *ACS Chemical Biology, 5*(7), 691–700.

Dutton, J. L., Renda, R. F., Waine, C., Clark, R. J., Daly, N. L., Jennings, C. V., et al. (2004). Conserved structural and sequence elements implicated in the processing of gene-encoded circular proteins. *Journal of Biological Chemistry, 279*(45), 46858–46867.

von Eggelkraut-Gottanka, R., Klose, A., Beck-Sickinger, A. G., & Beyermann, M. (2003). Peptide (alpha)thioester formation using standard Fmoc-chemistry. *Tetrahedron Letters, 44*(17), 3551–3554.

Eliasen, R., Daly, N. L., Wulff, B. S., Andresen, T. L., Conde-Frieboes, K. W., & Craik, D. J. (2012). Design, synthesis, structural and functional characterization of novel melanocortin agonists based on the cyclotide kalata B1. *Journal of Biological Chemistry, 287*(48), 40493–40501.

Felizmenio-Quimio, M. E., Daly, N. L., & Craik, D. J. (2001). Circular proteins in plants: solution structure of a novel macrocyclic trypsin inhibitor from *Momordica cochinchinensis*. *Journal of Biological Chemistry, 276*(25), 22875–22882.

Futaki, S., Sogawa, K., Maruyama, J., Asahara, T., & Niwa, M. (1997). Preparation of peptide thioesters using Fmoc-solid-phase peptide synthesis and its application to the construction of a template-assembled synthetic protein (TASP). *Tetrahedron Letters, 38*(35), 6237–6240.

Garcia, A. E., & Camarero, J. A. (2010). Biological activities of natural and engineered cyclotides, a novel molecular scaffold for peptide-based therapeutics. *Current Molecular Pharmacology, 3*(3), 153–163.

Gillon, A. D., Saska, I., Jennings, C. V., Guarino, R. F., Craik, D. J., & Anderson, M. A. (2008). Biosynthesis of circular proteins in plants. *Plant Journal, 53*(3), 505–515.

Gould, A., Li, Y., Majumder, S., Garcia, A. E., Carlsson, P., Shekhtman, A., et al. (2012). Recombinant production of rhesus theta-defensin-1 (RTD-1) using a bacterial expression system. *Molecular BioSystems, 8*(4), 1359–1365.

Gray, K., Elghadban, S., Thongyoo, P., Owen, K. A., Szabo, R., Bugge, T. H., et al. (2014). Potent and specific inhibition of the biological activity of the type-II transmembrane serine protease matriptase by the cyclic microprotein MCoTI-II. *Thrombosis and Haemostasis, 112*(2), 402–411.

Gruber, C. W., Elliott, A. G., Ireland, D. C., Delprete, P. G., Dessein, S., Goransson, U., et al. (2008). Distribution and evolution of circular miniproteins in flowering plants. *Plant Cell, 20*(9), 2471–2483.

Gunasekera, S., Aboye, T. L., Madian, W. A., El-Seedi, H. R., & Goransson, U. (2013). Making ends meet: microwave-accelerated synthesis of cyclic and disulfide rich proteins via in situ thioesterification and native chemical ligation. *International Journal of Peptide Research and Therapeutics, 19*(1), 43–54.

Gunasekera, S., Foley, F. M., Clark, R. J., Sando, L., Fabri, L. J., Craik, D. J., et al. (2008). Engineering stabilized vascular endothelial growth factor-A antagonists: synthesis, structural characterization, and bioactivity of grafted analogues of cyclotides. *Journal of Medicinal Chemistry, 51*(24), 7697–7704.

Heitz, A., Hernandez, J. F., Gagnon, J., Hong, T. T., Pham, T. T., Nguyen, T. M., et al. (2001). Solution structure of the squash trypsin inhibitor MCoTI-II. A new family for cyclic knottins. *Biochemistry, 40*(27), 7973–7983.

Hernandez, J. F., Gagnon, J., Chiche, L., Nguyen, T. M., Andrieu, J. P., Heitz, A., et al. (2000). Squash trypsin inhibitors from *Momordica cochinchinensis* exhibit an atypical macrocyclic structure. *Biochemistry, 39*(19), 5722–5730.

Huang, Y. H., Colgrave, M. L., Clark, R. J., Kotze, A. C., & Craik, D. J. (2010). Lysine-scanning mutagenesis reveals an amendable face of the cyclotide kalata B1 for the optimization of nematocidal activity. *Journal of Biological Chemistry, 285*(14), 10797–10805.

Ingenito, R., Bianchi, E., Fattori, D., & Pessi, A. (1999). Solid phase synthesis of peptide C-terminal thioesters by Fmoc/t-Bu chemistry Source. *Journal of the American Chemical Society, 121*(49), 11369–11374.

Iwai, H., Zuger, S., Jin, J., & Tam, P. H. (2006). Highly efficient protein trans-splicing by a naturally split DnaE intein from *Nostoc punctiforme*. *FEBS Letters, 580*(7), 1853–1858.

Jagadish, K., Borra, R., Lacey, V., Majumder, S., Shekhtman, A., Wang, L., et al. (2013). Expression of fluorescent cyclotides using protein trans-splicing for easy monitoring of cyclotide-protein interactions. *Angewandte Chemie, International Edition in English, 52*(11), 3126–3131.

Jagadish, K., Gould, A., Borra, R., Majumder, S., Mushtaq, Z., Shekhtman, A., et al. (2015). *Angewandte Chemie, International Edition in English, 54*(29), 8390–8394.

Ji, Y., Majumder, S., Millard, M., Borra, R., Bi, T., Elnagar, A. Y., et al. (2013). In vivo activation of the p53 Tumor Suppressor pathway by an engineered cyclotide. *Journal of the American Chemical Society, 135*(31), 11623–11633.

Jia, X., Kwon, S., Wang, C. I., Huang, Y. H., Chan, L. Y., Tan, C. C., et al. (2014). Semi-enzymatic cyclization of disulfide-rich peptides using Sortase A. *Journal of Biological Chemistry, 289*(10), 6627–6638.

Kang, J., Richardson, J. P., & Macmillan, D. (2009). 3-Mercaptopropionic acid-mediated synthesis of peptide and protein thioesters. *Chemical Communications (Cambridge), 4*, 407–409.

Katoh, T., Goto, Y., Reza, M. S., & Suga, H. (2011). Ribosomal synthesis of backbone macrocyclic peptides. *Chemical Communications (Cambridge), 47*(36), 9946–9958.

Kenner, G. W., McDermott, J. R., & Sheppard, R. C. (1971). The safety catch principle in solid phase peptide synthesis. *Chemical Communications*, 636–637.

Kimura, R., & Camarero, J. A. (2005). Expressed protein ligation: a new tool for the biosynthesis of cyclic polypeptides. *Protein & Peptide Letters, 12*(8), 789–794.

Kimura, R. H., Tran, A. T., & Camarero, J. A. (2006). Biosynthesis of the cyclotide kalata B1 by using protein splicing. *Angewandte Chemie, International Edition, 45*(6), 973–976.

Koehbach, J., O'Brien, M., Muttenthaler, M., Miazzo, M., Akcan, M., Elliott, A. G., et al. (2013). Oxytocic plant cyclotides as templates for peptide G protein-coupled receptor ligand design. *Proceedings of the National Academy of Sciences of the United States of America, 110*(52), 21183–21188.

Kritzer, J. A., Hamamichi, S., McCaffery, J. M., Santagata, S., Naumann, T. A., Caldwell, K. A., et al. (2009). Rapid selection of cyclic peptides that reduce alpha-synuclein toxicity in yeast and animal models. *Nature Chemical Biology, 5*(9), 655–663.

Marglin, A., & Merrifield, R. B. (1970). Chemical synthesis of peptides and proteins. *Annual Review of Biochemistry, 39*, 841–866.

Marraffini, L. A., DeDent, A. C., & Schneewind, O. (2006). Sortases and the art of anchoring proteins to the envelopes of gram-positive bacteria. *Microbiology and Molecular Biology Reviews, 70*(1), 192–221.

Marx, U. C., Korsinczky, M. L., Schirra, H. J., Jones, A., Condie, B., Otvos, L., Jr., et al. (2003). Enzymatic cyclization of a potent Bowman-Birk protease inhibitor, sunflower trypsin inhibitor-1, and solution structure of an acyclic precursor peptide. *Journal of Biological Chemistry, 278*(24), 21782–21789.

Mylne, J. S., Chan, L. Y., Chanson, A. H., Daly, N. L., Schaefer, H., Bailey, T. L., et al. (2012). Cyclic peptides arising by evolutionary parallelism via asparaginyl-endopeptidase-mediated biosynthesis. *Plant Cell, 24*(7), 2765–2778.

Mylne, J. S., Colgrave, M. L., Daly, N. L., Chanson, A. H., Elliott, A. G., McCallum, E. J., et al. (2011). Albumins and their processing machinery are hijacked for cyclic peptides in sunflower. *Nature Chemical Biology,* 7(5), 257–259.

Nguyen, G. K., Wang, S., Qiu, Y., Hemu, X., Lian, Y., & Tam, J. P. (2014). Butelase 1 is an Asx-specific ligase enabling peptide macrocyclization and synthesis. *Nature Chemical Biology, 10*(9), 732–738.

Nguyen, G. K., Zhang, S., Nguyen, N. T., Nguyen, P. Q., Chiu, M. S., Hardjojo, A., et al. (2011). Discovery and characterization of novel cyclotides originated from chimeric precursors consisting of albumin-1 chain a and cyclotide domains in the Fabaceae family. *Journal of Biological Chemistry, 286*(27), 24275–24287.

Parthasarathy, R., Subramanian, S., & Boder, E. T. (2007). Sortase A as a novel molecular "stapler" for sequence-specific protein conjugation. *Bioconjugate Chemistry, 18*(2), 469–476.

Perler, F. B. (2002). InBase: the intein database. *Nucleic Acids Research, 30*(1), 383–384.

Popp, M. W., & Ploegh, H. L. (2011). Making and breaking peptide bonds: protein engineering using sortase. *Angewandte Chemie, International Edition in English, 50*(22), 5024–5032.

Poth, A. G., Colgrave, M. L., Lyons, R. E., Daly, N. L., & Craik, D. J. (2011). Discovery of an unusual biosynthetic origin for circular proteins in legumes. *Proceedings of the National Academy of Sciences of the United States of America, 108*(25), 1027–1032.

Poth, A. G., Colgrave, M. L., Philip, R., Kerenga, B., Daly, N. L., Anderson, M. A., et al. (2010). Discovery of cyclotides in the fabaceae plant family provides new insights into the cyclization, evolution, and distribution of circular proteins. *ACS Chemical Biology, 6*(4), 345–355.

Poth, A. G., Mylne, J. S., Grassl, J., Lyons, R. E., Millar, A. H., Colgrave, M. L., et al. (2012). Cyclotides associate with leaf vasculature and are the products of a novel precursor in petunia (Solanaceae). *Journal of Biological Chemistry, 287*(32), 27033–27046.

Puttamadappa, S. S., Jagadish, K., Shekhtman, A., & Camarero, J. A. (2010). Backbone dynamics of cyclotide MCoTI-I free and complexed with trypsin. *Angewandte Chemie, International Edition in English, 49*(39), 7030–7034.

Puttamadappa, S. S., Jagadish, K., Shekhtman, A., & Camarero, J. A. (2011). Erratum in: backbone dynamics of cyclotide MCoTI-I free and complexed with trypsin. *Angewandte Chemie, International Edition in English, 50*(31), 6948–6949.

Reiss, S., Sieber, M., Oberle, V., Wentzel, A., Spangenberg, P., Claus, R., et al. (2006). Inhibition of platelet aggregation by grafting RGD and KGD sequences on the structural scaffold of small disulfide-rich proteins. *Platelets, 17*(3), 153–157.

Richardson, J. P., Chan, C. H., Blanc, J., Saadi, M., & Macmillan, D. (2010). Exploring neoglycoprotein assembly through native chemical ligation using neoglycopeptide thioesters prepared via N→S acyl transfer. *Organic & Biomolecular Chemistry, 8*(6), 1351–1360.

Rosengren, K. J., Daly, N. L., Plan, M. R., Waine, C., & Craik, D. J. (2003). Twists, knots, and rings in proteins. Structural definition of the cyclotide framework. *Journal of Biological Chemistry, 278*(10), 8606–8616.

Saether, O., Craik, D. J., Campbell, I. D., Sletten, K., Juul, J., & Norman, D. G. (1995). Elucidation of the primary and three-dimensional structure of the uterotonic polypeptide kalata B1. *Biochemistry, 34*(13), 4147–4158.

Sancheti, H., & Camarero, J. A. (2009). "Splicing up" drug discovery. Cell-based expression and screening of genetically-encoded libraries of backbone-cyclized polypeptides. *Advanced Drug Delivery Reviews, 61*(11), 908–917.

Sando, L., Henriques, S. T., Foley, F., Simonsen, S. M., Daly, N. L., Hall, K. N., et al. (2011). A synthetic mirror image of kalata B1 reveals that cyclotide activity is independent of a protein receptor. *ChemBioChem, 12*(16), 2456–2462.

Saska, I., Gillon, A. D., Hatsugai, N., Dietzgen, R. G., Hara-Nishimura, I., Anderson, M. A., et al. (2007). An asparaginyl endopeptidase mediates in vivo protein backbone cyclization. *Journal of Biological Chemistry, 282*(40), 29721–29728.

Scott, C. P., Abel-Santos, E., Jones, A. D., & Benkovic, S. J. (2001). Structural requirements for the biosynthesis of backbone cyclic peptide libraries. *Chemical Biology, 8*(8), 801–815.

Scott, C. P., Abel-Santos, E., Wall, M., Wahnon, D., & Benkovic, S. J. (1999). Production of cyclic peptides and proteins in vivo. *Proceedings of the National Academy of Sciences of the United States of America, 96*(24), 13638–13643.

Shin, Y., Winans, K. A., Backes, B. J., Kent, S. B. H., Ellman, J. A., & Bertozzi, C. R. (1999). Fmoc-based synthesis of peptide-^{a}thioesters: application to the total chemical synthesis of a glycoprotein by native chemical ligation. *Journal of the American Chemical Society, 121*, 11684–11689.

Simonsen, S. M., Sando, L., Rosengren, K. J., Wang, C. K., Colgrave, M. L., Daly, N. L., et al. (2008). Alanine scanning mutagenesis of the prototypic cyclotide reveals a cluster of residues essential for bioactivity. *Journal of Biological Chemistry, 283*(15), 9805–9813.

Stanger, K., Maurer, T., Kaluarachchi, H., Coons, M., Franke, Y., & Hannoush, R. N. (2014). Backbone cyclization of a recombinant cystine-knot peptide by engineered Sortase A. *FEBS Letters, 588*(23), 4487–4496.

Tam, J. P., & Lu, Y. A. (1997). Synthesis of large cyclic cystine-knot peptide by orthogonal coupling strategy using unprotected peptide precursor. *Tetrathedron Letters, 38*(32), 5599–5602.

Tam, J. P., & Lu, Y. A. (1998). A biomimetic strategy in the synthesis and fragmentation of a cyclic protein. *Protein Science,* 7(7), 1583–1592.

Tam, J. P., Lu, Y. A., Yang, J. L., & Chiu, K. W. (1999). An unusual structural motif of antimicrobial peptides containing end-to-end macrocycle and cystine-knot disulfides. *Proceedings of the National Academy of Sciences of the United States of America, 96*(16), 8913–8918.

Tavassoli, A., & Benkovic, S. J. (2007). Split-intein mediated circular ligation used in the synthesis of cyclic peptide libraries in *E. coli*. *Nature Protocols, 2*(5), 1126–1133.

Thongyoo, P., Bonomelli, C., Leatherbarrow, R. J., & Tate, E. W. (2009). Potent inhibitors of beta-tryptase and human leukocyte elastase based on the MCoTI-II scaffold. *Journal of Medicinal Chemistry, 52*(20), 6197–6200.

Thongyoo, P., Roque-Rosell, N., Leatherbarrow, R. J., & Tate, E. W. (2008). Chemical and biomimetic total syntheses of natural and engineered MCoTI cyclotides. *Organic & Biomolecular Chemistry, 6*(8), 1462–1470.

Thongyoo, P., Tate, E. W., & Leatherbarrow, R. J. (2006). Total synthesis of the macrocyclic cysteine knot microprotein MCoTI-II. *Chemical Communications (Cambridge), 27*, 2848–2850.

Tsukiji, S., & Nagamune, T. (2009). Sortase-mediated ligation: a gift from gram-positive bacteria to protein engineering. *ChemBioChem, 10*(5), 787–798.

Werle, M., Kafedjiiski, K., Kolmar, H., & Bernkop-Schnurch, A. (2007). Evaluation and improvement of the properties of the novel cystine-knot microprotein McoEeTI for oral administration. *International Journal of Pharmaceutics, 332*(1–2), 72–79.

Wong, C. T., Rowlands, D. K., Wong, C. H., Lo, T. W., Nguyen, G. K., Li, H. Y., et al. (2012). Orally active peptidic bradykinin B1 receptor antagonists engineered from a cyclotide scaffold for inflammatory pain treatment. *Angewandte Chemie, International Edition in English, 51*(23), 5620–5624.

Young, T. S., Young, D. D., Ahmad, I., Louis, J. M., Benkovic, S. J., & Schultz, P. G. (2011). Evolution of cyclic peptide protease inhibitors. *Proceedings of the National Academy of Sciences of the United States of America, 108*(27), 11052–11056.

Zettler, J., Schutz, V., & Mootz, H. D. (2009). The naturally split Npu DnaE intein exhibits an extraordinarily high rate in the protein trans-splicing reaction. *FEBS Letters, 583*(5), 909–914.

Zhang, J., Hua, Z., Huang, Z., Chen, Q., Long, Q., Craik, D. J., et al. (2014). Two Blast-independent tools, CyPerl and CyExcel, for harvesting hundreds of novel cyclotides and analogues from plant genomes and protein databases. *Planta*. http://dx.doi.org/10.1007/s00425-014-2229-5.

CHAPTER TEN

Cyclotides in a Biotechnological Context: Opportunities and Challenges

Mark A. Jackson, Edward K. Gilding[1]
Institute for Molecular Bioscience, The University of Queensland, Brisbane, QLD, Australia
[1]Corresponding author: E-mail: e.gilding@imb.uq.edu.au

Contents

Advances in Botanical Research, Volume 76
ISSN 0065-2296
http://dx.doi.org/10.1016/bs.abr.2015.09.010

Abstract

Plant biotechnology offers us the ability to augment and improve upon biological systems to produce value added crops, medicines, or specialized products. The application of cyclotides, either naturally occurring or engineered, requires their production at industrial-scale yields. Here we discuss the most relevant applications of cyclotides within a plant biotechnological context and present biological, regulatory, and market factors that should be considered when embarking on a cyclotide production project *in planta*. It is evident that only a select few plants appear to have evolved to produce cyclotides, as introduced in Chapters 1–4, where they putatively function as highly abundant defense molecules. Their high abundance and stability suggest that they are of primary importance to the species that contain them. Plant biotechnology could be used to harness this functionality and to engineer novel designer cyclotides for expression in plants. We finish with a brief discussion of what the future may hold for cyclotide-based biotechnology.

1. CYCLOTIDES IN A BIOTECHNOLOGICAL CONTEXT

1.1 Insecticidal Applications

Plants are continually evolving and adapting defense response to pest and disease. For crop species, breeding varieties that harbor defense traits to resist infestations is essential, as is the tailored application of pesticides. Associated with these approaches, however, is the constant threat of insects developing resistance and environmental concerns of prolonged pesticide use. Intensive research into developing novel, environmentally friendly alternatives is underway, with cyclotides presenting a promising solution.

In nature, cyclotides are thought to accumulate in plant tissues to deter herbivore feeding. Evidence for this was first presented in a pioneering study by Jennings, West, Waine, Craik, & Anderson (2001), who tested this hypothesis by feeding *Helicoverpa punctigera* larvae a diet containing purified kalata B1, the prototypic cyclotide from *Oldenlandia affinis*. After 16 days of feeding, the surviving larvae reared on the peptide-containing media was found to weigh on average only 3.3 mg (±1.1 mg) compared to controls that had reached 284 mg (±96.1 mg). This phenomenon appears not to be kalata B1 specific, with similar bioactivity found with the cyclotide termed cter M found within the evolutionarily distant *Clitoria ternatea* of the Fabaceae family (Poth, Colgrave, Lyons, Daly, & Craik, 2011; Poth, Colgrave, Philip, et al., 2011).

Insights into the insecticidal mode of action of kalata B1 were revealed using electron microscopy techniques (Barbeta, Marshall, Gillon, Craik,

& Anderson, 2008). Inspection of larvae that had been fed kalata B1 in their diet revealed ruptured epithelium cells of the midgut, a response similar to that observed when larvae ingest the *Bacillus thuringiensis* (Bt) delta-endotoxin, a commercially used insecticide (Pardo-Lopez, Soberon, & Bravo, 2013). Unlike Bt toxins, however, which assemble at membrane bound receptors to cause membrane pore formation, a receptor for kalata B1 has never been identified. Rather, membrane-binding studies point to a unique receptor-independent cyclotide—lipid interaction driven only by hydrophobic forces (Henriques & Craik, 2012; Wang, Wacklin, & Craik, 2012), with the critical residues required present within a hydrophobic patch of kalata B1 (Simonsen et al., 2008). Importantly, this receptor-independent mode of action is advantageous for the deployment of cyclotides as insecticides as it would limit the development of insect resistance.

1.1.1 Cyclotides as Insecticidal Sprays

Overuse and reliance on insecticides has led to concerns for the environment, human health, and for the continued development of insect resistance. Cyclotides, being natural, plant-derived, and already present in the ecosystem represent an ideal solution. This may be through their development as an insecticidal spray or through the genetic manipulation of crop species to express insecticidal proteins *in planta*. Cost-effective production, formulation, and product stability are vital qualities for insecticidal spray products. In cyclotide-producing plants, cyclotides can accumulate to very high yields (up to 2 mg/g FW), often in nutrient poor growth conditions (Gran, 1973b). At these levels and with simple solvent-based extraction procedures production costs could be kept to a minimum. Furthermore, cyclotides have been shown to remain intact even after boiling, enzymatic and acid treatments (Colgrave & Craik, 2004), thus storage and stability at room temperature is feasible.

Estimates of cyclotide diversity just within the Rubiaceae are estimated to be in the tens of thousands, with these cyclotide permutations possibly representing uniquely active molecules (Gruber et al., 2008). This is intriguing and may provide for synergistic and broadening of insecticide activity or even for variants with as yet unidentified modes of action. Further testing of these variants singly or in combination against a wide range of insects would be astute.

1.1.2 Cyclotides as Nematicides

Cyclotides have attracted interest for their deployment as nematicidal agents for use in livestock and agricultural applications. Initial functional studies of the kalata cyclotides revealed variants with anthelmintic activity against two economically important intestinal parasites that infect sheep, namely *Haemonchus contortus* and *Trichostrongylus colubriformis* (Colgrave, Kotze, Huang, et al., 2008). Further investigations revealed that this activity extends to include the medically important human and canine hookworms (Colgrave et al., 2009). More potent cyclotides have since been identified with cycloviolacin O2 from *Viola odorata* being the most potent, attributed to a high percentage of basic residues in this peptide (Colgrave, Kotze, Ireland, Wang, & Craik, 2008). Structure activity studies using a library of synthetic kalata B1 peptide variants indicated that kalata B1 contains an amenable face where residues substituted with a lysine result in improved anthelmintic activity (Huang, Colgrave, Clark, Kotze, & Craik, 2010). A triple lysine substitution within this face resulted in a 13- and 6-fold improvement in IC_{50} values relative to native kalata B1 against *H. contortus* and *T. colubriformis* nematodes respectively. This amenable face is separate from the previously defined hydrophobic and hydrophilic patches, which are known to be essential for membrane binding and structure stabilization respectively. In another study, Colgrave, Kotze, Huang, et al. (2008) set out to determine the mode of anthelmintic action and specifically whether ingestion of the peptide was a requirement for toxicity. To accomplish this, the researchers first tracked the location of a biotinylated kalata B1 variant in *H. contortus* by using a fluorescein-conjugated avidin probe. After incubation for 24 h, fluorescence was visible in all internal organs and around the periphery of the worm's surface. Ingestion was ruled out as a requirement for toxicity as worms that were treated with ivermectin, which prevents feeding, also displayed toxicity effects upon exposure to kalata B1. It was suggested that the membrane pore formation ability of kalata B1 might provide for the passage of kalata B1 into the internal organs of the worm.

In agriculture, parasitic nematodes are particularly devastating pests, which left untreated, have devastating effects on root phenotype and ultimately yield of susceptible crop species. Because root nematodes exist belowground, they are particularly difficult to treat once infection takes hold. For this reason, preplanting treatments and ground preparation controls are essential. For many decades, fumigation with broad-acting chemical fumigants has provided some control; however, many are now deemed unacceptable for environmental reasons. In many cyclotide-producing species, cyclotide variants are produced at high concentration in belowground tissue such as roots, which suggest that they may play a native role in inhibiting nematode infection. Although no published data are available, a recent patent application filed by Pioneer Hi-Bred identifies a number of cyclotides extracted from *C. ternatea* seed with potent activity against model *Caenorhabditis elegans* L1 larvae (Altier, Hu, Siehl, Wei, & Wu, 2014). Further research is required to determine if this potency extends to pathogenic root nematodes and if field application is feasible.

1.1.3 Genetic Engineering for Insect Resistance

Since the beginning of time, farmers have knowingly or unknowingly selected crop varieties that harbor genes for pest resistance and higher yields. Today large-scale breeding programs continue this tradition; however, with continued selection comes a narrowing of the gene pool available. Inevitably this will lead to crop varieties more susceptible to disease outbreaks. Fortunately, advances in plant biotechnology offer a solution through genetic engineering. Sometimes referred to as precision breeding, genetic engineering allows the insertion of a gene of interest from any source into a recipient plant's genome. In this way novel approaches to improving yield, nutrient acquisition and pest resistance can be approached, otherwise impossible to achieve through conventional plant breeding.

Cotton has been a major beneficiary of GM technology with 70% of the hectares planted for the worldwide cotton crop being transgenic as of 2013 (gmo-compass.org, 2013). In developed nations, this number is close to 100% with the major GM trait being insect resistance, induced through expression of a Bt toxin gene, naturally found in the soil bacterium *B. thuringiensis* (cottonaustralia.com.au, 2015). Importantly, for cotton this technology has significantly reduced the need for spraying of insecticides, a plus for the environment and farm gate income. Disturbingly, however, since the initial introduction of the Bt gene into cotton, examples of larvae displaying resistance have surfaced particularly in India. To counter this second- and third-generation Bt genes have been introduced (Tabashnik, Brevault, & Carriere, 2013). The need to keep ahead of resistance illustrates the requirement of the continued development of novel gene-encoded insecticides.

Thus far cyclotides have not been found in any major crop species of interest, the closest being *Petunia* hybrids which are cultivated for flowers (Poth et al., 2012). Therefore transfer of cyclotide precursor genes into cotton or other agricultural important species will require genetic transformation techniques. Although genetic transformation is feasible for most commercially important crops, this is not without difficulty, especially for selecting superior lines that provide reliable and sustained transgene expression. In addition, the precise mechanism of cyclotide biosynthesis *in planta* is not fully understood. What has been established is that in addition to the precursor cyclotide gene a set of accessory proteins is required for efficient peptide cyclization. Thus far attempts to transfer cyclotide biosynthesis capacity into nonnative producer plants such as tobacco or *Arabidopsis* have resulted only in inefficient and nonprecise cyclotide processing. Further research is required to understand the requirement for accessory proteins, no doubt present within native cyclotide-producing plants.

1.2 Pharmaceutical Applications

It has been estimated that the average cost of bringing a new pharmaceutical to the market is $2 billion US dollars (www.pfizer.com). Most leads fall over and the further down the development pipeline they are the more costly. It is thus imperative to tailor screening programs to weed out likely failures early. In this light, peptide-based drugs have traditionally been seen as unfavorable for progression due to their instability, lack of oral bioavailability and cost of manufacture. On the other hand, peptide therapeutics has enormous potential, if only these bottlenecks could be overcome. When compared to small molecule drugs, they provide for highly specific interactions with their targets resulting in fewer side effects. When compared to larger protein-based drugs they are generally cheaper to produce and less likely to induce any unwanted autoimmune response. Most peptide drugs currently on the market require intravenous administration, as oral routes require resistance to digestive enzymes and an ability to permeate through cellular membranes. Strategies to improve peptide stability and oral bioavailability are being actively sought and include the use of nonnatural amino acids, pegylation, conjugation to carriers, and backbone cyclization. Here we describe the use of plant cyclotide frameworks for stabilizing bioactive epitopes.

1.2.1 Cyclotide Grafts for Extracellular Targets

As described in Chapter 7, cyclotides are considered ideal scaffolds for which bioactive peptide sequences can be inserted for stabilization. In nature, many hundreds of cyclotides have been discovered each containing six conserved cysteine residues that flank variable backbone loop sequences. This large natural diversity of sequences provides confidence that the framework can accept the insertion of foreign peptide sequences. Additionally being small in size, grafting studies can be achieved at lab scale almost routinely using solid-phase peptide synthesis techniques.

Initial grafting studies have focused on developing peptides for interaction with extracellular targets found within human circulatory system. For these applications, peptides need not traverse the cellular membranes, simplifying peptide design.

1.2.1.1 Controlling Angiogenesis

Balancing the chemical signals that controls angiogenesis (blood vessel formation) is crucial for maintaining physiological homeostasis. In tumor tissue, this balance is lost with blood vessel formation essential for feeding tumor growth. Antiangiogenesis therapeutics are thus of considerable interest for the treatment of many human cancers with several now FDA approved (for review, see Chakrabarti, Barrow, Kanwar, Ramana, & Kanwar, 2014). Currently the most widely used is bevacizumab (Avastin®), a monoclonal antibody that specifically recognizes and attaches to the vascular endothelial growth factor (VEGF) protein, slowing angiogenesis progression. The development of peptides as antiangiogenesis agents has gained traction (Rosca et al., 2011) due to their capacity for highly specific inhibition. Although many promising peptides have been discovered, stability in vivo remains a concern. In an attempt to overcome this, a study by Gunasekera et al. (2008) reported the use of the kalata B1 cyclotide framework to stabilize an arginine-rich antiangiogenesis epitope that is an antagonist of VEGF. This epitope termed polyR (RRKRRR) was inserted into loops 2, 3, 5, and 6 of kalata B1 and tested in VEGF competition assays. The best performing peptide (loop 3 graft) showed significant inhibitory activity with an IC_{50} of 12 μM. Furthermore, this graft showed enhanced stability in serum over its linear counterpart.

For some human conditions, such as ischemic heart disease, rheumatoid arthritis, and in wound healing, vascularization is limited, and strategies to promote angiogenesis are warranted. In a study by Chan et al. (2011), several small peptides known to promote angiogenic activity, but by themselves lacked stability were grafted into the *Momordica cochinchinensis* trypsin inhibitor-II (MCoTI-II) and sunflower trypsin inhibitor (SFTI-1) cyclic peptide scaffolds. These novel grafted compounds showed remarkable ability to induce blood vessel growth and displayed higher stability in serum than linear peptide counterparts.

1.2.1.2 Antiobesity

The melanocortin-4 receptor gene (MCR4) plays a key role in energy homeostasis with misregulation leading to severe obesity. In mice models, stimulation of MCR4 leads to reduced food intake and reduced weight gain, while the opposite is true when MCR4 is inhibited (Samama, Rumennik, & Grippo, 2003). In humans, endogenous agonists of MCR4 include the adrenocorticotropic and α-, β-, and γ-melanocyte-stimulating hormones (MSH) with all sharing the tetrapeptide sequence (His-Phe-Arg-Trp). Several peptide analogues of this sequence have been tested in rodent feeding trials with some promising results obtained (Fani, Bak, Delhanty, van Rossum, & van den Akker, 2014). In one study, a single injection of a linear tetrapeptide compound reduced the overnight food intake by 24% when compared to controls (Ye et al., 2005). In another study, a library of 16 backbone cyclized pentapeptides containing analogues of the tetrapeptide sequence were prepared and tested for oral delivery and efficacy in a rat feeding trial (Hess et al., 2008). Over 12 days of oral delivery, one analogue demonstrated a significant reduction in body weight gain. Grafting into a cyclotide framework has more recently been investigated as a means to stabilize and potentially allow oral delivery of the (His-Phe-Arg-Trp) peptide. In the initial study, Eliasen et al. (2012) prepared four different grafts, each into loop 6 of the kalata B1 cyclotide and tested these for affinity to the MCR4 receptor. Excitingly one graft displayed a higher affinity to MCR4 than the native agonist α-MSH with only weak affinity detected to other MCR isoforms. Unfortunately in this case, however, higher affinity did not translate to higher functional activity with a 150-fold less potent EC_{50} calculated compared to that of α-MSH. Stability of the grafted peptide was, however, vastly improved with chymotrypsin cleavage assays showing complete degradation of α-MSH within 10 min compared with 80% of the grafted peptide remaining intact after 5 h of incubation.

1.2.1.3 Multiple Sclerosis

Multiple sclerosis is a debilitating and incurable autoimmune disease that affects the central nervous system. Most treatment options aim to suppress immune function, which although effective at reducing relapse brings many unwanted side effects. Research by Wang et al. (2014) set out to test if a cyclotide framework could be used to present epitopes of the myelin oligodendrocyte glycoprotein (MOG) known to possess immunoregulatory potential. Of the 17 grafts tested, 9 adopted a well-defined fold and 7 were chosen for testing in a mouse model of experimental autoimmune encephalomyelitis. Mice were vaccinated over a 3-day period, and disease progression was monitored. One of the grafts tested, termed MOG3, demonstrated a significant reduction in both the incidence and severity of disease.

Importantly, MOG3 was shown to be relatively stable in human serum, hydrochloric acid, and pancreatin, when compared to the linear epitope.

1.2.1.4 Pain

A further application of the cyclotide scaffold for grafting has been in the development of oral therapeutics for pain management. For this, Wong et al. (2012) used the kalata B1 scaffold to stabilize epitopes known to antagonize bradykinin, a potent inducer of pain. Two grafted cyclotides were prepared containing 8- and 9-amino-acid length exchanges within loop 6 of kalata B1. These grafts were shown to be highly stable in human serum with over 90% of grafted peptides remaining intact after 6 h of incubation. When tested in a mouse model of visceral pain, the best performing cyclic graft (ckb-kal) reduced the pain response (writhing action) by 49% and 42% respectively for peptide administered by injection (1 mg/kg) or orally (10 mg/kg). In contrast, reduction in pain response could only be detected with the linear control peptide if administered by injection. This research highlights the potential of the cyclotide scaffold to improve bioavailability upon oral delivery.

1.2.1.5 Immunosuppressants

Currently there are a number of immunosuppressant drugs on the market used for treating autoimmune diseases and for preventing organ rejection after surgery. Many, however, bring unwanted side effects, are costly to produce and difficult to administer given the often lifelong requirement for patients to comply. Newer, less toxic and cheaper alternatives are thus being actively sort. Somewhat surprisingly the prototypic cyclotide kalata B1 found in abundance within *O. affinis* leaf tissue was recently demonstrated to function in immune suppression (Gründemann, Koehbach, Huber, & Gruber, 2012). Structure activity studies and T cell receptor signaling analysis has revealed an active cyclotide face that is essential for immunosuppression and a mode of action that involves the downregulation of the autocrine growth factor interleukin 2 (Gründemann et al., 2013). These findings open up opportunities for the fine-tuning of cyclotides into therapeutic stable peptides for the treatment of autoimmune disease or as preventatives of organ rejection.

1.2.2 Cyclotide Grafts for Intracellular Targets

Intracellular therapeutic targets have long been the mainstay of small molecule drugs, which are small enough to easily diffuse through cellular membranes. In some instances, however, small molecule drugs can lack the specificity required for therapeutic benefit. In contrast, peptides or larger proteins can provide for highly specific interactions, but generally lack the ability for cell penetration. Strategies to overcome this limitation would represent a game changer for peptide therapeutics.

The cyclotides MCoTI-I and the related MCoTI-II from the trypsin inhibitor subfamily of cyclotides have recently garnered strong interest for their capacity for cell penetration (Contreras, Elnagar, Hamm-Alvarez, & Camarero, 2011). This feature when combined with serum stability and capacity for grafting opens up opportunities to target intracellular disease processes. In a pioneering example of this, Ji et al. (2013) used the MCoTI-I framework to deliver a grafted peptide that specifically inhibits intracellular oncogenic protein-mediated downregulation of the p53 tumor suppressor. Unlike numerous other small peptides that have shown promise, the grafted peptides prepared in this study provide a sustained effect. When tested in an in vivo tumor model, the cyclotide grafts significantly reduced tumor growth coinciding with a reactivation of the tumor suppressor p53 pathway.

With this success aside their remains much room for improvement of the internalization efficiency of both McoTI-I and II. Recently, research by Huang, Chaousis, Cheneval, Craik, and Henriques (2015) addressed this by assembling and testing multiple MCoTI-II variants, which differed in distribution of basic residues, inclusion of additional cell-penetrating peptide sequences and overall positive charge. The reengineered peptides were conjugated with an Alexa fluor 488 molecule so that internalization efficiency into cells could be monitored. The best performing variant showed a 2.4-fold improvement in internalization putting it on par with the cell-penetrating ability of TAT, a peptide derived from the human immunodeficiency virus transactivator of transcription factor, which is considered the gold standard of cell-penetrating peptides.

1.3 Other Potential Areas of Exploration

1.3.1 Antifouling and Biofilm Formation

Cycloviolacins have been found to prevent marine fouling when applied to surfaces exposed to marine environments (Goransson, Sjogren, Svangard, Claeson, & Bohlin, 2004). Such activity is welcome as a means to reduce the costs of maintenance and removal of marine surfaces. The idea is that cyclotides with antifouling effects could be added to paints and other coatings to reduce the establishment of barnacles. Commercial interest to protect the use of two cycloviolacins in antifouling applications has been sought; however, the implementation of this technology remains elusive due to the large amounts of cyclotide needed (Bohlin, Sjogren, Claeson, Goransson, & Svangard, 2008; Goransson et al., 2004). In this case, large-scale heterologous production in engineered plants could offer a solution to the need for large amounts of cyclotide, thus illustrating the importance of plant biotechnology to aid in maturation of cyclotide-based technologies.

2. CYCLOTIDE PRODUCTION SYSTEMS

Plant biotechnology has been utilized to introduce various useful genic sequences to protect and increase crop yield or to produce value added products. Stable transformation of plant hosts with genes encoding desired peptides or enzymes is an option where a protocol for stable transformation exists. Where vertical transmission is not desirable or the expressed genes will need to be adapted quickly from year to year, transient expression is an alternative option. In the case of the latter, the advantage of speed and flexibility in using plants was demonstrated by the culture of virus-like particles (VLPs) in plants to be used as influenza vaccines that are effectively immunogenic (D'Aoust et al., 2008; Landry et al., 2010). This speed and flexibility in VLP production is readily applicable at the rapid and adaptable production of novel cyclotides in transient systems. Furthermore, transient expression mitigates the possibility of vertical transmission and therefore alleviates concerns over transgene escape. Additional factors to consider for the development of a plant-based cyclotide production system are the regulatory framework that oversees transgenic plant biotechnology and public acceptance of the technology. Typically, however, it is noted that medicines for human use produced by biotechnological pathways are more easily accepted by consumers (WHO, 2014). Cyclotides produced in plants for medicines should similarly be accepted by consumers.

2.1 Strategies and Goals of Cyclotide Production *In Planta*

Complete treatise on genetic transformation of economically important plants are available as books or reviews, and the reader is encouraged to seek these out ahead of planning a project involving the expression of cyclotides in host plants. For illustrative purposes, we will discuss use of *Nicotiana benthamiana* as a noncyclotide-producing host plant and *Petunia* × *hybrida* as an example of a transformable cyclotide-producing plant. Large-scale production of protein products in *N. benthamiana* has been demonstrated, and *P.* × *hybrida* produces relatively few cyclotides, thus minimizing issues with isolation of heterologously produced cyclotides (D'Aoust et al., 2008; Poth et al., 2012). Current dogma regarding cyclotide production is that it is essential for coexpression of a cyclization-competent artificially optimized biosynthetic enzyme (AEP) alongside the desired cyclotide precursor gene. This would be necessary to convert a nonproducing host plant into a cyclotide-containing plant (Figure 1). Below we suggest four different strategies to produce cyclotides in plants considering different end uses and taking into account minimization of biosafety risks and provisions for flexibility of the cyclotide production pipeline. It is worth mentioning that like all other schemes to express mammalian

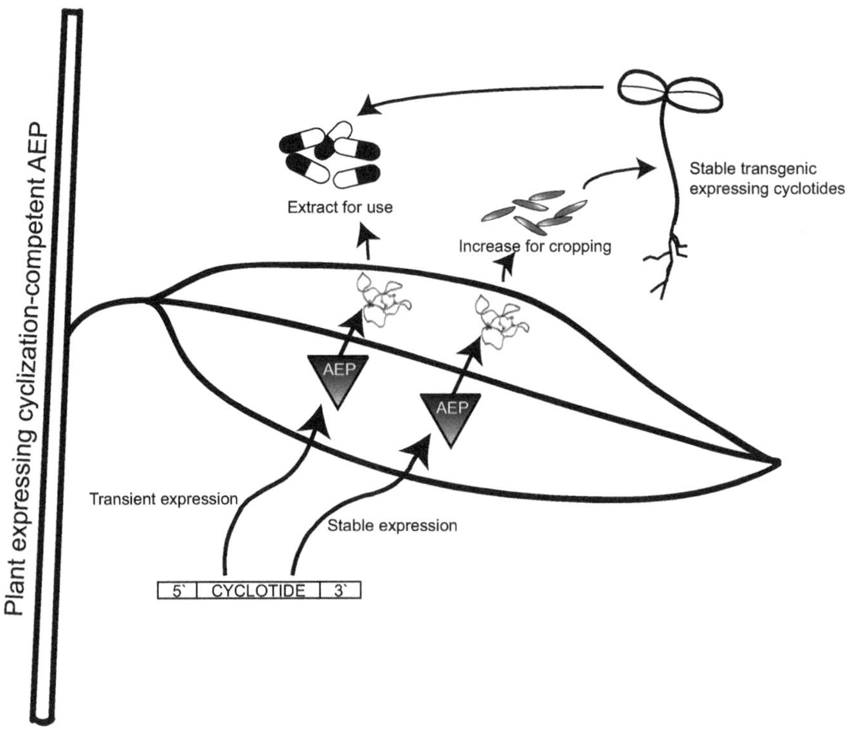

Figure 1 Schematic of the overall strategy to produce cyclotides efficiently in a plant. Assuming the plant is not a native cyclotide producer, the introduction of a cyclization-competent artificially optimized biosynthetic enzyme (AEP) in the host plant genome is first achieved. Subsequently, transient expression experiments may be used to rapidly test gene sequences encoding precursors in the plant system, or to rapidly produce peptides that must keep pace with evolving biological applications (e.g., antivirals). Stable transformation is an ultimate goal or alternative pathway to obtain a finished product, depicted as pills.

pharmaceuticals in plants, expression of cyclotide drugs in plants has the advantage of not being at risk of being a vector for mammal viruses and other mammalian pathogens (Ma, Drake, Chargelegue, Obregon, & Prada, 2005).

1. **Cyclotides for use as seasonal vaccines or as antagonist against rapidly evolving pathogens**: We propose that a combination of transient and stable transformation would be desirable to express pharmacologically important molecules against rapidly evolving targets (e.g., if cyclotide targets are viruses or other pathogens). In this scheme, host plants may be a natural cyclotide-producing species such as *P.* × *hybrida*, which carries an inbuilt capacity for peptide cyclization (Poth et al., 2012). Alternatively, the host plant may be a nonnatural cyclotide producer such as *N. benthamiana*, which carries advantages as a tried and tested system for transient expression of a wide range of recombinant products *in planta*. For the latter, an initial stable transformation would be required to insert an efficient cyclization-competent AEP gene before transient expression of cyclotide precursor genes of interest as the endogenous cyclotide-processing machinery is inadequate (Gillon et al., 2008).
2. **Cyclotides for field-grown crop protection**: Crop plants would be stably transformed with a cyclization-enhanced AEP together with precursor genes for protective cyclotides. Transgene expression would be controlled by use of tissue-specific or pathogen-inducible gene promoters (Xu, McElroy, Thornburg, & Wu, 1993). This application presents regulatory and biosafety issues depending on where the crop is intended to be grown and risk mitigation against gene flow from the crop to wild relatives is possible. See Section 1.1.3 above.
3. **Cyclotides for mammalian pharmacological targets**: Similarly to the second strategy above, we propose that stable transformation of host plants with an appropriate AEP and desired cyclotides be performed, but with minimization of biosafety risk through restricting cultivation of engineered cyclotide producer plants to glasshouses or growth chambers. Alternatively, plant cell suspension cultures or transgenic hairy root cultures could be developed which allows a controlled and sterile environment for therapeutic peptide production. Indeed the first FDA approved recombinant therapeutic produced from plants utilizes a carrot suspension cell production system (Grabowski, Golembo, & Shaaltiel, 2014).
4. **Altering cyclotide profiles of cyclotide-producing plants to favor those of therapeutic or agricultural value**: This strategy is useful if one or several cyclotides are identified as of value and that production

of these target genes can be enhanced. In this case, endogenous expression levels may be manipulated by selective hormone treatments and growth conditions. Attempts to alter cyclotide expression by induction in *O. affinis* resulted in no alteration in cyclotide expression levels (Mylne, Wang, van der Weerden, & Craik, 2010). Other cyclotide-producing plants may behave differently, however. Alternatively, stable transformation with the desired cyclotide-encoding genes driven by high-expression or inducible promoters is applicable. Downregulation of less desirable endogenous cyclotides may be accomplished through genome editing or RNAi approaches.

2.2 Expression Considerations

Rates of success, efficiency of production, and the time required to obtain plants producing a desired product is heavily dependent upon the nature of the host plant and the molecule or phenotype being hosted. Here, we give a broad overview of considerations that should be analyzed before embarking on a venture to produce cyclotides heterologously, and highlight past trials and successes in plant production of peptide products.

Irrespective of the expression platform chosen, plant-based therapeutic production should be conducted in a controlled manner. This is important, firstly to satisfy pharmaceutical regulatory bodies that demand uniform batch-to-batch consistency, and secondly to minimize biosafety hazards such as contamination of the food supply or outcrossing to weedy relatives. The preferred containment strategy is to limit expression to transient only or by induction with an agent that is not commonly encountered by plants. An example of the latter is ethanol-inducible promoters, where expression of the desired product can be restricted until a short period ahead of harvesting the plant biomass (Caddick et al., 1998). Other systems available include the dexamethasone-inducible glucocorticoid receptor system and the in-plant activation cassette (Dugdale et al., 2014; Schena, Lloyd, & Davis, 1991). Of particular interest has been the development of transient expression approaches that utilize leaf tissue and viral-based replication vectors which are delivered to plant cells using *Agrobacterium* (Lindbo, 2007). Variations in this technology have underpinned the formation of numerous plant biotechnology companies established to use plants as a vehicle for therapeutic protein production. Yield obtained are largely product dependent with proteolytic stability as a common determinant. The reporter protein GFP that is intrinsically stable has been reported to accumulate to an astonishing 3–5 mg of GFP per gram fresh weight (Lindbo, 2007). Importantly, these transient approaches eliminate the need for laborious stable transformations and production timeframes are short with product yields most optimal 4–6 days after initiation. One limitation of these systems is, however, host cell dependency of the viral replication machinery, with most commercialized expression systems optimized for function in *N. benthamiana*.

Stable transformation protocols have been established for many of the economically important plant species. Although an initial laborious and potentially time-consuming process, the establishment of high-yielding stable transgenic lines is advantageous as it ensures a constant supply of recombinant protein with a capacity to scale up as needs require. Many different plant species have been tested as production hosts, including high leaf biomass crops such as lettuce to high seed-yielding crops such as rice and wheat (Chen & Lai, 2014; Stöger et al., 2000). Each tissue type targeted and plant species have unique advantages and disadvantages that need to be assessed on a case-by-case basis. Tobacco, for example, is advantageous in that it produces a large biomass (tens of tons per hectare per year), can be grown in a controlled environment and is not part of the food chain (Sheen, 1983). On the other hand, tobacco leaf is low in protein content and contains a variety of endogenous proteases that may limit product stability and yield. Furthermore tobacco alkaloids may ultimately impact negatively on regulatory acceptance. In contrast to leaf, plant seed has a natural function to store endogenous proteins required for later germination. By hijacking this process various recombinant proteins have been produced and remain effective in seed for some time (Lamphear et al., 2002; Schünmann, Coia, & Waterhouse, 2002).

Strategies to prevent unwanted proteolysis of recombinant therapeutics include the coexpression of suitable protease inhibitors, targeted downregulation of endogenous proteases and directed the accumulation of protein into subcellular organelles devoid of harmful proteases. Subcellular targeting is also important where posttranslational modifications are required or when the recombinant protein is toxic if allowed to accumulate within the cells' cytoplasm. For cyclotide production, targeting the precursor protein through the endomembrane system is key to the successful maturation of cyclotides arising from the precursor as it is observed that every full-length precursor carries a predicted signal peptide. It is presently unknown if there are regulatory sequences within precursors that may work best with some hosts and not others, or if certain precursors target nascent cyclotides to discrete vacuole types where they interact with specific processing enzymes. By virtue of passing through the ER, cyclotide precursors occupy the same compartments as protein disulfide isomerases (PDIs) that are theorized to catalyze the formation of the cyclotide's disulfide bonds (Gruber et al., 2007). Gruber et al. tested the ability of recombinant *O. affinis* PDI to catalyze disulfide bond formation and act as a chaperone in vitro using heterologous assays. Definitive proof of the roles one or more PDIs may play in cyclotide maturation in vivo remains to be produced and could be addressed by RNAi or mutant analysis. Given the ability of some cyclic cyclotide formation in noncyclotide-producing plants and the observation that disulfide bond formation is not required for cyclization of precursors in vitro, it is likely that most host plants will contain adequate oxidative and chaperonin activity (Nguyen et al., 2014).

Cyclotides for use in plant protection applications should be ideally geared for expression only within tissues prone to pest infestation. This would limit any fitness loss that is sometimes seen as a result of constitutive expression of foreign peptides. This could be implemented similarly to that of antimicrobial peptides (AMPs), where expression is driven by pathogen-activated promoter sequences such as the wound inducible-3 (win3) promoter from poplar (Hollick & Gordon, 1993, 1995). Other promoters similar to win3 may be mined from existing microarray and short-read expression profile data from plant material challenged with pathogen types of interest. Interestingly, in *O. affinis*, cyclotides are constitutively expressed at very high levels with no evidence of induction by various signals or environmental variables (Mylne et al., 2010). Thus, these observations indicate that high levels of cyclotide expression may not be detrimental to plant fitness, unlike AMPs, and that cyclotides may be involved in a plant's innate immunity against disease. It should be noted, however, that cyclotide-producing plants might have also evolved mechanisms to tolerate this level of accumulation. Unintended toxicity of heterologously expressed cyclotides has not been reported in literature to date; however, given their membrane disrupting potential it is conceivable that certain compartments and plant species may not tolerate cyclotides well (Schöpke, Hasan, Kraft, Otto, & Hiller, 1993).

Research into the efficiency of various plant AEPs to cyclize cyclotide precursors in nonnative hosts has revealed two broad findings: plant AEPs have some measure of cyclization ability regardless of origin, and that some plants are better than others at efficient maturation of cyclotides (Gillon et al., 2008). The inference of these findings is that specific cyclization-competent AEPs may be the key for developing cyclotide therapeutic or plant protection platforms in a variety of host species. Detailed mechanistic studies between cyclization-efficient and nonefficient AEPs would be informative.

2.2.1 Plant-Based Pharma—A Short History

Vaccines based on viral coat proteins or bacterial proteins have been developed and tested for efficacy since 1992 (Mason, Lam, & Arntzen, 1992). The first plant-produced vaccine, effective against the Norwalk virus, measured through antibody production in humans, was reported in the landmark study by Tacket et al. (2000). While immunogenicity against the target was observed in 95% of trial participants after a single exposure to the transgenic potato making the immunogenic protein, it was noted that serum antibody production increased only modestly. Expression of the B subunit of cholera toxin by Arakawa, Chong, & Langridge (1998) in potato is an example of a bacterial vaccination target that showed some effect in laboratory mice challenged with the toxin. These early studies showed that further work would be required to create an effective vaccine in plants.

Subsequent research into the production of VLPs against influenza demonstrate that transient expression of influenza VLPs could be completed within 3 weeks of identification of new pandemic influenza strains (Landry et al., 2010). Landry et al. also propose that the short lead time needed to respond to emerging disease threats when using a plant-based transient expression system outperforms the currently used egg-based influenza vaccine production system. The relevance of these VLP production examples to cyclotide production is that transient expression systems of producing cyclotide-based medicines could also benefit from the option of dynamic flexibility in responding to emerging pathogen resistance while offering plug-and-play simplicity.

In recent years, the use of plants as production vehicles for therapeutics has gained traction. In May of 2012, the first plant expression-based drug termed Elelyso (taliglucerase alfa) was approved by the FDA for use in enzyme replacement therapy for Gaucher's disease. To accomplish this, the company responsible Protalix Biotherapeutics developed a platform technology that enabled the active ingredient to be produced in either transgenic carrot or tobacco cells. Start-up infrastructure costs was minimal with the company using disposable bioreactor bags for plant cell growth. Although Elelyso is currently only approved for intravenous dosing, the company is in the process of developing an orally available form as well as a number of other plant-expressed therapeutics that are in different stages of clinical trials (www.protalix.com).

Other plant biotechnology companies are focusing their effort on producing and storing therapeutic proteins in plant seed, which provides an environment where high yields of protein can be accommodated, and can remain stable for a number of years. A number of biotechnology companies have formed to work in this space, many using different plant hosts for varying reasons. Ventria Bioscience has had most success using rice as a production platform with two products currently in clinical trials for the treatment of HIV-associated chronic inflammation and inflammatory bowel disease respectively (www.ventria.com). Other companies such as Soymeds are targeting soybean with an aim to deliver therapeutics as soy milk (www.soymeds.net). With continued improvements in plant-based expression technology and an increase in consumer confidence, no doubt this field of plant biofactory production will continue to attract commercial and academic interest.

3. MARKET ANALYSIS OF CYCLOTIDES IN PLANT BIOTECHNOLOGY

As introduced earlier in this chapter, there are two applications for plant-produced cyclotides. Broadly, these are medicines and pesticides. Here, we discuss some considerations before pioneering work into cyclotide production in plant systems for those application areas.

3.1 Cyclotides for Medicines

Cyclotides as medicines have promise in that the cyclotide scaffold offers a means to host and preserve peptide drug epitopes within the human body. Whereas, linear and nonstabilized peptides may have a half-life of minutes in the bloodstream, cyclotides are known to persist for hours in serum to potentially give an effective therapeutic period comparable to small molecule-based pharmaceuticals (Chan et al., 2011; Gunasekera et al., 2008; Werle & Bernkop-Schnurch, 2006). Although no cyclotide-based drug has been commercialized to date, engineered cyclotide-based drugs have been developed and tested in laboratory settings. Real-world use of kalata tea as an uterotonic along with the isolation of vasopressin—receptor agonist activity due to kalata B7 within kalata tea illustrates the ease of preparation and utility of such drugs (Gran, 1973a; Koehbach et al., 2013).

Market risks with cyclotide-based medicines are something to be considered before embarking on a program to develop a cyclotide-based drug. The strong headwinds that exist in conventional drug-development pipelines affect peptide-based drugs as well. The big advantage that mitigates the common risk of peptide-based drugs as being short half-life pharmaceuticals lies within the stable cyclotide scaffold. Use of plants to produce pharmaceutically relevant molecules adds a layer of complexity, but not necessarily risk. The multitude of transformable plant species and flexibility of gene technology means that programs can easily find a "best-fit" expression system. While research into grafting drug targets into cyclotides and expressing them in plants has been underway in several labs for about 13 years, progress and time to maturation of the technology are comparable to the production of vaccines in plants. Plant-based production of antigens that were to form the basis of future vaccines began in 1995, with Phase II clinical trials showing effective immune responses in humans just being completed in 2010 (Landry et al., 2010). Thus, it is plausible that as of 2015, we are in the second half of the journey from drug concept to a fully vetted cyclotide pharmaceutical product produced in plants. To summarize, risks to cyclotide-based plant-produced pharmaceuticals are similar to those faced by the community that is working toward plant-based vaccines. We foresee further development of cyclotide-based pharmaceuticals, particularly with recent developments in defining the *in planta* cyclization mechanism.

3.2 Cyclotides as Insecticides or Cyclotides in Plant Protection

Du Pont was first to enclose the use of a cyclotide as an insecticide within a patent (WO2005118628-A3) that was filed in 2005. In the case of the Du Pont patent, the cyclotide sequences that were protected are from the Violaceae and included cycloviolacin O13 and O17 (Rafael, Lu, McCutchen & Clement, 2006). More recently, two patent applications filed 2 months apart have come to light. The first of these two patent applications was filed by Innovate Ag (WO2012113017-A1) for the use of *C. ternatea* ethanolic extracts as insecticides and modifiers of insect behavior for use against lepidopteran larval infestations (Mensah, 2012). This patent is the first move into commercialization of cyclotide-based insecticides applied exogenously to crops. The latter patent application (US 8735147 B2) encloses the use of cyclotides from *C. ternatea* to control nematodes in plants by heterologous expression of these peptides in crop plants (Altier et al., 2014). In addition to these patent applications, laboratory assays have already shown the insecticidal activity of kalata B2 and cter M against *Helicoverpa*, highlighting their utility in crop protection (Jennings et al., 2001; Poth, Colgrave, Lyons, et al., 2011). These patent documents highlight that intellectual property surrounding cyclotides in plant protection has commercial interest and investments into discoveries showing novel use of cyclotides will need to be assessed against existing intellectual property claims.

Biological risks associated with utilizing cyclotides as modulators of plant protection are multifold, but offer the advantage of reducing small molecule pesticide use that is known to have various impacts on the environment. In the case of cyclotides as insecticides, there is some concern about endogenous expression of these molecules to protect plants and consumption of parts containing these cyclotides. While the well-touted stability of the cyclotide scaffold makes them attractive for further exploration, it also opens up the possibility that these molecules will persist in environments outside of their desired distribution. Release of cyclotides into the environment from agriculture also requires further assessment in light of a report that kalata B1, B2, and cycloviolacin O2 are toxic against algae, duckweed, lettuce, and bacteria in laboratory conditions (Ovesen, Brandt, et al., 2011). Cyclotide sorption to naturally occurring mineral species was significant, suggesting mobility could be controlled (Ovesen, Nielsen, & Hansen, 2011). How cyclotides are handled and degraded in both producer plants and the environment requires further study.

Finally, we have the risks associated with public acceptance of transgenic crops, and in particular, crops that produce natural pesticides entering the food supply chain. Recently it was shown that sweet potato (*Ipomoea batatas*) is effectively a transgenic crop having gained either at least one fragment encoding multiple ORFs around the time of domestication (Kyndt et al., 2015). Kyndt et al. suggest that this finding challenges issues with consumer perception of transgenic foods considering sweet potato has been cultivated and consumed for generations by many cultures. Corn is a major edible that has also hosted insecticidal proteins as protection against lepidopteran pests. The endogenous production of Bt toxin in maize has led to much debate around policy and implementation of regulations restricting the cultivation of such crops. These restrictions are regardless of evidence that Bt maize is safe and being cultivated across an area exceeding a million hectares combined over 17 years (Romeis, McLean, & Shelton, 2013).

3.3 Advantages of Cyclotide Production in Plants

As introduced above, the production of cyclotides *in planta* is expected to be a relatively economical method to produce large quantities of these special molecules. In a basic research setting, it may cost hundreds to over a 1000 US dollars to synthesize a quantity of a cyclotide in the laboratory using fluorenylmethyloxycarbonyl chloride (FMOC) or tert-butyloxycarbonyl (BOC) solid-phase peptide syntheses. Cyclotides with the bracelet topology have proven to be very difficult to fold into a cystine knot in vitro, thus adding cost due to low yields and unwelcome technical difficulty (Tam & Lu, 1998). Thus, production in plants is the more attractive alternative to produce quantities of cyclotides at a fraction of the cost.

In addition, a plant expression platform would be essentially plug-and-play in nature, thus allowing production pipelines to be adaptable to consumer needs. While it might take several months to a year to produce a plant crop that produces cyclotides of choice, it is expected that the repeatability of the method that has power. The current aim of those developing plant production systems is to ensure that a plant capable of producing cyclotides will almost always produce that cyclotide despite minor engineered variations in sequence or environmental conditions.

3.4 Consumer Market Value of Cyclotides

It is difficult to assess the true value of a technology before it has been implemented; however, we discuss the various applications and global costs/trade in similar existing assets that current fulfill such applications.

Per capita expenditure on pharmaceuticals varies from country to country, with 2012 values ranging from 70 USD in Mexico to 1010 USD in the USA (OECD, 2014). These values are not adjusted for relative cost to per capita income or cost of living, thus the economic burden upon individuals of pharmaceutical expenditure is not captured in these numbers. In the USA, this equates to a total expenditure of approximately 316 billion USD for 2012. Clearly, this alone is a sizeable market. The cost of production *in planta*, being comparatively lower than chemical synthesis in a laboratory, also means that the economic burden of these pharmaceuticals will be lessened in countries sensitive to increases in medical costs.

Agricultural applications of cyclotides could possibly include a portion of the market value of pesticides. The pesticide market was valued at approximately 39 billion USD in 2007, and is expected to be worth upward of 75 billion USD in 2019 (Grube, Donaldson, Kiely, & Wu, 2011; PRNewswire, 2014). These data illustrate that the pesticide market is one of growth, creating an environment ripe for innovation and the gain of market share by new technologies.

It is hoped by those performing cyclotide research that cyclotides will earn a place among pharmaceutical or pesticide industries.

4. FUTURE PROSPECTS FOR CYCLOTIDES AND PLANT BIOTECHNOLOGY

4.1 Cyclotides for the Developing World

It is an unfortunate reality that the people who most often need crucial medicines are those who just can not afford it. Cyclotide-based therapeutics could potentially fill this need due to capacity for cost-effective plant-based production and cheaper storage cost (e.g., no refrigeration required). Numerous studies have identified peptides that specifically interfere with progression of diseases prevalent in the developing world (Kanodia et al., 2014; Lhouvum, Bhuyar, & Trivedi, 2015). Unfortunately for most, a lack of oral bioavailability and serum stability means a requirement for dosing by injection and refrigerated storage. If these peptides could be cyclized within a cyclotide framework and expressed within local crops, then this would allow the end users access and control of life-saving medicine. Unfortunately, funding opportunities for this type of research are limited with philanthropist being the largest contributor.

The insecticidal nature of cyclotides could also be harnessed in developing nations to increase crop productivity. In the simplest case, cyclotide-producing crops could be grown in paddocks, harvested, and applied as ethanolic extracts to act as potent insecticides. Although not tested, cyclotide variants could also play a role in preventing malaria by providing long-lasting antimosquito activity once coating onto mosquito nets. Currently pyrethroid insecticides are used for this purpose as they pose low risk to human health and are relatively stable; however, the rise of pyrethroid resistance is limiting effectiveness (Lindblade et al., 2015).

4.2 Genome Editing and Cyclotide Production

In recent years, there has been a burst of new technologies that allow precise genome editing, including targeted gene knockouts or site-specific transgene integration. The prerequisite for this has been the development of designer nucleases that allow for site-specific double-stranded genomic breaks (DSBs) (for a recent review see Osakabe & Osakabe, 2015). Once detected DSBs must be quickly repaired which occurs commonly by the nonhomologous end-joining (NHEJ) pathway or less commonly through homologous recombination (HR). NHEJ repair is particularly error prone, which often results in small insertions or deletions, which if occurring within coding sequence may disrupt the reading frame, effectively knocking out gene function. On the contrary, if left and right flanking homologous sequence to the DSB is provided repair may occur via the more precise HR. In this way, user-directed nucleotide changes from single bases to whole genes can be introduced into the plant genome.

Two applications of genome-editing technology for cyclotide biotechnology are envisaged. First, endogenously encoded AEP or other interfering protease genes from a target plant may be knocked out. In this way, inefficient cyclization by maladapted endogenous AEPs can be removed, thus reducing competition of the substrate for preferred cyclization AEP(s). *Arabidopsis* and *N. benthamiana* both misprocess kalata B1 precursors and *Arabidopsis* mutants lacking expression of all four AEPs encoded in the *Arabidopsis* genome are not lethal (Gillon et al., 2008; Kuroyanagi et al., 2005). Together this suggests that we may edit, by NHEJ, the protease population expressed in a plant to suit our needs. Second, we may target an easily transformable plant (i.e., *Petunia*) that naturally produces cyclotides to express edited versions of cyclotides in their native framework by HR. In this second strategy, the modified cyclotide gene will be regulated in an identical fashion to the wild-type cyclotide gene in theory.

4.3 Artificially Optimized Biosynthetic Enzymes

The apparent convergent evolution of cyclotides, the varied structure of cyclotide precursors, and the recruitment of AEPs to cyclize nascent cyclotides suggest that there may be plasticity within the cyclotide—AEP system. Improvements could be achieved by tailoring AEPs to cyclize substrates of divergent sequence. The technical input for directed evolution of AEPs may be significant, as only a handful of AEPs have been successfully expressed recombinantly (Bernath-Levin et al., 2015; Zhao et al., 2014). Additionally, a high-throughput method to screen for transpeptidation of substrates would be essential. Theoretically, a Förster resonance energy transfer (FRET)-based assay would be ideal. In this assay, a pair of substrates would be utilized, one with a FRET fluorophore pair and another labeled with a quenching moiety. Upon AEP-mediated transpeptidation, the joining of the FRET excitation fluorophore to the quenching label would result in a loss of a FRET signal and an increase of straight emission fluorophore signal. Such an assay, carried out with a cell sorter, could operate as a high-throughput screen to identify AEP clones that have qualities of interest.

ACKNOWLEDGMENTS

Work in our laboratory on cyclotides is supported by a grant from the Australian Research Council (DP20121234).

REFERENCES

Altier, D. J., Hu, X., Siehl, D. L., Wei, J. Z., & Wu, G. (2014). *Polynucleotides encoding cyclotides having nematocidal activity*. Google Patents.

Arakawa, T., Chong, D. K. X., & Langridge, W. H. R. (1998). Efficacy of a food plant-based oral cholera toxin B subunit vaccine. *Nature Biotechnology, 16*(3), 292–297.

Barbeta, B. L., Marshall, A. T., Gillon, A. D., Craik, D. J., & Anderson, M. A. (2008). Plant cyclotides disrupt epithelial cells in the midgut of lepidopteran larvae. *Proceedings of the National Academy of Sciences of the United States of America, 105*(4), 1221–1225. http://dx.doi.org/10.1073/pnas.0710338104.

Bernath-Levin, K., Nelson, C., Elliott, A. G., Jayasena, A. S., Millar, A. H., Craik, D. J., et al. (2015). Peptide macrocyclization by a bifunctional endoprotease. *Chemistry & Biology, 22*(5), 571–582. http://dx.doi.org/10.1016/j.chembiol.2015.04.010.

Bohlin, L., Sjogren, M., Claeson, P., Goransson, U., & Svangard, E. (2008). *On-growth inhibiting compounds*. Google Patents.

Caddick, M. X., Greenland, A. J., Jepson, I., Krause, K. P., Qu, N., Riddell, K. V., et al. (1998). An ethanol inducible gene switch for plants used to manipulate carbon metabolism. *Nature Biotechnology, 16*(2), 177–180. http://dx.doi.org/10.1038/nbt0298-177.

Chakrabarti, S., Barrow, C. J., Kanwar, R. K., Ramana, V., & Kanwar, J. R. (2014). Current protein-based anti-angiogenic therapeutics. *Mini-Reviews in Medicinal Chemistry, 14*(3), 291–312.

Chan, L. Y., Gunasekera, S., Henriques, S. T., Worth, N. F., Le, S. J., Clark, R. J., et al. (2011). Engineering pro-angiogenic peptides using stable, disulfide-rich cyclic scaffolds. *Blood, 118*(25), 6709–6717. http://dx.doi.org/10.1182/blood-2011-06-359141.

Chen, Q., & Lai, H. (2014). Gene delivery into plant cells for recombinant protein production. *BioMed Research International, 2015*.

Colgrave, M. L., & Craik, D. J. (2004). Thermal, chemical, and enzymatic stability of the cyclotide kalata B1: the importance of the cyclic cystine knot. *Biochemistry, 43*(20), 5965–5975. http://dx.doi.org/10.1021/bi049711q.

Colgrave, M. L., Kotze, A. C., Huang, Y. H., O'Grady, J., Simonsen, S. M., & Craik, D. J. (2008). Cyclotides: natural, circular plant peptides that possess significant activity against gastrointestinal nematode parasites of sheep. *Biochemistry, 47*(20), 5581–5589. http://dx.doi.org/10.1021/bi800223y.

Colgrave, M. L., Kotze, A. C., Ireland, D. C., Wang, C. K., & Craik, D. J. (2008). The anthelmintic activity of the cyclotides: natural variants with enhanced activity. *Chembiochem, 9*(12), 1939–1945. http://dx.doi.org/10.1002/cbic.200800174.

Colgrave, M. L., Kotze, A. C., Kopp, S., McCarthy, J. S., Coleman, G. T., & Craik, D. J. (2009). Anthelmintic activity of cyclotides: in vitro studies with canine and human hookworms. *Acta Tropica, 109*(2), 163–166. http://dx.doi.org/10.1016/j.actatropica.2008.11.003.

Contreras, J., Elnagar, A. Y., Hamm-Alvarez, S. F., & Camarero, J. A. (2011). Cellular uptake of cyclotide MCoTI-I follows multiple endocytic pathways. *Journal of Controlled Release, 155*(2), 134–143. http://dx.doi.org/10.1016/j.jconrel.2011.08.030.

cottonaustralia.com.au. (2015). *Cotton and biotechnology*. Retrieved from 30.04.15: http://cottonaustralia.com.au/cotton-library/fact-sheets/cotton-fact-file-biotechnology.

D'Aoust, M.-A., Lavoie, P.-O., Couture, M. M. J., Trépanier, S., Guay, J.-M., Dargis, M., et al. (2008). Influenza virus-like particles produced by transient expression in *Nicotiana benthamiana* induce a protective immune response against a lethal viral challenge in mice. *Plant Biotechnology Journal, 6*(9), 930–940. http://dx.doi.org/10.1111/j.1467-7652.2008.00384.x.

Dugdale, B., Mortimer, C. L., Kato, M., James, T. A., Harding, R. M., & Dale, J. L. (2014). Design and construction of an in-plant activation cassette for transgene expression and recombinant protein production in plants. *Nature Protocols, 9*(5), 1010–1027. http://dx.doi.org/10.1038/nprot.2014.068.

Eliasen, R., Daly, N. L., Wulff, B. S., Andresen, T. L., Conde-Frieboes, K. W., & Craik, D. J. (2012). Design, synthesis, structural and functional characterization of novel melanocortin agonists based on the cyclotide kalata B1. *The Journal of Biological Chemistry, 287*(48), 40493–40501. http://dx.doi.org/10.1074/jbc.M112.395442.

Fani, L., Bak, S., Delhanty, P., van Rossum, E. F. C., & van den Akker, E. L. T. (2014). The melanocortin-4 receptor as target for obesity treatment: a systematic review of emerging pharmacological therapeutic options. *International Journal of Obesity, 38*(2), 163–169. http://dx.doi.org/10.1038/ijo.2013.80.

Gillon, A. D., Saska, I., Jennings, C. V., Guarino, R. F., Craik, D. J., & Anderson, M. A. (2008). Biosynthesis of circular proteins in plants. *The Plant Journal, 53*(3), 505–515. http://dx.doi.org/10.1111/j.1365-313X.2007.03357.x.

gmo-compass.org. (2013). *Genetically modified plants: Global cultivation on 174 million hectares.* Retrieved from 30.04.15: http://www.gmo-compass.org/eng/agri_biotechnology/gmo_planting/257.global_gm_planting_2013.html.

Goransson, U., Sjogren, M., Svangard, E., Claeson, P., & Bohlin, L. (2004). Reversible antifouling effect of the cyclotide cycloviolacin O2 against barnacles. *Journal of Natural Products, 67*(8), 1287–1290. http://dx.doi.org/10.1021/np0499719.

Grabowski, G. A., Golembo, M., & Shaaltiel, Y. (2014). Taliglucerase alfa: an enzyme replacement therapy using plant cell expression technology. *Molecular Genetics and Metabolism, 112*(1), 1–8. http://dx.doi.org/10.1016/j.ymgme.2014.02.011.

Gran, L. (1973a). On the effect of a polypeptide isolated from "Kalata-Kalata" (*Oldenlandia affinis* DC) on the oestrogen dominated uterus. *Acta Pharmacologica et Toxicologica (Copenh), 33*(5), 400–408.

Gran, L. (1973b). Oxytocic principles of *Oldenlandia affinis. Lloydia, 36*(2), 174–178.

Grube, A., Donaldson, D., Kiely, T., & Wu, L. (2011). *Pesticides industry sales and usage 2006 and 2007 market estimates* (p. 41). Washington, DC: U.S. Environmental Protection Agency.

Gruber, C. W., Cemazar, M., Clark, R. J., Horibe, T., Renda, R. F., Anderson, M. A., et al. (2007). A novel plant protein-disulfide isomerase involved in the oxidative folding of cystine knot defense proteins. *The Journal of Biological Chemistry, 282*(28), 20435–20446. http://dx.doi.org/10.1074/jbc.M700018200.

Gruber, C. W., Elliott, A. G., Ireland, D. C., Delprete, P. G., Dessein, S., Göransson, U., et al. (2008). Distribution and evolution of circular miniproteins in flowering plants. *The Plant Cell, 20*(9), 2471–2483. http://dx.doi.org/10.1105/tpc.108.062331.

Gründemann, C., Koehbach, J., Huber, R., & Gruber, C. W. (2012). Do plant cyclotides have potential as immunosuppressant peptides? *Journal of Natural Products, 75*(2), 167–174. http://dx.doi.org/10.1021/np200722w.

Gründemann, C., Thell, K., Lengen, K., Garcia-Käufer, M., Huang, Y.-H., Huber, R., et al. (2013). Cyclotides suppress human T-lymphocyte Proliferation by an interleukin 2-dependent mechanism. *PLoS One, 8*(6), e68016. http://dx.doi.org/10.1371/journal.pone.0068016.

Gunasekera, S., Foley, F. M., Clark, R. J., Sando, L., Fabri, L. J., Craik, D. J., et al. (2008). Engineering stabilized vascular endothelial growth factor-A antagonists: synthesis, structural characterization, and bioactivity of grafted analogues of cyclotides. *Journal of Medicinal Chemistry, 51*(24), 7697–7704. http://dx.doi.org/10.1021/jm800704e.

Henriques, S. T., & Craik, D. J. (2012). Importance of the cell membrane on the mechanism of action of cyclotides. *ACS Chemical Biology, 7*(4), 626–636. http://dx.doi.org/10.1021/cb200395f.

Hess, S., Linde, Y., Ovadia, O., Safrai, E., Shalev, D. E., Swed, A., et al. (2008). Backbone cyclic peptidomimetic melanocortin-4 receptor agonist as a novel orally administrated drug lead for treating obesity. *Journal of Medicinal Chemistry, 51*(4), 1026–1034. http://dx.doi.org/10.1021/jm701093y.

Hollick, J., & Gordon, M. (1993). A poplar tree proteinase inhibitor-like gene promoter is responsive to wounding in transgenic tobacco. *Plant Molecular Biology, 22*(4), 561–572. http://dx.doi.org/10.1007/BF00047398.

Hollick, J. B., & Gordon, M. P. (1995). Transgenic analysis of a hybrid poplar wound-inducible promoter reveals developmental patterns of expression similar to that of storage protein genes. *Plant Physiology, 109*(1), 73–85.

Huang, Y. H., Chaousis, S., Cheneval, O., Craik, D. J., & Henriques, S. (2015). Optimization of the cyclotide framework to improve cell penetration properties. *Frontiers in Pharmacology, 6*, 7.

Huang, Y. H., Colgrave, M. L., Clark, R. J., Kotze, A. C., & Craik, D. J. (2010). Lysine-scanning mutagenesis reveals an amendable face of the cyclotide kalata B1 for the optimization of nematocidal activity. *The Journal of Biological Chemistry, 285*(14), 10797–10805. http://dx.doi.org/10.1074/jbc.M109.089854.

Jennings, C., West, J., Waine, C., Craik, D., & Anderson, M. (2001). Biosynthesis and insecticidal properties of plant cyclotides: the cyclic knotted proteins from *Oldenlandia affinis*. *Proceedings of the National Academy of Sciences of the United States of America, 98*(19), 10614–10619. http://dx.doi.org/10.1073/pnas.191366898.

Ji, Y. B., Majumder, S., Millard, M., Bora, R., Bi, T., Elnagar, A. Y., et al. (2013). Vivo Activation of the p53 Tumor Suppressor Pathway by an Engineered Cyclotide. *Journal of the American Chemical Society, 135*(31), 11623–11633.

Kanodia, S., Kumar, G., Rizzi, L., Pedretti, A., Hodder, A. N., Romeo, S., et al. (2014). Synthetic peptides derived from the C-terminal 6 kDa region of *Plasmodium falciparum* SERA5 inhibit the enzyme activity and malaria parasite development. *Biochimica et Biophysica Acta-General Subjects, 1840*(9), 2765–2775. http://dx.doi.org/10.1016/j.bbagen.2014.04.013.

Koehbach, J., O'Brien, M., Muttenthaler, M., Miazzo, M., Akcan, M., Elliott, A. G., et al. (2013). Oxytocic plant cyclotides as templates for peptide G protein-coupled receptor ligand design. *Proceedings of the National Academy of Sciences of the United States of America, 110*(52), 21183–21188. http://dx.doi.org/10.1073/pnas.1311183110.

Kuroyanagi, M., Yamada, K., Hatsugai, N., Kondo, M., Nishimura, M., & Hara-Nishimura, I. (2005). Vacuolar processing enzyme is essential for mycotoxin-induced cell death in *Arabidopsis thaliana*. *The Journal of Biological Chemistry, 280*(38), 32914–32920. http://dx.doi.org/10.1074/jbc.M504476200.

Kyndt, T., Quispe, D., Zhai, H., Jarret, R., Ghislain, M., Liu, Q., et al. (2015). The genome of cultivated sweet potato contains *Agrobacterium* T-DNAs with expressed genes: an example of a naturally transgenic food crop. *Proceedings of the National Academy of Sciences of the United States of America*. http://dx.doi.org/10.1073/pnas.1419685112.

Lamphear, B. J., Streatfield, S. J., Jilka, J. M., Brooks, C. A., Barker, D. K., Turner, D. D., et al. (2002). Delivery of subunit vaccines in maize seed. *Journal of Controlled Release, 85*(1–3), 169–180. http://dx.doi.org/10.1016/S0168-3659(02)00282-1.

Landry, N., Ward, B. J., Trépanier, S., Montomoli, E., Dargis, M., Lapini, G., et al. (2010). Preclinical and clinical development of plant-made virus-like particle vaccine against avian H5N1 influenza. *PLoS One, 5*(12), e15559. http://dx.doi.org/10.1371/journal.pone.0015559.

Lhouvum, K., Bhuyar, K. S., & Trivedi, V. (2015). Molecular modeling and correlation of PFI1625c-peptide models of bioactive peptides with antimalarial properties. *Medicinal Chemistry Research, 24*(4), 1527–1533. http://dx.doi.org/10.1007/s00044-014-1232-5.

Lindblade, K. A., Mwandama, D., Mzilahowa, T., Steinhardt, L., Gimnig, J., Shah, M., et al. (2015). A cohort study of the effectiveness of insecticide-treated bed nets to prevent malaria in an area of moderate pyrethroid resistance, Malawi. *Malaria Journal, 14*, 15. http://dx.doi.org/10.1186/s12936-015-0554-1.

Lindbo, J. A. (2007). TRBO: a high-efficiency tobacco mosaic virus RNA-based overexpression vector. *Plant Physiology, 145*(4), 1232–1240. http://dx.doi.org/10.1104/pp.107.106377.

Ma, J. K. C., Drake, P. M. W., Chargelegue, D., Obregon, P., & Prada, A. (2005). Antibody processing and engineering in plants, and new strategies for vaccine production. *Vaccine, 23*(15), 1814–1818. http://dx.doi.org/10.1016/j.vaccine.2004.11.011.

Mason, H. S., Lam, D. M., & Arntzen, C. J. (1992). Expression of hepatitis B surface antigen in transgenic plants. *Proceedings of the National Academy of Sciences of the United States of America, 89*(24), 11745–11749.

Mensah, R. (2012). *Control of insect pests.* Google Patents.

Mylne, J. S., Wang, C. K., van der Weerden, N. L., & Craik, D. J. (2010). Cyclotides are a component of the innate defense of *Oldenlandia affinis*. *Biopolymers, 94*(5), 635–646. http://dx.doi.org/10.1002/bip.21419.

Nguyen, G. K., Wang, S., Qiu, Y., Hemu, X., Lian, Y., & Tam, J. P. (2014). Butelase 1 is an Asx-specific ligase enabling peptide macrocyclization and synthesis. *Nature Chemical Biology, 10*(9), 732–738. http://dx.doi.org/10.1038/nchembio.1586.

OECD. (2014). *Public expenditure on health 2014/1* (Vol. 3).

Osakabe, Y., & Osakabe, K. (2015). Genome editing with engineered nucleases in plants. *Plant and Cell Physiology, 56*(3), 389–400.

Ovesen, R. G., Brandt, K. K., Göransson, U., Nielsen, J., Hansen, H. C. B., & Cedergreen, N. (2011). Biomedicine in the environment: cyclotides constitute potent natural toxins in plants and soil bacteria. *Environmental Toxicology and Chemistry, 30*(5), 1190–1196. http://dx.doi.org/10.1002/etc.496.

Ovesen, R. G., Nielsen, J., & Hansen, H. C. (2011). Biomedicine in the environment: sorption of the cyclotide kalata B2 to montmorillonite, goethite, and humic acid. *Environmental Toxicology and Chemistry, 30*(8), 1785–1792. http://dx.doi.org/10.1002/etc.579.

Pardo-Lopez, L., Soberon, M., & Bravo, A. (2013). *Bacillus thuringiensis* insecticidal three-domain Cry toxins: mode of action, insect resistance and consequences for crop protection. *FEMS Microbiology Reviews, 37*(1), 3–22. http://dx.doi.org/10.1111/j.1574-6976.2012.00341.x.

Poth, A. G., Colgrave, M. L., Lyons, R. E., Daly, N. L., & Craik, D. J. (2011). Discovery of an unusual biosynthetic origin for circular proteins in legumes. *Proceedings of the National Academy of Sciences of the United States of America, 108*(25), 10127–10132. http://dx.doi.org/10.1073/pnas.1103660108.

Poth, A. G., Colgrave, M. L., Philip, R., Kerenga, B., Daly, N. L., Anderson, M. A., et al. (2011). Discovery of cyclotides in the Fabaceae plant family provides new insights into the cyclization, evolution, and distribution of circular proteins. *ACS Chemical Biology, 6*(4), 345–355. http://dx.doi.org/10.1021/cb100388j.

Poth, A. G., Mylne, J. S., Grassl, J., Lyons, R. E., Millar, A. H., Colgrave, M. L., et al. (2012). Cyclotides associate with leaf vasculature and are the products of a novel precursor in petunia (Solanaceae). *The Journal of Biological Chemistry, 287*(32), 27033–27046. http://dx.doi.org/10.1074/jbc.M112.370841.

PRNewswire. (2014). *Reportbuyer.com has added a new market research report: Global pesticides market – segmented by type, application area and geography – trends and forecasts (2014–2020) – sustainability, regulation & competition.* (Online: PR Newswire).

Rafael, H., Lu, L. A., McCutchen, B. F., Presnail, J. K., & Clement, W. K. W. (2006). *Nucleic acid molecules encoding plant cyclotide polypeptides and methods of use*. Google Patents.
Romeis, J., McLean, M. A., & Shelton, A. M. (2013). When bad science makes good headlines: *Bt* maize and regulatory bans. *Nature Biotechnology, 31*(5), 386–387. http://dx.doi.org/10.1038/nbt.2578.
Rosca, E. V., Koskimaki, J. E., Rivera, C. G., Pandey, N. B., Tamiz, A. P., & Popel, A. S. (2011). Anti-angiogenic peptides for cancer therapeutics. *Current Pharmaceutical Biotechnology, 12*(8), 1101–1116.
Samama, P., Rumennik, L., & Grippo, J. F. (2003). The melanocortin receptor MCR4 controls fat consumption. *Regulatory Peptides, 113*(1–3), 85–88. http://dx.doi.org/10.1016/s0167-0115(02)00299-9.
Schena, M., Lloyd, A. M., & Davis, R. W. (1991). A steroid-inducible gene expression system for plant cells. *Proceedings of the National Academy of Sciences of the United States of America, 88*(23), 10421–10425.
Schöpke, T., Hasan, A. M. I., Kraft, R., Otto, A., & Hiller, K. (1993). Hämolyttisch aktive Komponenten aus *Viola tricolor* L. und *Viola arvensis* Murray. *Scientia Pharmaceutica, 61*, 145–153.
Schünmann, P. D., Coia, G., & Waterhouse, P. (2002). Biopharming the SimpliRED™ HIV diagnostic reagent in barley, potato and tobacco. *Molecular Breeding, 9*(2), 113–121. http://dx.doi.org/10.1023/A:1026752805494.
Sheen, S. (1983). Biomass and chemical composition of tobacco plants under high density growth. *Beiträge zur Tabakforschung/Contributions to Tobacco Research, 12*(1), 35–42.
Simonsen, S. M., Sando, L., Rosengren, K. J., Wang, C. K., Colgrave, M. L., Daly, N. L., et al. (2008). Alanine scanning mutagenesis of the prototypic cyclotide reveals a cluster of residues essential for bioactivity. *The Journal of Biological Chemistry, 283*(15), 9805–9813. http://dx.doi.org/10.1074/jbc.M709303200.
Stöger, E., Vaquero, C., Torres, E., Sack, M., Nicholson, L., Drossard, J., et al. (2000). Cereal crops as viable production and storage systems for pharmaceutical scFv antibodies. *Plant Molecular Biology, 42*(4), 583–590. http://dx.doi.org/10.1023/A:1006301519427.
Tabashnik, B. E., Brevault, T., & Carriere, Y. (2013). Insect resistance to Bt crops: lessons from the first billion acres. *Nature Biotechnology, 31*(6), 510–521. http://dx.doi.org/10.1038/nbt.2597.
Tacket, C. O., Mason, H. S., Losonsky, G., Estes, M. K., Levine, M. M., & Arntzen, C. J. (2000). Human immune responses to a novel norwalk virus vaccine delivered in transgenic potatoes. *The Journal of Infectious Diseases, 182*(1), 302–305. http://dx.doi.org/10.1086/315653.
Tam, J. P., & Lu, Y.-A. (1998). A biomimetic strategy in the synthesis and fragmentation of cyclic protein. *Protein Science, 7*(7), 1583–1592. http://dx.doi.org/10.1002/pro.5560070712.
Wang, C. K., Gruber, C. W., Cemazar, M., Siatskas, C., Tagore, P., Payne, N., et al. (2014). Molecular grafting onto a stable framework yields novel cyclic peptides for the treatment of multiple sclerosis. *ACS Chemical Biology, 9*(1), 156–163. http://dx.doi.org/10.1021/cb400548s.
Wang, C. K., Wacklin, H. P., & Craik, D. J. (2012). Cyclotides insert into lipid bilayers to form membrane pores and destabilize the membrane through hydrophobic and phosphoethanolamine-specific interactions. *The Journal of Biological Chemistry, 287*(52), 43884–43898. http://dx.doi.org/10.1074/jbc.M112.421198.
Werle, M., & Bernkop-Schnurch, A. (2006). Strategies to improve plasma half life time of peptide and protein drugs. *Amino Acids, 30*(4), 351–367. http://dx.doi.org/10.1007/s00726-005-0289-3.
WHO (Producer). (2014). *Frequently asked questions on genetically modified foods* (Online document) Retrieved from 28.04.15: http://www.who.int/foodsafety/areas_work/food-technology/Frequently_asked_questions_on_gm_foods.pdf.

Wong, C. T. T., Rowlands, D. K., Wong, C., Lo, T. W. C., Nguyen, G. K. T., Li, H., & Tam, J. P. (2012). Orally Active Peptidic Bradykinin B-1 Receptor Antagonists Engineered from a Cyclotide Scaffold for Inflammatory Pain Treatment, *Angewandte Chemie-International Edition, 51*(23), 5620–5624.

Xu, D., McElroy, D., Thornburg, R. W., & Wu, R. (1993). Systemic induction of a potato pin2 promoter by wounding, methyl jasmonate, and abscisic acid in transgenic rice plants. *Plant Molecular Biology, 22*(4), 573–588.

Ye, Z. X., MacNeil, T., Weinberg, D. H., Kalyani, R. N., Tang, R., Strack, A. M., et al. (2005). Structure-activity relationship of linear tetrapeptides Tic-DPhe-Arg-Trp-NH_2 at the human melanocortin-4 receptor and effects on feeding behaviors in rat. *Peptides, 26*(10), 2017–2025. http://dx.doi.org/10.1016/j.peptides.2004.12.035.

Zhao, L., Hua, T., Crowley, C., Ru, H., Ni, X., Shaw, N., et al. (2014). Structural analysis of asparaginyl endopeptidase reveals the activation mechanism and a reversible intermediate maturation stage. *Cell Research, 24*(3), 344–358. http://dx.doi.org/10.1038/cr.2014.4.

SUBJECT INDEX

Note: Page numbers followed by "f" indicate figures, and "t" indicate tables.

C

D

E

J

K

L

T

U

V

AUTHOR INDEX

Note: Page numbers followed by "f" indicate figures and "t" indicate tables.

A

B

D

E

F

G

H

L

M

N

O

P

Q

R

T

V

W

X

Y

Z

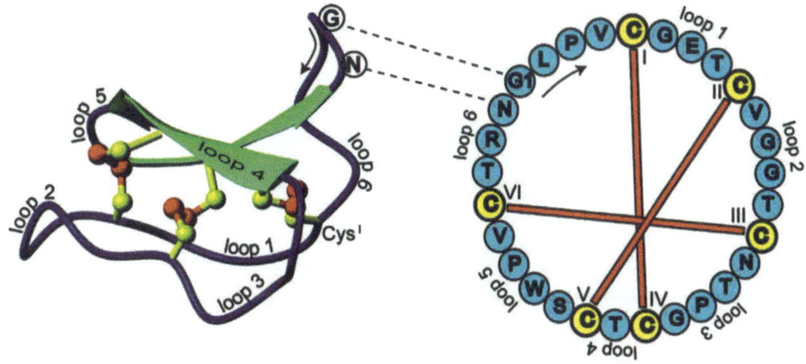

David J. Craik, Figure 1 ***Sequence and structure of kalata B1.*** On the left is the NMR structure (PDB ID code 1NB1). On the right is the amino acid sequence, with the circular backbone shown using the one-letter amino acid code. Arrows indicate the direction of the peptide chain from N to C direction. Backbone loops between successive cysteine residues are numbered loops 1 to 6. Cysteine residues are numbered I to VI.

Ulf Göransson *et al.*, Figure 1 ***Origin of cyclotides in Violaceae.*** The species that started the cyclotide story in Violaceae, *Viola arvensis* (A), through the serendipitous discovery of violapeptide I. The varv peptides A-H confined the first "library" of cyclotides (Claeson et al., 1998; Göransson et al., 1999). Varv peptide A is most likely identical to the violapeptide I, which sequence was not complete when reported. (B) *Viola odorata*, the main source of cycloviolacin O2 (cyO2) and other cyclotides that formed the bracelet subfamily (Craik et al., 1999). The sequence alignment below exemplifies cyclotide sequences: the prototype kalata B1, varv peptide A, and cycloviolacin O2. Brackets indicate the three disulfide bonds and the cyclic backbone.

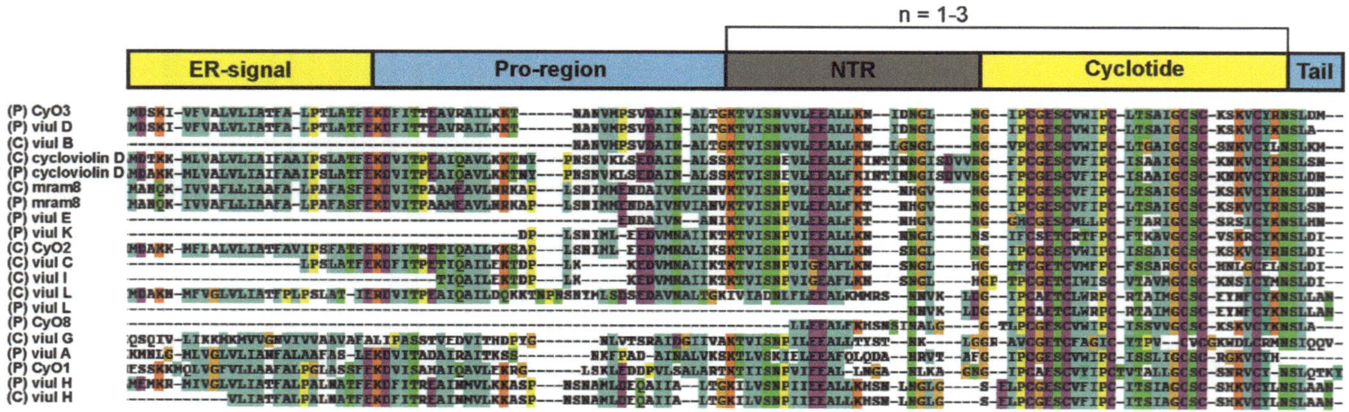

Ulf Göransson *et al.*, Figure 3 ***Example of structures of the cyclotide precursors.*** Peptide and alignment of sequences predicted from two *Viola uliginosa* transcripts: the native plant (P) and callus from in vitro culture (C) (Slazak, Sliwinska, et al., 2015). Similarly to precursors produced by plants from other families, five regions can be distinguished: ER-signal, pro region, N-terminal repeat (NTR), the cyclotide sequence, and a short C-terminal tail (characteristic for Violaceae).

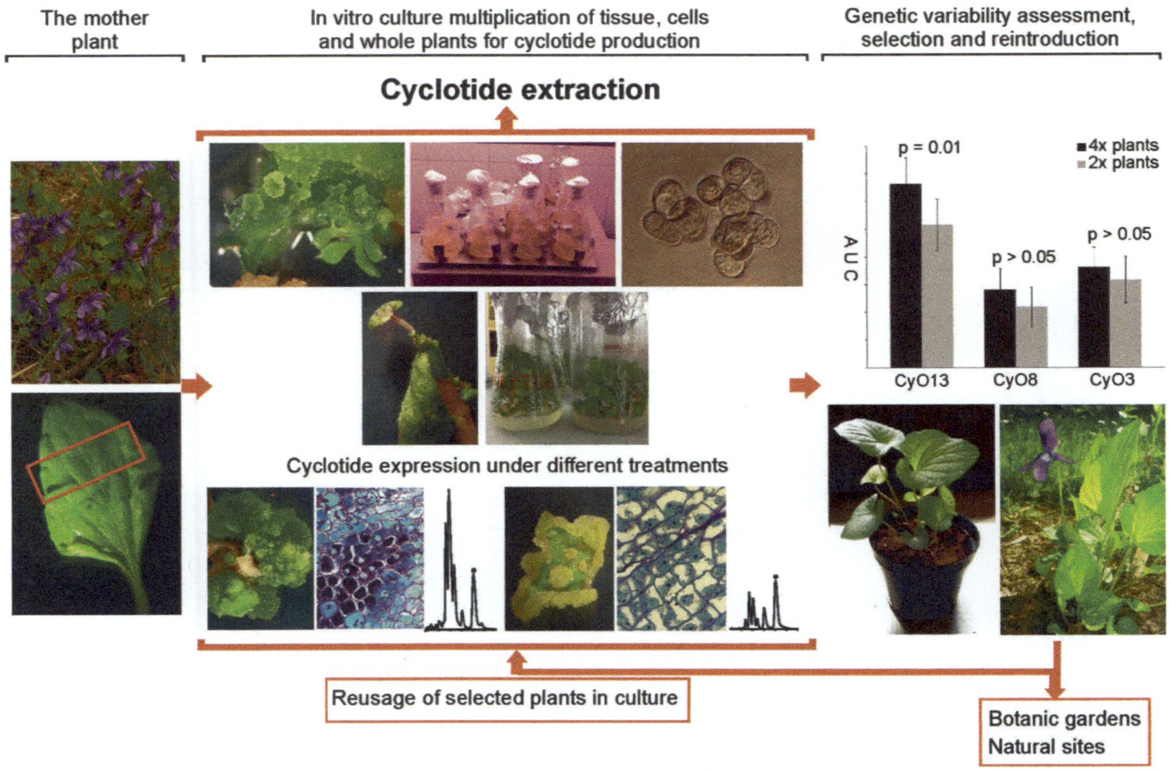

Ulf Göransson *et al.*, Figure 5 ***In vitro cultures for micropropagation of* Viola uliginosa *and cyclotide production.*** By using in vitro culture techniques, multiplication of maternal plant, its tissues and cells are achieved in relatively short time. Regenerated plantlets can serve as a source of material for reintroduction, whereas tissues, especially fast-growing suspension cultures, can be utilized for cyclotide production. Culture conditions induce somaclonal variation leading to changes in genomes and cause genetic variation of regenerated plants. The changes may be beneficial as demonstrated for *V. uliginosa*, where tetraploid (4×)-regenerated plants produced more cyclotides than plants with standard genome size (diploid, 2×). The selected specimens can be reused in *in vitro* culture as initial material to obtain higher yields (Slazak, Jacobsson, et al., 2015; Slazak, Sliwinska, et al., 2015). *Parts of Figure 5 reproduced from Slazak, Jacobsson, et al. (2015), Slazak, Sliwinska, et al. (2015) with permissions from Elsevier and Springer.*

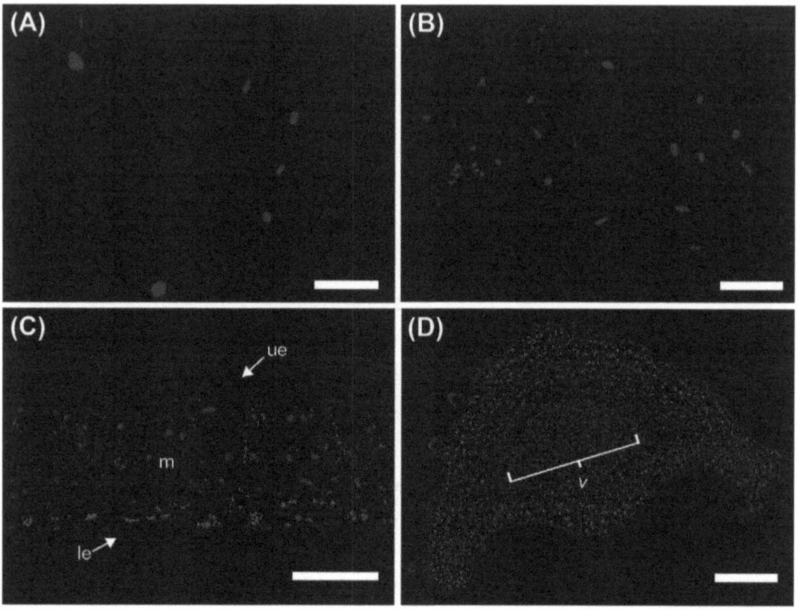

Ulf Göransson *et al.*, Figure 6 Immunohistochemistry techniques in depicting cyclotides localization in plant tissues and organs. (A) and (B), negative controls proving antibodies specificity: *Nicotiana benthamiana* (A), plant known for not producing cyclotides; *Viola odorata* leaf cross section (B) stained without using primary antibody against CyO2. Only blue autofluorescence of cell walls, nuclei, and chloroplasts (stained with DAPI) could be seen. (C) and (D), *V. odorata* leaf (C) and petiole (D) cross sections with stained cyclotides (red fluorescence). In the leaf (C) big clumps of cyclotides were found in lower epidermis region (le) and smaller amounts in leaf mezofile (m) and upper epidermis (ue). In the petiole (D) cyclotides are evenly distributed and found in the vascular bundle (v). The clumps of cyclotides are most probably an artifact of the fixation procedure (proteins and peptides, e.g., very concentrated cyclotides, may cross-link with paraformaldehyde and glutaraldehyde and become insoluble in water (Mannich & Krösche, 1912; Richards & Knowles, 1968)). Bars represent 100 μm in (A–C), and 250 μm in (D).

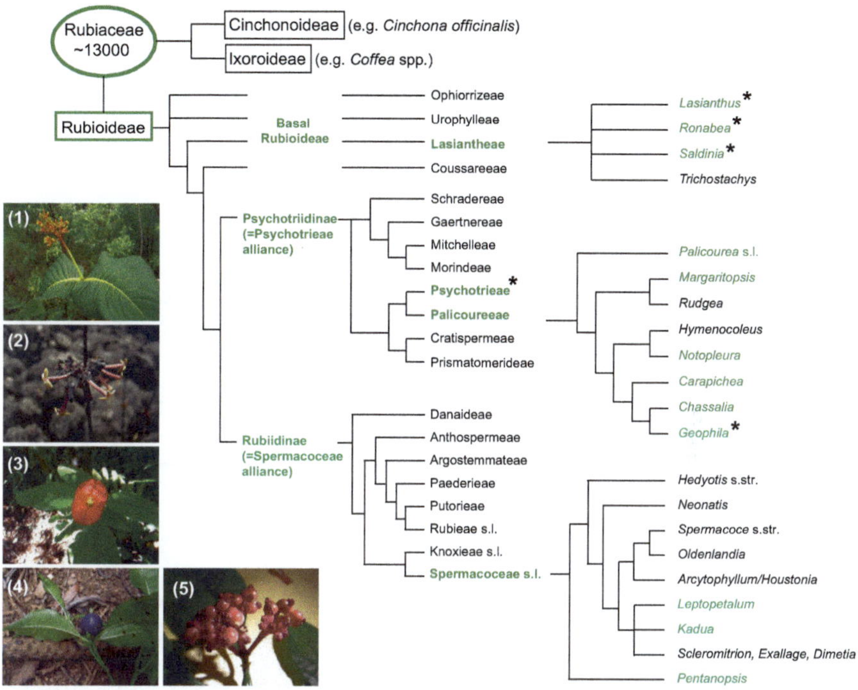

Johannes Koehbach and Christian W. Gruber, Figure 2 ***Phylogenetic distribution of cyclotide-containing plants in the Rubiaceae.*** The presence or absence of cyclotide-containing species within the genera listed (right column, italic font) and tribe (middle column, bold font) is indicated with green colored font. Cyclotides have been discovered within the tribes Spermacoceae (supertribe Rubiidinae), Psychotrieae and Palicoureeae (Psychotriidinae), as well as Lasiantheae (Basal Rubioideae). Genera and tribes wherein cyclotide-containing species have been identified according to Gruber et al. (2008), but for which no cyclotide sequence has been reported hitherto are marked with an asterisk. Phylogeny of Rubiaceae follows the literature (Bremer & Eriksson, 2009; Neupane et al., 2015; Robbrecht & Manen, 2006). s.l.—sensu lato, s.str.—sensu. stricto. Illustrations of the floral diversity and colorful fruits of cyclotide-containing species are shown; (1) *Palicourea rigida* Kunth, (2) *Kadua centranthoides* Hook. & Arn., (3) *Psychotria poeppigiana* Müll. Arg., (4) *Psychotria suterella* Müll. Arg., (5) *Chassalia discolor* K. Schum.

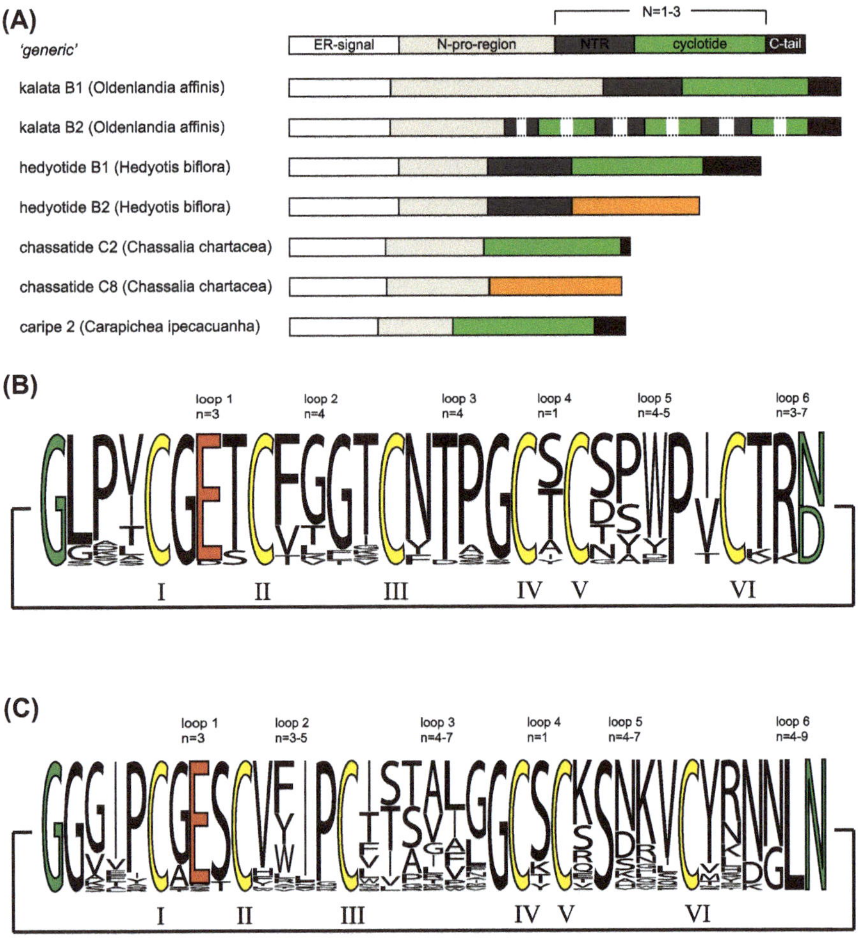

Johannes Koehbach and Christian W. Gruber, Figure 3 ***Rubiaceae cyclotide precursors and mature peptides.*** (A) An overview of the architecture of cyclotide precursor proteins among different species is shown. The typical cyclotide precursor architecture is depicted as "generic" and its domains are color-coded, i.e., ER-signal (white), N-pro-region (light gray), N-terminal repeats (NTR, dark gray), mature cyclotide domain (green), and C-terminal tail sequence (black). Orange indicates cyclotide domains encoding for linear cyclotides. Rubiaceae precursor proteins may contain up to three cyclotide domains encoding the same or different cyclotide sequences (illustrated as N = 1–3). Sequence logos of all Rubiaceae cyclotide sequences as presented in Table 1 are shown for Moebius-type (B) and bracelet-type cyclotides (C). For illustration purposes "hybrid cyclotides," i.e., kalata B8, B9, and Oak7 cyclotide have been excluded. Conserved cysteines (yellow), glutamic acid (red), asparagine/aspartic acid, and glycine residues (green) at the predicted cyclization site are highlighted and loop sizes are given.

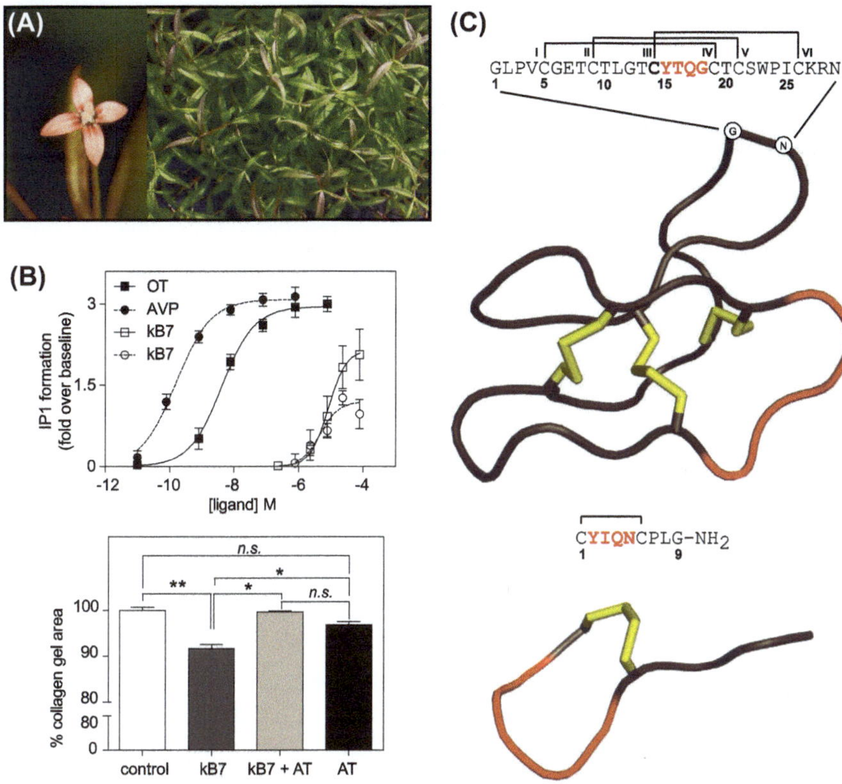

Johannes Koehbach and Christian W. Gruber, Figure 4 ***Bioactivity of Rubiaceae cyclotides.*** The African plant *Oldenlandia affinis* (photograph by David Wilson, James Cook University, Australia) (A) contains a number of different cyclotides, known as kalata peptides (Table 1) and is used in traditional medicine for its uterotonic properties. Recent pharmacological characterization of this activity revealed the oxytocin (OTR) and vasopressin V_{1a} ($V_{1a}R$) receptors as molecular targets of the cyclotide kalata B7 (kB7). Functional receptor activation showed kB7 (blank symbols) to be a partial agonist at both receptors (OTR—solid lines, $V_{1a}R$—dashed lines) (B, upper panel). Oxytocin (OT, filled squares) and vasopressin (AVP, filled circles) were used as control peptides. The contractile effects of kB7 were studied using a collagen-gel assay (B, lower panel). *(Figure has been adapted and modified from Koehbach, O'Brien, et al. (2013).)* Sequence alignment and NMR structural studies revealed a high similarity of kalata B7 (PDB ID: 2M9O) with human oxytocin (PDB ID: 2MGO) (C). Structures are shown as ribbon diagrams with disulphide bonds highlighted in yellow. The "active" loop of kalata B7 (-YTQG-) showing sequence similarity to oxytocin (-YIQN-) is highlighted in red.

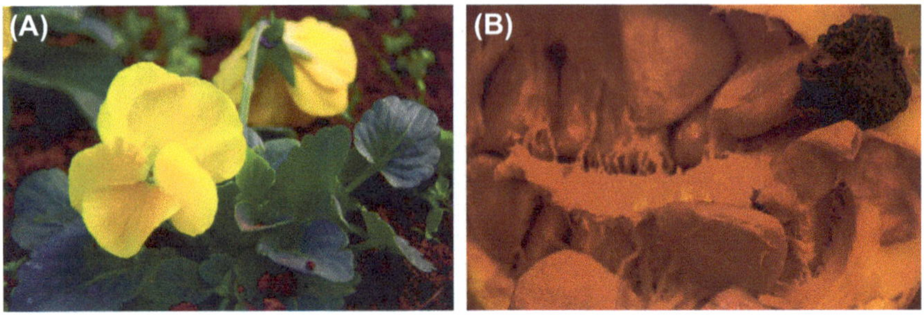

Ning-Hua Tan and Wen-Jun He, Figure 2 Two Chinese plants containing cyclotides, taken by Wen-Jun He. (A) The flowers of *Viola tricolor*; (B) the fruit of *Momordica cochinchinensis*.

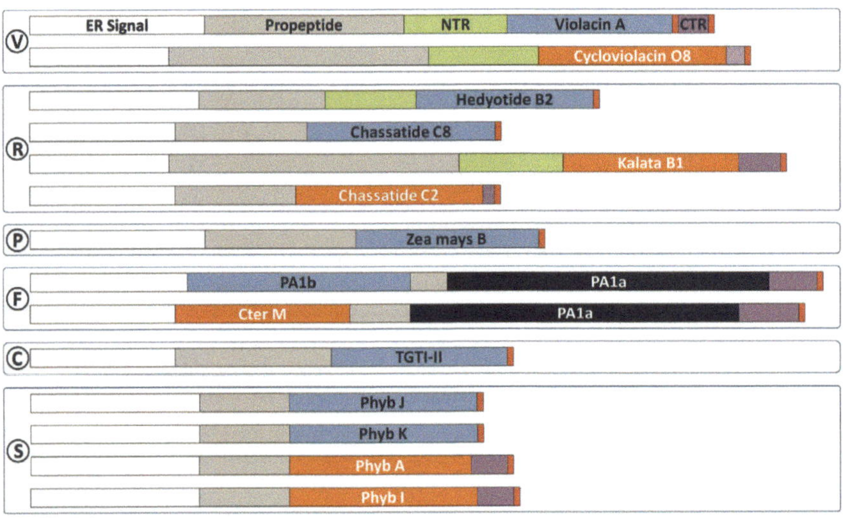

Michelle L. Colgrave, Figure 1 ***Variation in cyclotide and related knottin gene architecture among angiosperms.*** Signal sequence is shown in white boxes. Knottin and acyclotide domains are shown in light blue boxes. Cyclotide domains are shown in orange boxes. N-terminal prodomains are shown in green boxes. C-terminal repeats are shown in mauve boxes. *V*, Violaceae; *R*, Rubiaceae; *P*, Poaceae; *F*, Fabaceae; *C*, Cucurbitaceae; *S*, Solanaceae. *This research was originally published in Journal of Biological Chemistry. Poth et al. (2012), © the American Society for Biochemistry and Molecular Biology.*

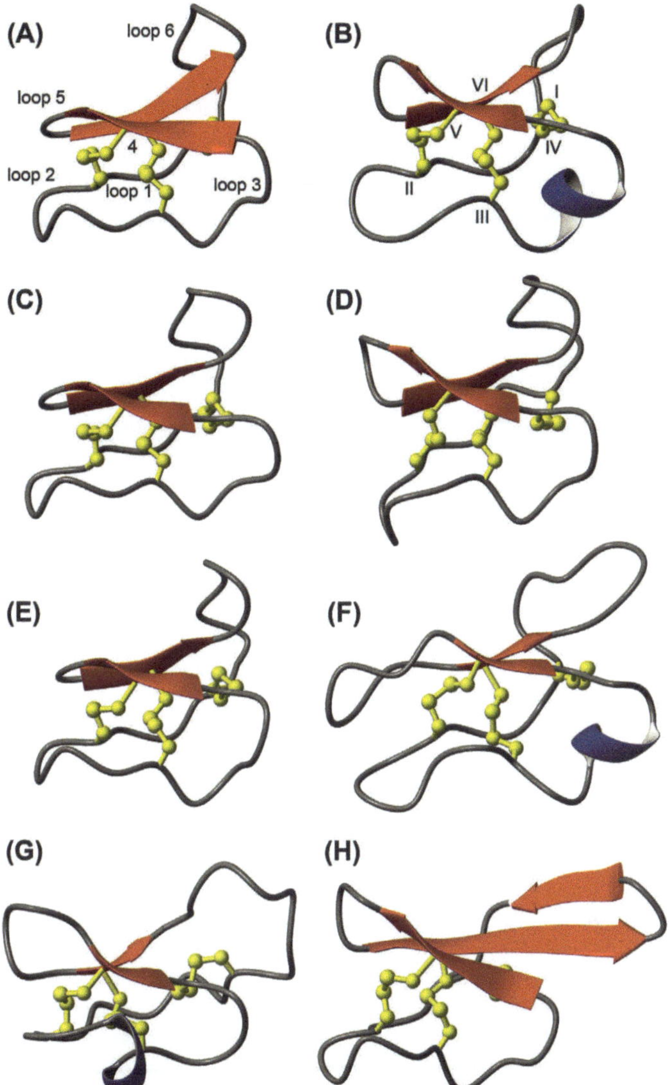

N.L. Daly and K.J. Rosengren, Figure 3 ***Representative structures highlighting the diversity of the cyclotide family.*** (A) Solution NMR structure of kalata B1 (PDB code 1NBI). (B) Solution NMR structure of cycloviolacin O1 (1NBJ). (C) Crystal structure of varv F (3E4H). (D) Solution NMR structure of kalata B8 (2B38). (E) Solution NMR structure of kalata B7 in DPC micelles and in complex with Mn^{2+} ions (2JWM). (F) Solution NMR structure of palicourein (1R1F). (G) Crystal structure of MCoTi-II in complex with trypsin (4GUX). (H) NMR solution structure of tricyclon A (1YP8). B-sheet secondary structure is shown in red and helical secondary structure in blue. Disulfides are shown in yellow in ball-and-stick representation.

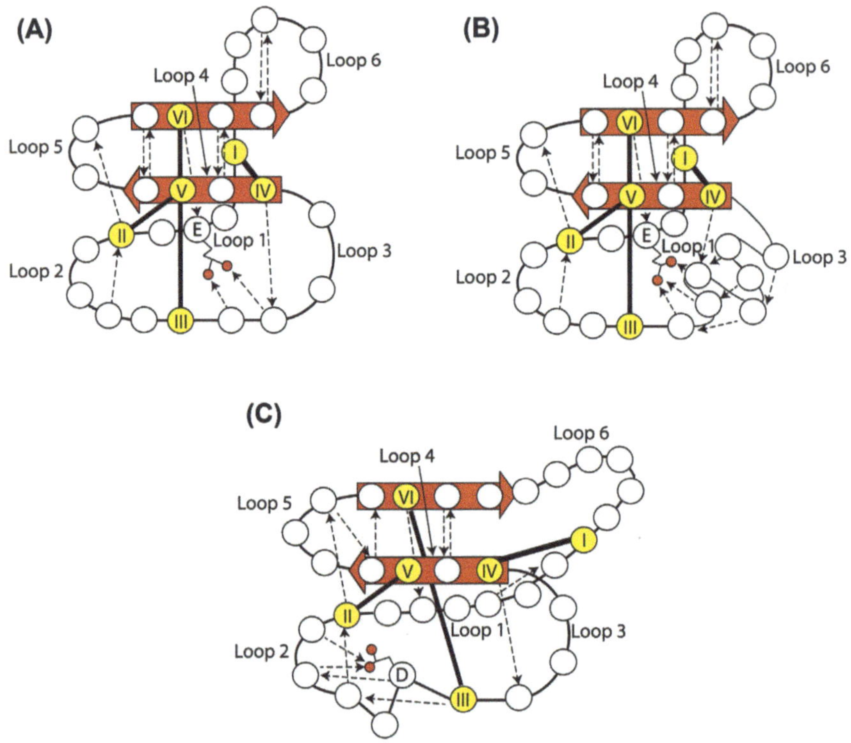

N.L. Daly and K.J. Rosengren, Figure 4 ***Conserved hydrogen bonding networks in cyclotides.*** Schematic illustration of the (A) Möbius, (B) bracelet, and (C) trypsin inhibitor folds, and key hydrogen bonds that stabilize the different folds. Backbone hydrogen bonds involving amide protons and carbonyls are shown as dashed arrows from the donator to the acceptor. In addition, backbone amide hydrogen bonds to key side chain carboxyl groups are highlighted. Hydrogen bonds were identified in crystal structures or by analysis of deuterium exchange and temperature coefficient data (Rosengren et al., 2003).

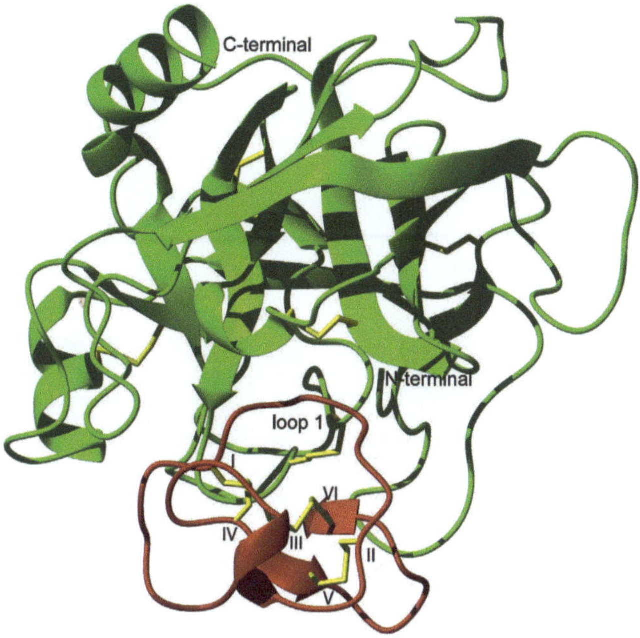

N.L. Daly and K.J. Rosengren, Figure 5 ***Three-dimensional structure of MCoTI-II bound to trypsin (PDB ID code 4GUX).*** Both proteins are show in ribbon style, with trypsin in green and MCoTI-II in red. Disulfide bonds are shown in yellow. The main interactions between the peptide and trypsin are associated with loop 1 of MCoTI-II. The diagram was made using Pymol (The PyMOL Molecular Graphics System, Schrödinger, LLC).

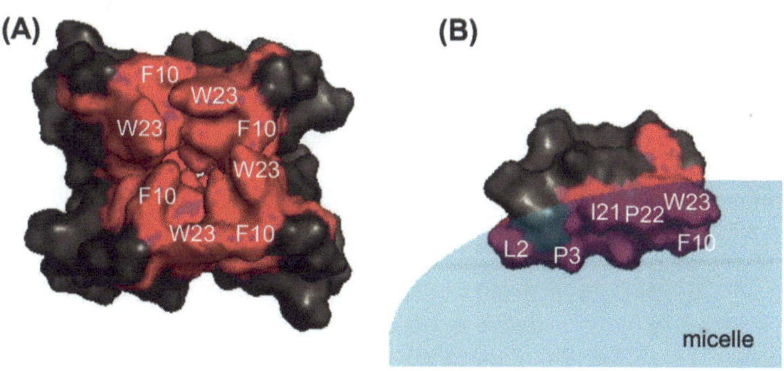

N.L. Daly and K.J. Rosengren, Figure 6 ***Aggregation and membrane interaction of kalata B2.*** (A) Proposed tetramer, which is a key oligomeric species of kalata B2 in aqueous buffer. The model is based on chemical shift changes occurring during aggregation, highlighted in pink, and the symmetry evident from the single set of resonances. An octamer is proposed to form by face-to-face association of two tetramers. Kalata B2 also binds and inserts into micelles according to the schematic view in (B), which is based on quenching of resonances using paramagnetic probes. Both interactions rely on the exposed hydrophobic patch.

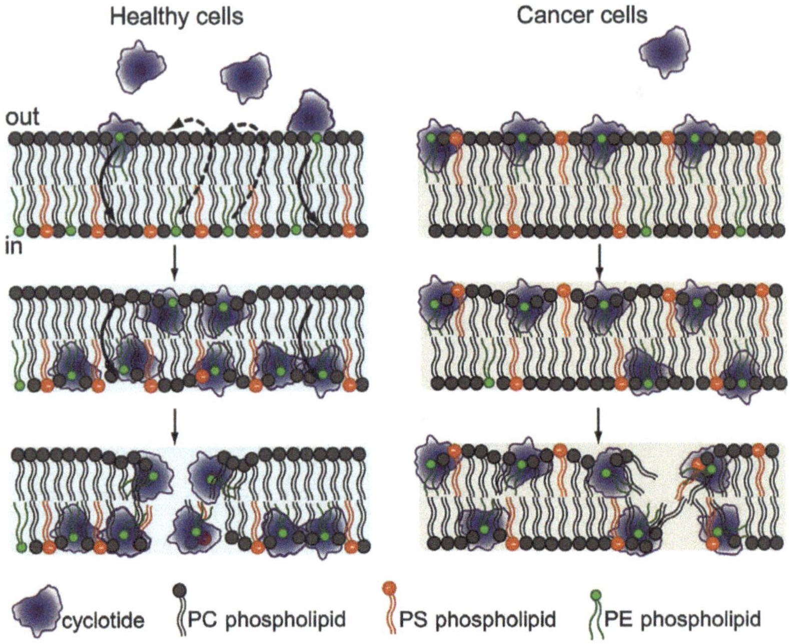

Georgianna K. Oguis *et al.*, Figure 5 *Mechanism of membrane binding of cyclotides.* The mechanism in which cyclotides target and disrupt membranes of human healthy and cancer cells is represented in a model showing phosphatidylcholine (PC), phosphatidylethanolamine (PE), and phosphatidylserine (PS) phospholipids. Trans-bilayer movement of PE phospholipids is indicated in dark arrows. To bind to the membrane of both healthy and cancer cells, cyclotides target PE phospholipids. Insertion and membrane disruption follow. Because cancer cells have more exposed PE phospholipids, cyclotides can target cancer cells more effectively than healthy cells. However, PE phospholipids in healthy cells are more dynamic than cancer cells; hence, peptide insertion is more favorable in healthy cells. *Reproduced with permission from ChemBiochem, 15, 1956 (2014). Copyright 1999–2015 John Wiley & Sons, Inc.*

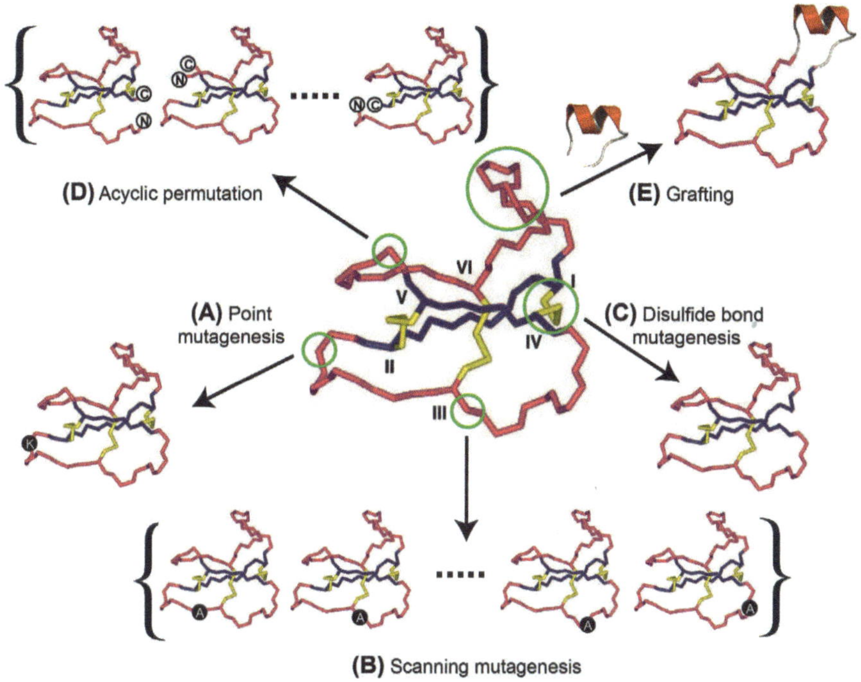

Georgianna K. Oguis *et al.*, Figure 6 ***Overview of structure–activity relationship studies of cyclotides.*** The prototypical cyclotide kalata B1 is shown in the middle with the types of mutagenesis and structure–activity relationship studies that have been done illustrated schematically. (A) Point mutagenesis—individual residues are replaced with another residue (lysine, for example) to investigate the effect on activity. (B) Scanning mutagenesis—a residue (alanine, for example) is systematically substituted to all positions within the sequence. (C) Disulfide bond mutagenesis—each one of the three disulfide bonds were successively removed to investigate their role in structure and folding. (D) Acyclic permutation—acyclic derivatives are made by breaking single peptide bond in the backbone to investigate the effect of cyclization on structure and folding. (E) Grafting—bioactive epitopes were grafted into the cyclotide framework to investigate the plasticity of the framework being used for development of therapeutic and diagnostic agents.

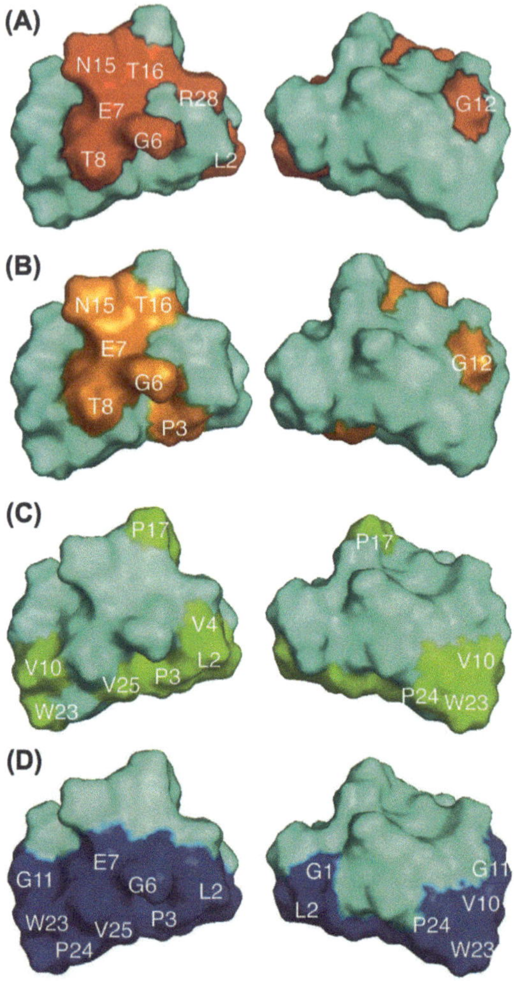

Georgianna K. Oguis *et al.*, Figure 7 ***Localization of residues critical for hemolytic and insecticidal activity of kalata B1 and the hydrophobic patch of kalata B1.*** The figures on the right are rotated 180° relative to those on the left. (A) Mutated residues giving <20% hemolysis at 60 μM peptide concentration are shown in red. The wild-type peptide displays 69% hemolysis at this concentration. Functionally essential residues without concomitant structural perturbation are seen to cluster on one face of the molecule around E7. (B) The surface position of residues is critical for insecticidal activity. Mutated residues giving <70% insecticidal activity at 230 ppm are shown in orange, relative to the wild-type peptide normalized to 100% activity. (C) The hydrophobic residues of kalata B1 are highlighted in green, showing a dominant hydrophobic patch on the lower section of the framework. (D) Membrane-penetrating residues as measured by NMR are shown in dark blue (Shenkarev et al., 2006), showing a correspondence with the dominant hydrophobic surface patch. *Reproduced with permission from Journal of Biological Chemistry, 283, 9805 (2008). Copyright 2015 American Society for Biochemistry and Molecular Biology.*

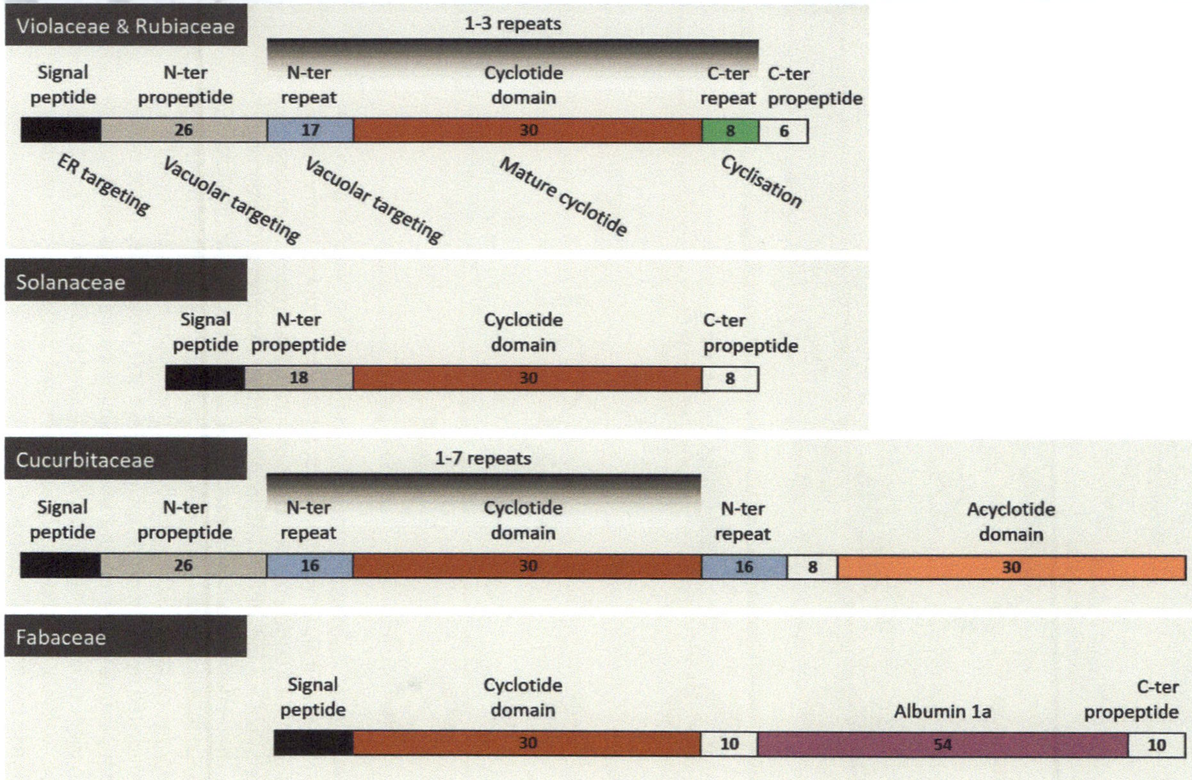

Thomas Shafee *et al.*, Figure 1 ***Schematic of cyclotide precursors. Violaceae and Rubiaceae***: Cyclotide precursors are composed of an endoplasmic reticulum (ER) signal peptide, an N-terminal propeptide and a region composed of an N-terminal repeat, a mature cyclotide domain and a C-terminal repeat. This region can be repeated up to three times and is followed by an additional short sequence at the C-terminus. ***Solanaceae***: Cyclotide transcripts from *Petunia hybrida* encode a precursor with an ER sequence and one N-terminal propeptide of 18 amino acids followed by a mature cyclotide domain and a C-terminal propeptide of 6 amino acids. ***Cucurbitaceae***: Cyclotide transcripts from *Momordica cochinchinensis* in the Cucurbitaceae family encode precursors with an ER signal sequence followed by an N-terminal propeptide of 2–9 amino acids and a region composed of an N-terminal repeat sequence of 16 amino acids and a mature cyclotide domain. This region is repeated up to seven times and is separated by 8 amino acids from a terminal acyclotide domain which is a typical noncyclic knottin. ***Fabaceae***: Cyclotide transcripts from the butterfly pea, *Clitoria ternatea*, from the Fabaceae family encode a precursor with an ER signal peptide followed immediately by a mature cyclotide domain that is flanked at the C-terminus by a linking peptide and the albumin-1a chain. The cyclotide domain replaces the PA1b subunit that is present in typical albumin-1 genes.

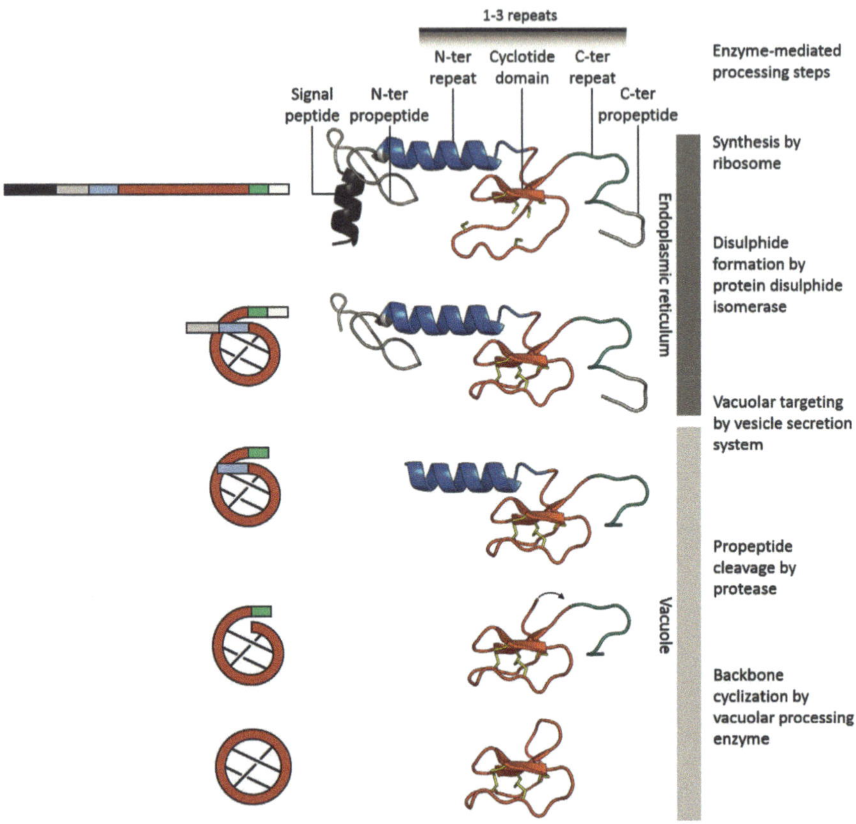

Thomas Shafee *et al.*, Figure 2 ***Cyclotide processing pathway.*** The cyclotide precursor is comprised of several parts: Firstly a hydrophobic signal peptide (thought to be α-helical (Nakamura & Matsuoka, 1993)) targets the precursor to the ER. The N-terminal propeptide and the helical N-terminal repeat contain vacuolar targeting sequences (Conlan et al., 2011) (PDB:1WN8). The cyclotide domain contains the cystine knot motif (Rosengren, Daly, Plan, Waine, & Craik, 2003) (PDB:1NB1). A final C-terminal propeptide of unknown structure is necessary for cyclization. The cyclotide domain folds in the ER. Disulphide bond formation is mediated by the oxidizing environment and protein disulphide isomerase. The first signal peptide is cleaved and the remaining protein is secreted to the vacuole, whereupon the N-terminal propeptide is also cleaved by an unknown protease. Finally, the protease, vacuolar processing enzyme performs backbone cyclization by transpeptidation to yield the final cyclic product.

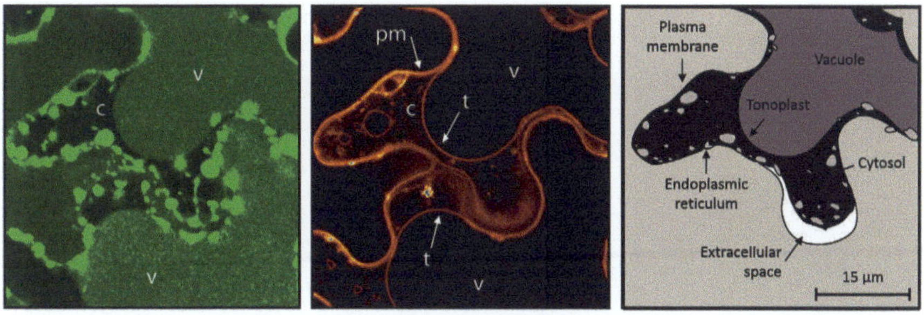

Thomas Shafee *et al.*, Figure 3 ***Vacuolar targeting by the signal peptide and* Oak1 *precursor.*** Transient expression of GFP linked to the *Oak1* cyclotide precursor from *Oldenlandia affinis* in leaf epidermal cells from *Nicotiana benthamiana*. Left panel: GFP fluorescence from GFP-Oak1 in the ER and Golgi on the periphery of the cell and GFP in the vacuole (v) after proteolytic processing and production of the kB1 cyclotide. Middle panel: Fluorescence from the membrane dye FM 5–95 highlighting the plasma membrane (pm), the tonoplast (t, vacuolar membrane) and other small membranous components. Right panel: Diagram illustrating the relevant subcellular compartments of one of the cells shown in the fluorescent micrographs. *Modified from Conlan et al. (2011).*

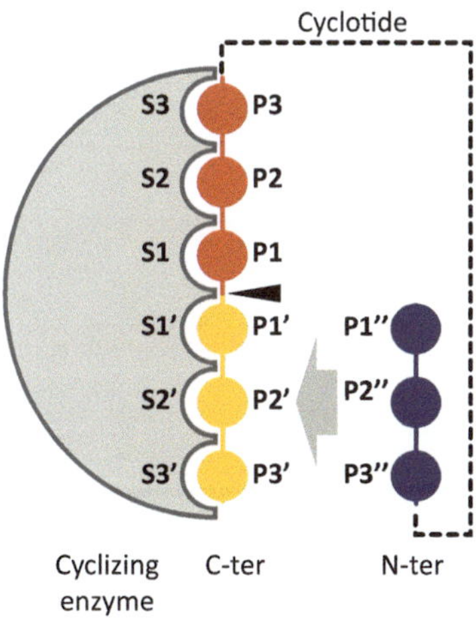

Thomas Shafee *et al.*, Figure 4 ***Enzyme binding sites.*** The cyclizing enzyme binds firstly to the C-terminus of the cyclotide precursor and subsequently to the N-terminus using the same binding site. The binding sites are labeled S1 to S3 before the cleavage site, and S1′ to S3′ after the cleavage site (indicated by a black triangle). The peptide residues that bind to these sites are similarly labeled P1 to P3 before and P1′ to P3′ after the cleavage site. Since the cyclotide N-terminus binds to the enzyme as a second substrate after cleavage, its residues are labeled P1″ to P3″.

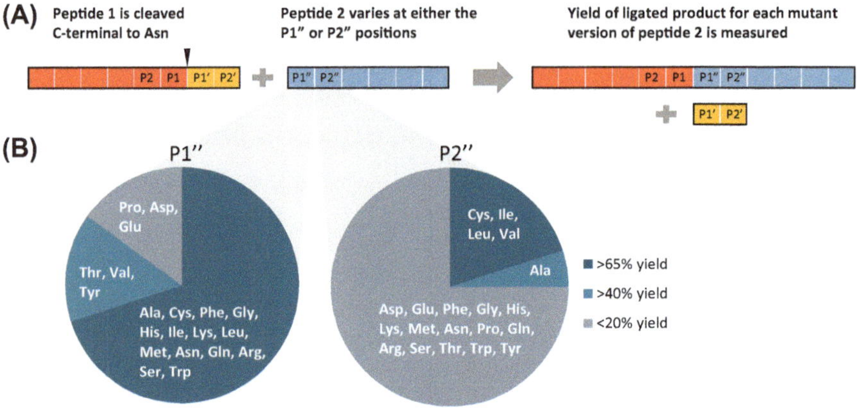

Thomas Shafee *et al.*, Figure 8 ***Substrate specificity of butelase 1 for the incoming N-terminus during transpeptidation.*** (A) Butelase 1 cleaves after a P1 Asn (black triangle) in an 8-residue model peptide and ligates the newly exposed C-terminus to the N-terminus of a second peptide. In the second peptide, either the P1″ or P2″ residue is substituted with each of the 20 amino acids and the effect on yield examined. (B) Yields of ligated product from model peptides with variations at the P1″ and P2″ positions. *Adapted from reference Nguyen et al.(2014).*

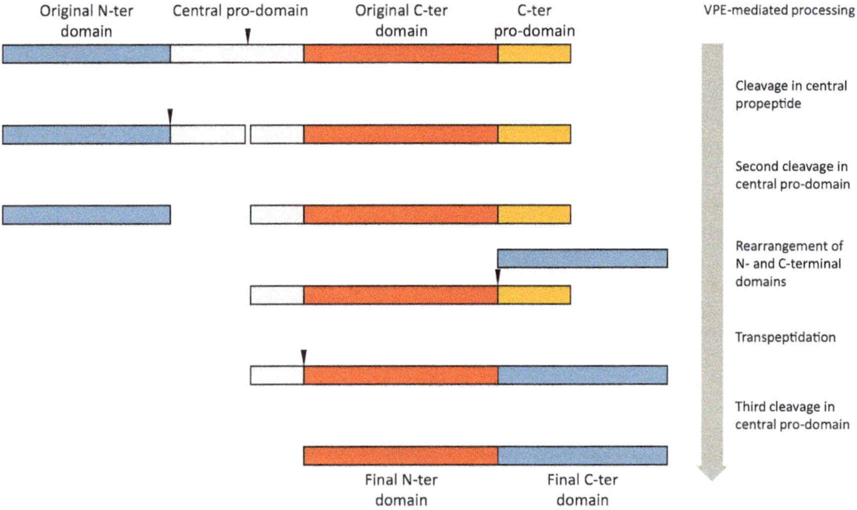

Thomas Shafee *et al.*, Figure 9 ***VPE-mediated ConA maturation.*** Schematic illustrating the hydrolysis and transpeptidation processing events required to produce mature ConA. Arrows indicate processing sites. *Figure adapted from Bowles et al. (1986) and Sheldon et al. (1996).*

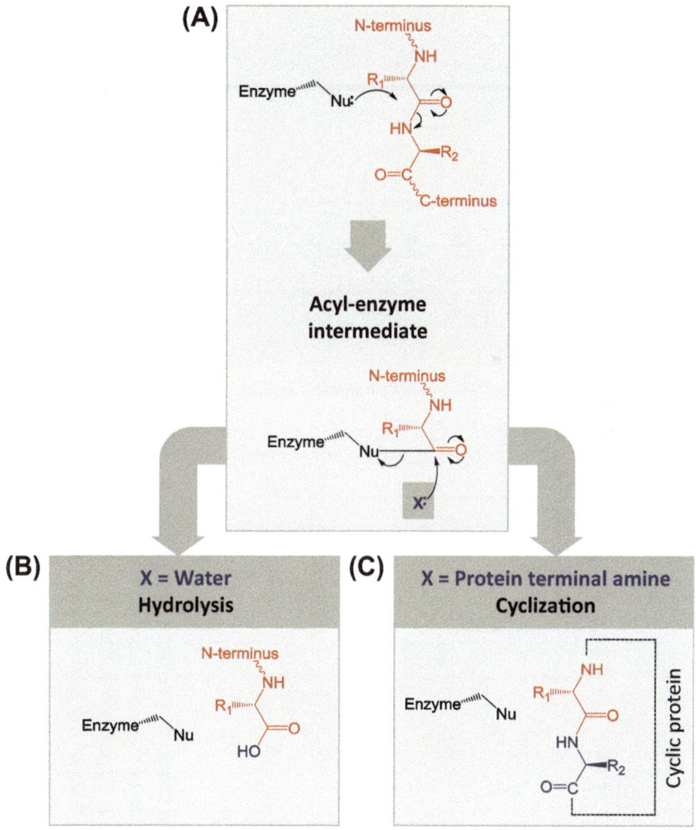

Thomas Shafee *et al.*, Figure 10 ***Chemical mechanism of cyclization by intramolecular transpeptidation.*** (A) The enzyme (black) contains a nucleophilic residue (Nu) in its active site which attacks the carbonyl of the substrate peptide (red) to form a covalent acyl-enzyme intermediate and eject the substrate's C-terminus. This intermediate is then attacked by a second nucleophile (X) to regenerate the free enzyme and release the final product. (B) In a typical protease, X is water and hydrolysis simply releases the remaining cleaved peptide. (C) Alternatively, the peptide's N-terminus enters the active site and the amine acts as the nucleophile X. This intramolecular transpeptidation yields a cyclic product.

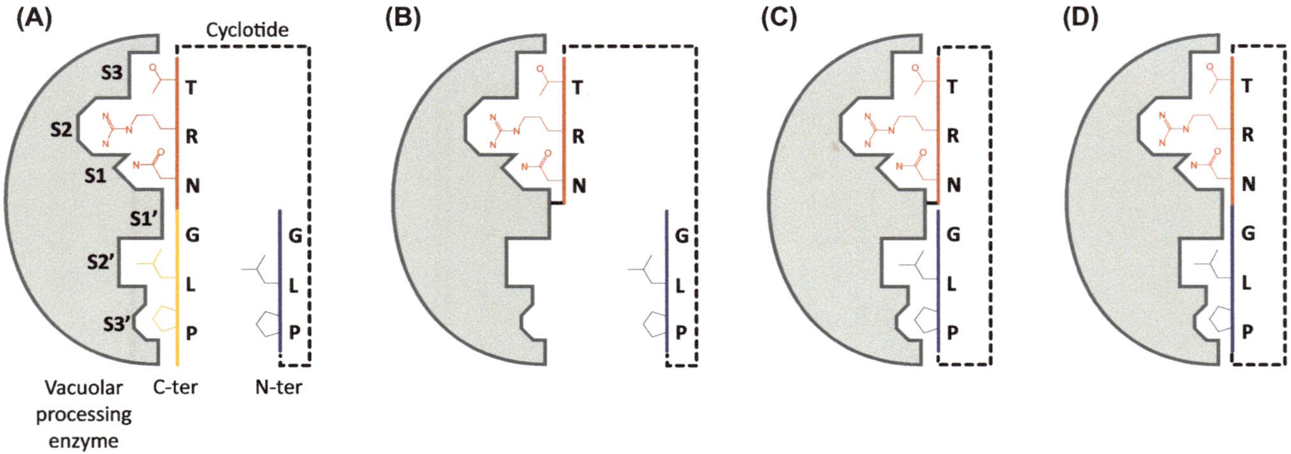

Thomas Shafee *et al.*, Figure 11 ***Proposed binding specificity for cyclization by VPE.*** (A) VPE specifically preferentially binds the substrate sequence TRNGLP in kB1. (B) Upon formation of the acyl-enzyme intermediate, the cleaved C-terminal peptide is released to expose the GLP binding pocket. (C) If the peptide N-terminus also contains the GLP motif, it binds to the exposed active site pocket. (D) The N-terminal amine outcompetes water as the nucleophile to resolve the acyl-enzyme intermediate into a cyclic product.

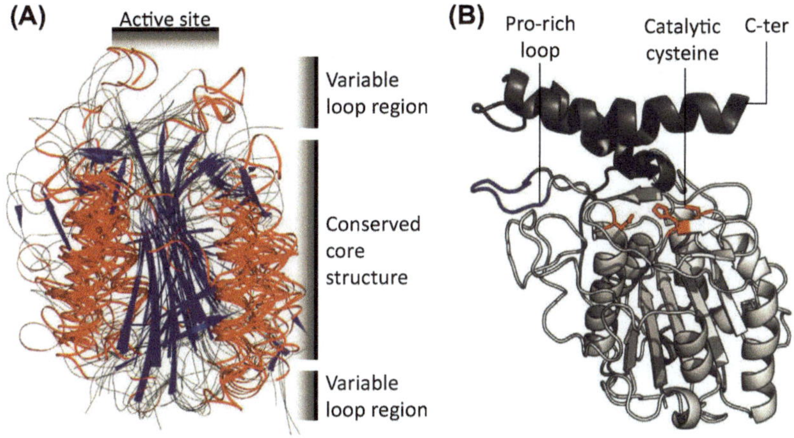

Thomas Shafee *et al.*, Figure 13 ***Structure of VPE and related enzymes.*** (A) Overlay of the known structures of CD clan proteases with α-helices in red, β-sheets in blue. (PDBs: 4AWA, 3H11, 4HQ0, 4JR0, 1CP3, 4F6P, 3V3K, 4HVA, 4JJ7, 4IEF, 3EEB, 3PA8). (B) Model of mature butelase 1 based on human pro-legumain (4FGU) with catalytic residues in red, VPE proline-rich loop in blue and C-terminal helices in gray. The exact orientation of the proline rich loop is unknown.

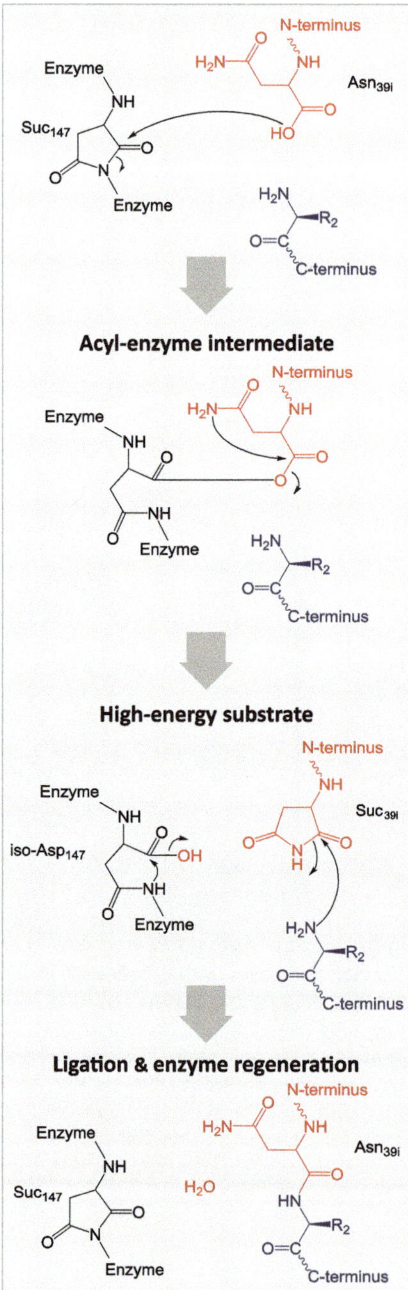

Thomas Shafee *et al.*, Figure 14 *Proposed condensation ligation mechanism for Legumain.* The enzyme (black) contains a high-energy succinimide (Suc_{147}) ring in its active site. This unnatural residue is the result of an intramolecular condensation of an aspartate residue. The carboxylic acid of the first substrate peptide (red) performs a nucleophilic attack on the succinimide ring to form an acyl-enzyme intermediate. This intermediate is then attacked by the substrate asparagine (Asn_{39i}) to release the substrate. The enzyme now contains an iso-aspartic acid (iso-Asp_{147}) and the substrate contains a high-energy succinimide ring (Suc_{39i}). Finally, the substrate's high-energy succinimide is attacked by the amine of the second substrate (blue) to form a new peptide bond and the enzyme regenerates its succinimide by intramolecular condensation. Overall, this condensation reaction yields a ligated peptide product.

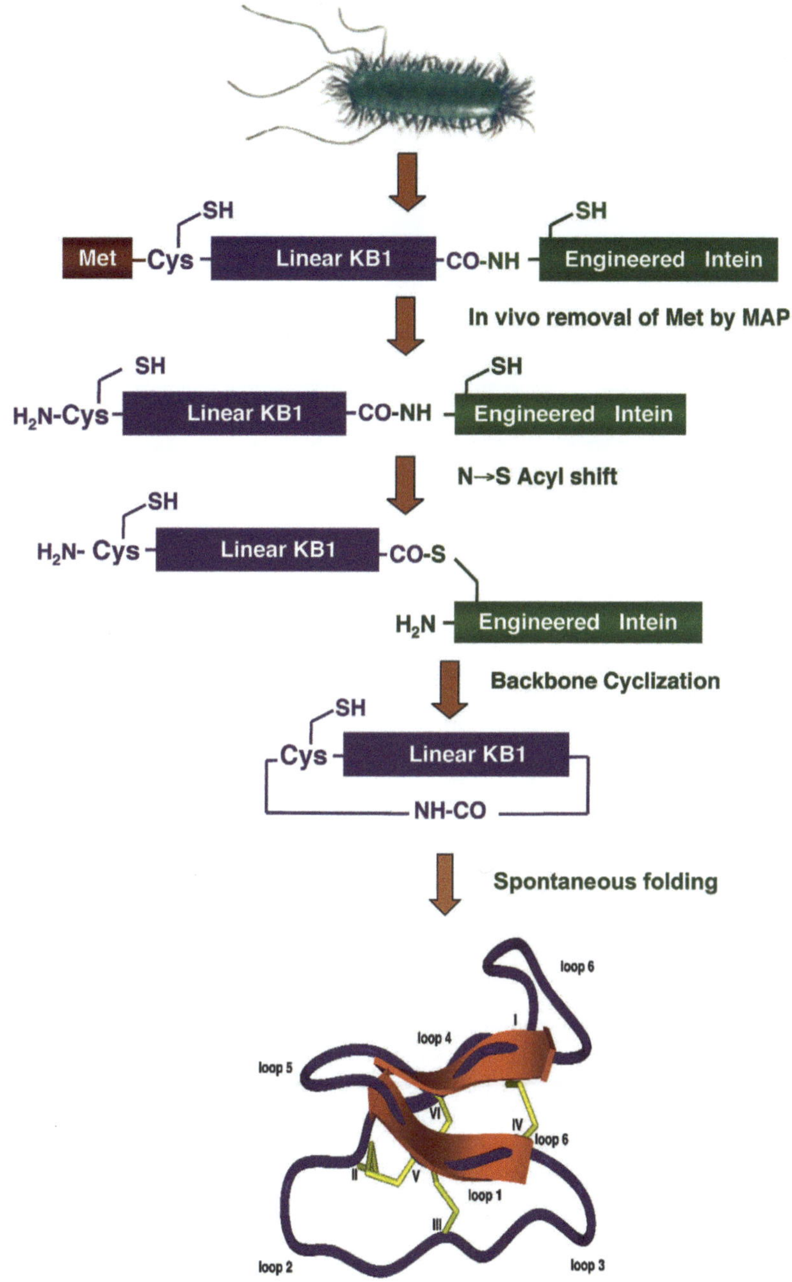

Yilong Li *et al.*, Figure 4 Intein-mediated backbone cyclization of a cyclotide linear precursor. The intein–cyclotide construct is recombinantly expressed in *E. coli* and can be cyclized and folded either in vitro (Kimura et al., 2006) or in vivo (Austin et al., 2009; Camarero et al., 2007). MAP, methionine aminopeptidase.

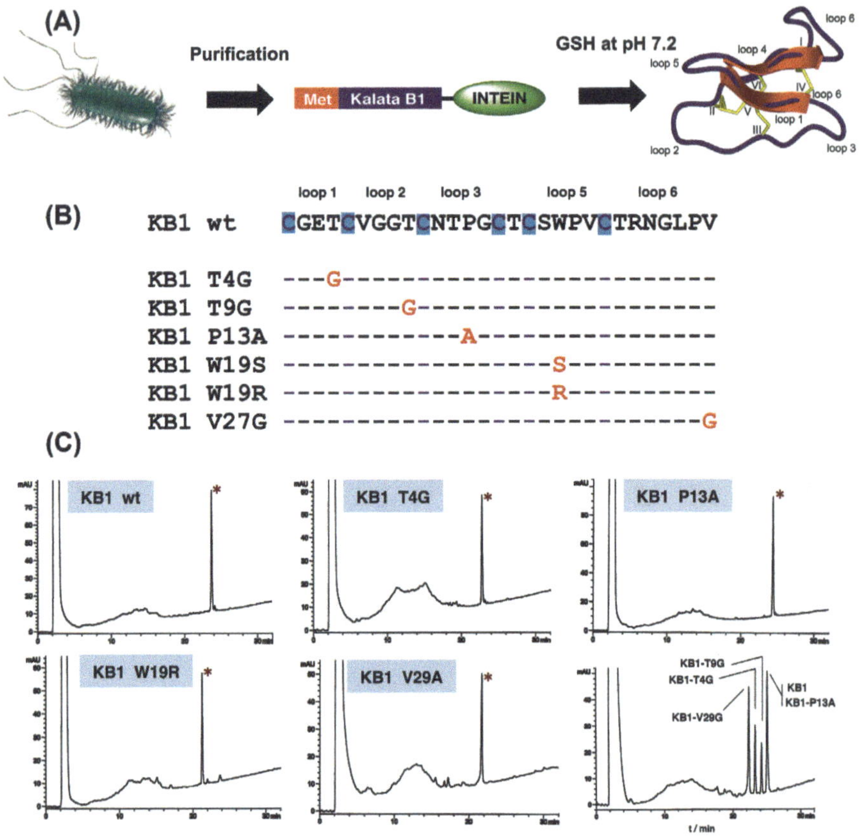

Yilong Li *et al.*, Figure 5 In vitro production of different kalata B1 cyclotides using an intein-mediated backbone cyclization (Kimura et al., 2006). (A) The intein–kalata constructs were expressed in *E. coli*, purified, and cyclized/folded in aqueous buffer at pH 7.2 containing reduced GSH for 24 h. (B) Sequences of all the katala B mutants cyclized using this approach. (C) Analytical high-performance liquid chromatography traces of the cyclization/folding crude after 18 h. The peak corresponding to the folded cyclotide is marked with an asterisk.

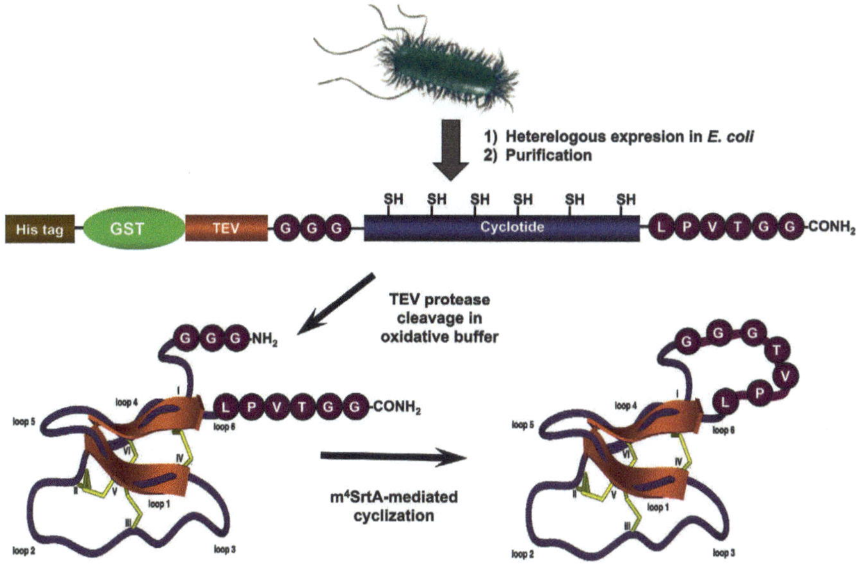

Yilong Li *et al.*, Figure 6 Strategy used for the biosynthesis of a cyclotide using SrtA-induced backbone cyclization (Stanger et al., 2014).

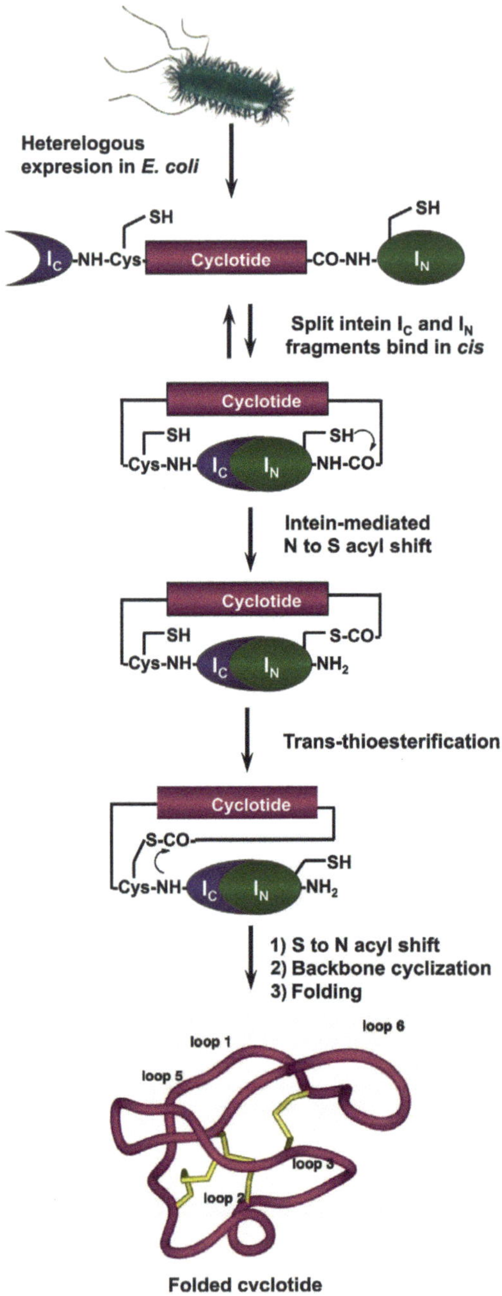

Yilong Li *et al.*, Figure 7 In-cell expression of native folded cyclotides using intein-mediated PTS (Jagadish et al., 2013).

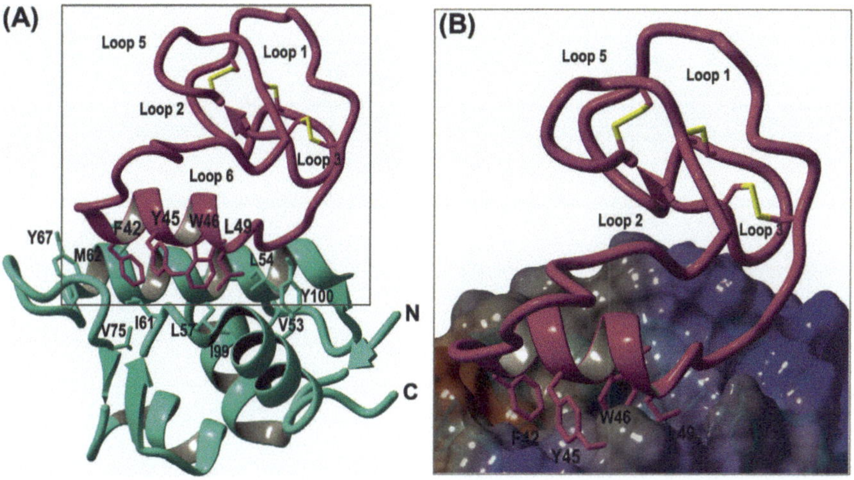

Yilong Li *et al.*, Figure 8 Solution structure of the complex formed by engineered cyclotide MCo-PMI and the p53-binding domain of Hdm2 (pdb code: 2M86) (Ji et al., 2013). (A) Ribbon representation of MCo-PMI (purple) and Hdm2 (cyan blue) complex. The side chains of Phe42, Trp46, and Leu49 in MCo-PMI and the Hdm2 residues shaping the hydrophobic binding pocket are shown as sticks in purple and cyan blue, respectively. (B) Close-up view of the binding interface within the Hdm2-MCo-PMI complex. The electrostatic potential at the molecular surface of Hdm2 is shown as positive in blue, negative in red, and noncharged in white.

Kolibri 360 GmbH
Ferdinand-Jühlke-Straße 7
99095 Erfurt, Deutschland
produktsicherheit@kolibri360.de